ARTIFICIAL NEURAL NETWORKS

Learning Algorithms, Performance Evaluation, and Applications

THE KLUWER INTERNATIONAL SERIES
IN ENGINEERING AND COMPUTER SCIENCE

ARTIFICIAL NEURAL NETWORKS
Learning Algorithms, Performance Evaluation, and Applications

by

N. B. Karayiannis
University of Houston

A. N. Venetsanopoulos
University of Toronto

KLUWER ACADEMIC PUBLISHERS
Boston/Dordrecht/London

Distributors for North America:
Kluwer Academic Publishers
101 Philip Drive
Assinippi Park
Norwell, Massachusetts 02061 USA

Distributors for all other countries:
Kluwer Academic Publishers Group
Distribution Centre
Post Office Box 322
3300 AH Dordrecht, THE NETHERLANDS

Library of Congress Cataloging-in-Publication Data

Karayiannis, N. B. (Nicolaos B.), 1960-
Artificial neural networks : learning algorithms, performance evaluation, and applications / N. B. Karayiannis, A. N. Venetsanopoulos.
p. cm. -- (The Kluwer international series in engineering and computer science ; SECS 0209)
Includes bibliographical references and index.
ISBN 0-7923-9297-3 (alk. paper)
1. Neural networks (Computer science) 2. Machine learning. 3. Algorithms. I. Venetsanopoulos, A. N. (Anastasios N.), 1941- . II. Title. III. Series.
QA76 . 87 . K37 1993
006 . 3--dc20 92-34614
CIP

Printed on acid-free paper.

Printed in the United States of America

To Vivian & Vaso

Ει γάρ ηδύνατο ὲκαστον τών οργάνων
κελευσθέν ή προαισθανόμενον
αποτελείν τό αυτού έργον
...
ουδέν άν έδει ούτε τοίς αρχιτέκτοσιν υπηρετών
ούτε τοίς δεσπόταις δούλων.

Αριστοτέλης, Πο*λιτικά* A4, 1253b34 - 1254a1.

If a tool could do its job
after obeying a command or its own feeling
...
neither the architects (experts) would require assistants
nor the masters slaves.

Aristotle, Politics A4, 1253b34 - 1254a1

Table of Contents

Acknowledgments xiii

1 Introduction 1

1.1 Overview 1

1.2 Book Organization 4

2 Neural Network Architectures and Learning Schemes 9

2.1 Introduction 9

2.2 Feed-forward Neural Networks 11

2.3 Feed-back Neural Networks 46

2.4 Self-organizing Neural Networks 67

2.5 Discussion 75

Appendix A2.1 76

Appendix A2.2 77

Appendix A2.3 79

Appendix A2.4 82

Appendix A2.5 84

3 ELEANNE: Efficient LEarning Algorithms for Neural NEtworks 87

3.1 Introduction 87

3.2 Recursive Least-squares Algorithms 90

3.3 Efficient Learning Algorithms for Single-layered Neural Networks 98

3.4 Efficient Learning Algorithms for Multi-layered Neural Networks 109

3.5 Computational Considerations 121

3.6 Experimental Results 126

3.7 Discussion 131
Appendix A3.1 132
Appendix A3.2 134
Appendix A3.3 135
Appendix A3.4 138

4 Fast Learning Algorithms for Neural Networks 141
4.1 Introduction 141
4.2 A Generalized Training Criterion 144
4.3 Fast Learning Algorithms for Single-layered Neural Networks 151
4.4 Fast Learning Algorithms for Multi-layered Neural Networks 161
4.5 Experimental Results 176
4.6 Discussion 184
Appendix A4.1 185
Appendix A4.2 187
Appendix A4.3 188
Appendix A4.4 192

5 ALADIN: Algorithms for Learning and Architecture DetermINation 195
5.1 Introduction 195
5.2 Training Criteria 198
5.3 Neural Networks with one Hidden Layer 199
5.4 Neural Networks with Multiple Hidden Layers . . 207
5.5 Experimental Results 209
5.6 Discussion 215
Appendix A5.1 216
Appendix A5.2 217

6 Performance Evaluation of Single-layered Neural Networks 219

6.1 Introduction 219

6.2 Optimal Least-squares Training of Single-layered Neural Networks 220

6.3 Capacity Considerations 234

6.4 Output Nonlinearities and Network Performance 247

6.5 Discussion 249

Appendix A6.1 250

Appendix A6.2 251

Appendix A6.3 254

Appendix A6.4 256

7 High-order Neural Networks and Networks with Composite Key Patterns 259

7.1 Introduction 259

7.2 High-order Neural Networks 260

7.3 Neural Networks with Composite Key Patterns . . 279

7.4 Capacity Considerations 287

7.5 Discussion 297

8 Applications of Neural Networks: A Case Study 299

8.1 Introduction 299

8.2 General Methodology for the Development of Neural Network Systems 300

8.3 Application of Neural Networks in Environmental Protection 303

8.4 Discussion 315

9 Applications of Neural Networks: A Review 317

9.1 Introduction 317

9.2 Optimization Problems 319

9.3 Image Compression 330

9.4 Recognition of Handwritten Signatures, Characters, and Digits 338

9.5 Text to Speech Conversion 347

9.6 Classification Applications 349

9.7 Medical Diagnosis 351

9.8 Prediction of Secondary Structures of Proteins . . 354

9.9 Weather Forecasting 358

9.10 Financial Predictions 360

9.11 Other Applications 368

10 Future Trends and Directions 371

References 375

Subject Index 413

Author Index 435

Acknowledgments

We are deeply indebted to the many people who have made the completion of this book possible. Dr. Nicholas DeClaris read an early version of the manuscript and motivated several improvements. Drs. Ben Jansen, Periklis Ktonas, Haluk Ogmen, and Erkki Oja carefully read portions of the manuscript and provided valuable suggestions. Particular thanks go to Buddy Marshall, who had the patience to read the entire manuscript. His interesting feed-back improved the manuscript in many ways. Demetri Androutsos, Gokalp Bayramoglu, Yezdi Kharas, Akbar H. Marand, Arun Padmanabhan, Pin-I Pai, Panagiotis Pandelis, and Costa S. Partsinevelos helped us by proofreading portions of the manuscript. Vivian Karayiannis, Arun Padmanabhan, Pin-I Pai, and Panagiotis Pandelis helped with the preparation of the final version of the manuscript. We would also like to thank Dr. Stuart Long, chairman of the Department of Electrical Engineering of the University of Houston, for providing the facilities used in the preparation of this book.

Finally, special thanks go to Vivian Karayiannis for spending the summer and several weekends at home. Her love, patience and encouragement made this book possible.

Nicolaos B. Karayiannis
Houston, Texas

Anastasios N. Venetsanopoulos
Toronto, Canada

ARTIFICIAL NEURAL NETWORKS
Learning Algorithms, Performance Evaluation, and Applications

Chapter 1

Introduction

1.1 Overview

We are living in a decade recently declared as the "Decade of the Brain". Neuroscientists may soon manage to work out a functional map of the brain, thanks to technologies that open windows on the mind. With the average human brain consisting of 15 billion neurons, roughly equal to the number of stars in our milky way, each receiving signals through as many as 10,000 synapses, it is quite a view. "The brain is the last and greatest biological frontier", says James Weston codiscoverer of DNA, considered to be the most complex piece of biological machinery on earth.

After many years of research by neuroanatomists and neurophysiologists, the overall organization of the brain is well understood, but many of its detailed neural mechanisms remain to be decoded. In order to understand the functioning of the brain, neurobiologists have taken a bottom-up approach of studying the stimulus-response characteristics of single neurons and networks of neurons, while psychologists have taken a top-down approach of studying brain functions from the cognitive and behavioral level. While these two approaches are gradually converging, it is generally accepted that it may take another fifty years before we achieve a solid microscopic, intermediate, and macroscopic understanding of brain.

Engineers and computer scientists, faced with the challenge of designing computational systems to solve practical problems, cannot wait that long. Over the last fifty years they have developed digital computers, which generally rely on a single central processing unit (CPU) to perform each processing task, one step at a time. More recently, they have focused on "parallel" processors (also called concurrent processors), which harness the power of numerous central processing units, operating simultaneously, to extend the capabilities of a single processor. However, parallel processing has become bogged down with the problem of communication among the individual units. Scientists have not yet discovered how to automate the partitioning task of real-world problems so that they can be processed in parallel.

Conventional digital computers are extremely well suited to executing sequences of instructions that have been precisely formulated for them. On the other hand, the human brain, which has taken millions of years to evolve to its present architecture, excels at such tasks as vision, speech, information retrieval, and complex spatial and temporal pattern recognition in the presence of noisy or distorted data, all of which are extremely difficult tasks for conventional computers.

How does the brain accomplish these tasks, given that its "processing elements" (neurons) are significantly slower than the processing elements of digital computers? The brain distributes its processing among billions of relatively simple neurons. It routinely handles communications among these neurons, each of which is constantly sending and receiving information. A number of brain mechanisms have been discovered, each one of which solves certain problems of simultaneous processing. These mechanisms include *association, generalization,* and *self-organization.* Neural networks in the brain process information according to noncomputational principles. They operate in continuous time as opposed to the discrete operations of digital computers.

Such observations led to the study and the development of artificial neurons and artificial neural networks. Instead of performing a program of instructions sequentially as in a Von Neumann computer, these networks were designed to explore many competing hypotheses simultaneously, using massively parallel architectures composed of many simple computational elements connected by links with variable weights.

In the 1940s McCulloch and Pitts showed that the neuron can be modeled as a single threshold device to perform logic functions (McCulloch and Pitts, 1943; Cowan, 1990). At about the same time, relationships between engineering principles, feedback, and brain function were considered by Wiener, while Ashly introduced his model of brain function and principle of homeostasis.

During the 1960s, more refined neuron models were introduced, including Rosenblatt's Perceptron, Widrow's Adaline (ADAptive LINear Element), and Steinbuck's learning matrix. The Perceptron attracted considerable attention, when it was first introduced, because of its conceptual simplicity (Rosenblatt, 1962). However, that excitement was short lived when Minsky and Papert proved that it can not be used for complex logic functions (Minsky and Papert, 1969). The fate of Adaline was different. Because of its linear and adaptive nature, it developed into a successful tool for adaptive signal processing, widely used in equalization, echo and noise cancellation, adaptive control and adaptive arrays (Widrow et al., 1975; Widrow and Stearns, 1985).

Today the term artificial neural network has come to mean any computing architecture that consists of massively parallel interconnection of simple neurons. The present impetus in neural network research is due to numerous contributions of Grossberg, Kohonen, Rumelhart, and Hopfield (Lippmann, 1987). Hopfield's contribution published in 1982 presented a model which consisted of a set of first-order (nonlinear) differential equations that minimize a certain "energy" function (Hopfield, 1982). Hopfield argued that there are many significant computational capabilities at the network level, which are nonexistent at the neuron level (Hopfield, 1982; Hopfield, 1984). The sensation generated by Hopfield's arguments was followed by the publication of the Error Back Propagation algorithm in 1986 by Rumelhart, Hinton, and Williams (Rumelhart et al., 1986). Although the invention of this algorithm is disputed by other researchers, Rumelhart et al. revitalized the neural network research by showing that the Error Back Propagation algorithm could train multi-layered neural networks to perform tasks where the Perceptron had failed.

Although neural networks can serve as models for further understanding brain functions, engineers are interested in neural networks for solving problems. As an engineering technique, neural networks can increase the speed of computation, because of their massively parallel nature. Due to their adaptive nature, they can adapt to changes in the data and learn the characteristics of input signals. Furthermore, due to their nonlinear nature, neural networks can perform filtering operations which are beyond the capabilities of conventional linear filtering techniques. Neural networks can be used for pattern classification by defining nonlinear regions in the feature space. They can also overcome some of the limitations of conventional computers and artificial intelligence (AI) due to their ability to learn and their parallel architecture.

Both the volume of publications and the number of professional society meetings focusing on neural networks are growing at an amazing rate. The neural network industry is expected to grow to some billions of dollars in annual U.S. sales by the year 2000 and then sustain an average annual growth rate of 40 % each year in the form of new software, chip designs, computer systems, and necessary support services. This monograph contributes to this goal by surveying the area, including theory and applications, and presenting a set of promising fast learning algorithms, essential for the efficient design of neural networks.

1.2 Book Organization

The recent interest in artificial neural networks has motivated the publication of numerous books, including selections of research papers and textbooks presenting the most popular neural architectures and learning schemes (Rumelhart and McClelland, 1986; Grossberg, 1987a; Grossberg, 1987b; Kohonen, 1987; Grossberg, 1988a; Anderson and Rosenfeld, 1988; Kohonen, 1989; Wasserman, 1989; Levine, 1990; Simpson, 1990; Hertz et al., 1991; Rogers and Kabrisky, 1991; Lau, 1992; Sanchez-Sinencio and Lau, 1992). This book presents some recent developments which can have a very significant impact on neural network research. The presentation of the material is based on the belief that the successful application of neural networks to real-world problems depends strongly on the knowledge of their learning properties and performance. Neural networks are introduced as trainable devices which have the unique ability to generalize. The pioneering work which appeared during the past decades is presented, together with the current developments in the field, through a comprehensive and unified review of the most popular neural network architectures and learning schemes, in addition to the selective review of the existing vast literature on artificial neural networks. This book can be read in different ways, depending on the background, specialization, and ultimate goals of the reader. A specialist will find well-defined and easily reproducible algorithms, which can perform the training of neural networks much faster than existing algorithms, along with the performance evaluation of various neural network architectures and training schemes. This book can also help a beginner interested in the development of neural network systems to acquire the necessary background in this field in an organized and comprehensive way.

Existing neural network architectures can be subdivided into three basic categories: feed-forward, feed-back, and self-organizing. Chapter 2 of this book presents several neural network architectures and learning schemes and serves as an introduction to the material presented in subsequent chapters. The neural network architectures and learning schemes presented in this chapter were selected on the basis of their popularity and their impact on neural network research. This chapter begins with the presentation of single- and multi-layered neural networks and their training. It also presents supervised learning algorithms based on the gradient descent method and several modified Hebbian rules proposed for the unsupervised training of single-layered feed-forward neural networks. Chapter 2 presents a general model for feed-back neural networks and focuses on specific feed-back models, with emphasis on the Hopfield model. It also describes alternative feed-back neural architectures and associated learning schemes, including Boltzmann machines and the Recurrent Back Propagation algorithm. Chapter 2 presents the Kohonen feature map and also describes a learning vector quantization algorithm proposed for fine tuning the map. Finally, this chapter presents an introduction to adaptive resonance theory (ART) and describes the ART 1 self-organizing neural architecture.

Among the recent developments in neural networks research contained in this book is a family of Efficient LEarning Algorithms for Neural NEtworks (ELEANNE) which can achieve much faster convergence than the existing learning algorithms. Chapter 3 presents the development of efficient learning algorithms for feed-forward neural networks of any architecture. It starts with the derivation of two recursive least-squares learning algorithms for single-layered neural networks. Chapter 3 presents the development of second-order leaning algorithms for single-layered neural networks with analog or binary output and discusses the development of learning algorithms for multi-layered neural networks. Finally, this chapter considers the computational complexity and presents experimental verification of the efficiency of the leaning algorithms proposed, as well as their comparison to algorithms available in the scientific literature.

Chapter 4 presents a generalized criterion for training neural networks. This criterion leads to a variety of fast learning algorithms for single- and multi-layered neural networks. The simplest algorithm devised on the basis of this generalized criterion is the Fast Delta Rule algorithm for single-layered neural networks. The gradient descent method in conjunction with this generalized criterion provides the Fast Back Propagation algorithm for multi-layered neural networks. Another set of fast algorithms with better convergence properties is derived on the basis of the same strategy that provided the family of efficient learning algorithms presented in chapter 3. Several experiments verify that the fast algorithms presented in this chapter perform the training of feed-forward neural networks much faster that the corresponding algorithms available in the literature.

Chapter 5 presents the development of a family of Algorithms for Learning and Architecture DetermINation (ALADIN). The learning algorithms presented in this chapter determine the minimal architecture of multi-layered feed-forward neural networks while performing their training by deactivating the redundant hidden units on the basis of a criterion measuring the effect of each hidden unit on the training process. In addition to the algorithms based on the quadratic error criterion frequently used for training neural networks, this chapter also presents the development of fast algorithms based on the generalized criterion which accelerates the training. It is experimentally verified that the training algorithms presented in this chapter select the simplest neural networks with the highest generalization ability.

Chapter 6 considers the performance of single-layered neural networks. This chapter begins with optimal least-squares training of single-layered neural networks with analog output. The outer-product rule is interpreted as an approximation of optimal least-squares training, and the efficiency of this approximation is analytically and experimentally evaluated. This chapter evaluates various capacity bounds proposed for single-layered neural networks trained using the outer-product rule and derives a capacity bound for optimally trained neural networks. The effect of the training scheme on the performance of single-layered neural networks is evaluated by experimentally comparing the performance of single-layered neural networks trained with respect to their analog and binary output.

Chapter 7 evaluates the efficiency of the outer-product rule, when it is applied to training neural networks of order higher than one. This evaluation is based on a comparison of the outer-product rule and optimal least-squares training. This analysis also reveals some important properties of high-order neural networks. The investigation of the architecture of higher-order neural networks leads to the development of neural networks with composite key patterns that are the essential generalization of high-order neural networks. The training, performance, and potential applications of neural networks with composite key patterns are also investigated.

Chapter 8 presents the general methodology for the development of neural network systems. This methodology provides the basis for the application of neural networks in decision making. The problem considered in this chapter is the evaluation of the relative contribution of various pollution sources to the pollution of a certain ecosystem.

Chapter 9 reviews applications of neural networks in optimization, image compression, character recognition, text to speech conversion, classification of sonar signals, medical diagnosis, prediction of secondary structures of proteins, weather forecasting, and financial predictions. Finally, this chapter presents a review of the extensive literature on neural network applications.

Future trends and directions are discussed in chapter 10 and an extensive list of references is presented at the end of the book.

Chapter 2

Neural Network Architectures and Learning Schemes

2.1 Introduction

Existing neural network architectures can be divided into three basic categories: *Feed-forward, Feed-back,* and *Self-organizing* neural networks. The most widely used neural architectures that can be classified into these three categories are shown in Figure 2.1. Although each of these categories is based on a different philosophy and obeys different principles, the characterization of a system by the general term "neural network" usually implies an ability to learn. *Learning* is the process by which a neural system acquires ability to carry out certain tasks by adjusting its internal parameters according to some learning scheme. Depending on the particular neural architecture considered, learning can be either *supervised* or *unsupervised.*

This chapter is organized as follows: Section 2.2 presents the most widely used feed-forward neural network architectures. This section describes the process of using artificial neurons as basic building elements for the development of single- and multi-layered, first and higher order neural networks. The learning schemes for feed-forward neural networks presented in this section include the Delta Rule based algorithms for single-layered neural networks and

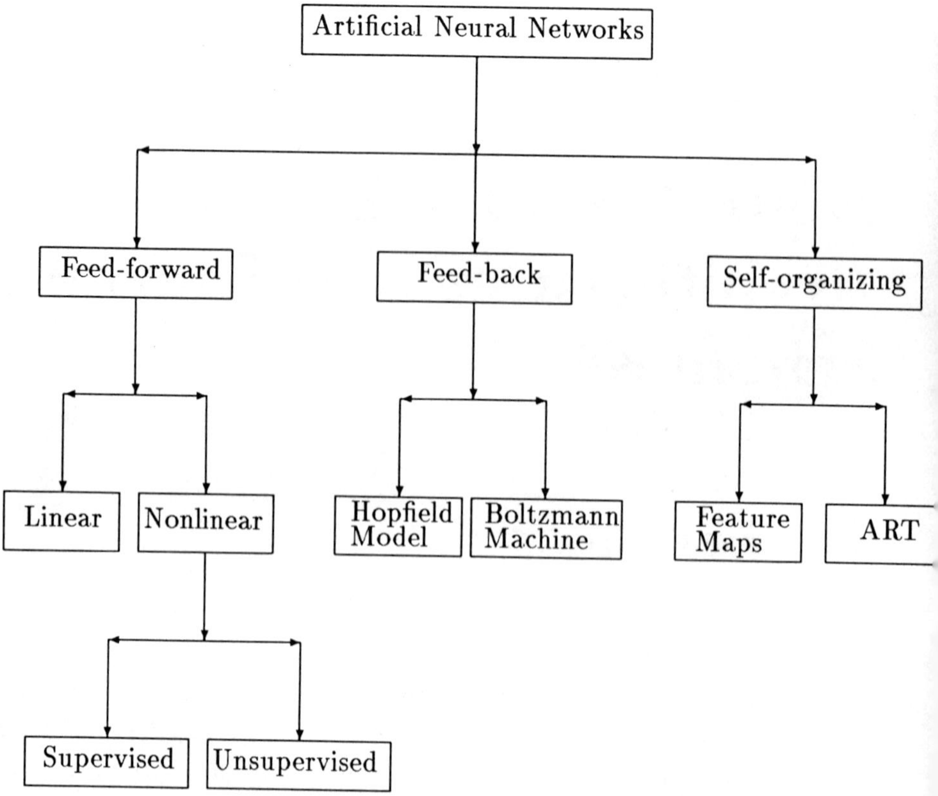

Figure 2.1: Neural network architectures

the Error Back Propagation algorithm for multi-layered neural networks. The remainder of this section presents unsupervised learning rules for single-layered neural networks which are based on the Hebbian learning principle. Section 2.3 presents some basic properties of feed-back neural networks, starting from a general feed-back neural network model proposed and studied by Cohen and Grossberg. Under certain simplifying assumptions, this model includes as special cases several feed-back neural network models which appeared independently in the literature, such as the McCulloch-Pitts model, the Brain-State-in-a-Box model, and the Hopfield model. In particular, this section focuses on the dynamics of the Hopfield model and its ability to function as a content addressable memory. This section also presents a learning algorithm for the Boltzmann machine, which is a special feed-back neural network architecture consisting of hidden and visible stochastic units. The remainder of this section presents the Recurrent Back Propagation algorithm, a gradient descent based algorithm for recurrent or feed-back neural networks. Section 2.4 presents Kohonen's feature maps. This class of pattern recognition systems differs from the ordinary neural network models but has some interesting self-organization properties. This section also presents a learning vector quantization (LVQ) algorithm, which is a supervised learning scheme used for the fine tuning of feature maps. The remainder of this section presents a general description of the adaptive resonance theory (ART) and some of the self-organizing neural network architectures based upon this theory.

2.2 Feed-forward Neural Networks

Feed-forward neural networks consist of one or more layers of nonlinear *processing elements* or *units*. The units belonging to neighboring layers are connected by sets of *synaptic weights*. Such neural architectures are called feed-forward since the output of each layer feeds the next layer of units. Feed-forward neural architectures appeared very early in the literature. The Perceptron (Rosenblatt, 1962) and the Adaline (Widrow, 1962; Widrow, 1987) are the earliest feed-forward neural architectures. Nevertheless, the potential of feed-forward neural networks was revealed quite recently, after

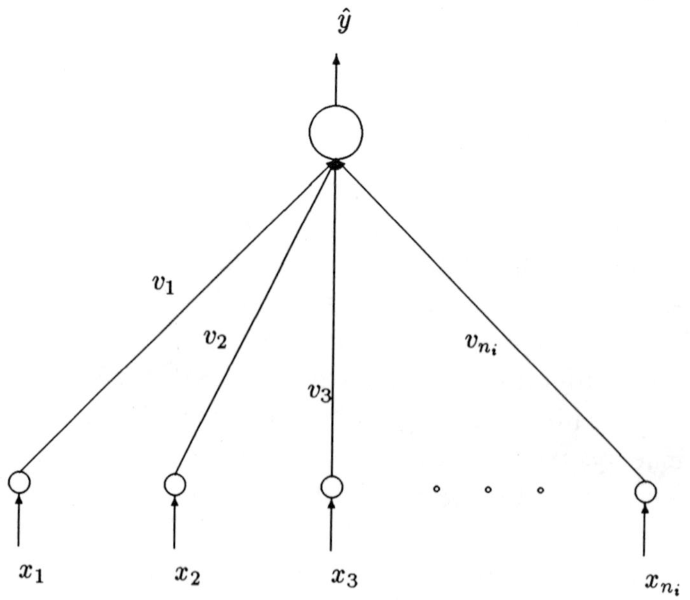

Figure 2.2: A single artificial neuron.

the appearance of multi-layered neural networks (Rumelhart et al., 1986; Lippmann, 1987; Widrow and Lehr, 1990). Such neural architectures include one or more layers of hidden units between the input and the output layer. A feed-forward neural network can be seen as a system transforming a set of input patterns into a set of output patterns. Such a network can be trained to provide a desired response to a given input. The network achieves such a behavior by adapting its synaptic weights during the learning phase on the basis of some learning rules. The training of feed-forward neural networks often requires the existence of a set of input and output patterns, called the *training set*. This kind of learning is called *supervised learning*.

2.2.1 Single Artificial Neurons

The function of a single artificial neuron is interpreted here as the equilibrium state of a dynamical model, in an attempt to establish a relationship between artificial and biological neurons (Kohonen, 1988). According to this interpretation, the activity of each neuron is determined by a set of continuous-valued input signals $x_i, i = 1, 2, \ldots, n_i$. These signals represent the momentary firing density of neural impulses delivered by another neuron to this input. It is assumed that $\hat{y}$ is the continuous-valued variable which represents the activity of the neuron. In addition, $\hat{y}$ is restricted to binary values which represent low or high activity. In order for this system to be consistent with the function of a biological neuron, it is assumed that the dynamic behavior of $\hat{y}$ is described by the following differential equation

$$\frac{d\hat{y}}{dt} = -\xi(\hat{y}) + I \tag{2.2.1}$$

In this model, I represents the integrated effect of all input currents to the cell membrane, while $\xi(.)$ is a nondecreasing nonlinear function. The term $-\xi(\hat{y})$ represents the decay of the neuron activity.

The simplest artificial neuron can be developed on the basis of the assumption that the input term I can be approximated by a linear function of the input signals $x_i, i = 1, 2, \ldots, n_i$, that is,

$$I = v_0 + \sum_{j=1}^{n_i} v_j x_j \tag{2.2.2}$$

where v_0 is a threshold and $v_j, j = 1, 2, \ldots, n_i$ are the synaptic weights which determine the effect of each particular input. If the input signals are held steady or are changing slowly, $\hat{y}$ approaches its asymptotic equilibrium, that is, $d\hat{y}/dt = 0$. At equilibrium, (2.2.1) provides $\xi(\hat{y}) = I$. Assuming that the input term I is given by (2.2.2),

$$\hat{y} = \sigma(v_0 + \sum_{j=1}^{n_i} v_j\, x_j) \tag{2.2.3}$$

where $\sigma(x) = \xi^{-1}(x)$. Figure 2.2 shows a typical representation of the single neuron described by (2.2.3). The function of this neuron

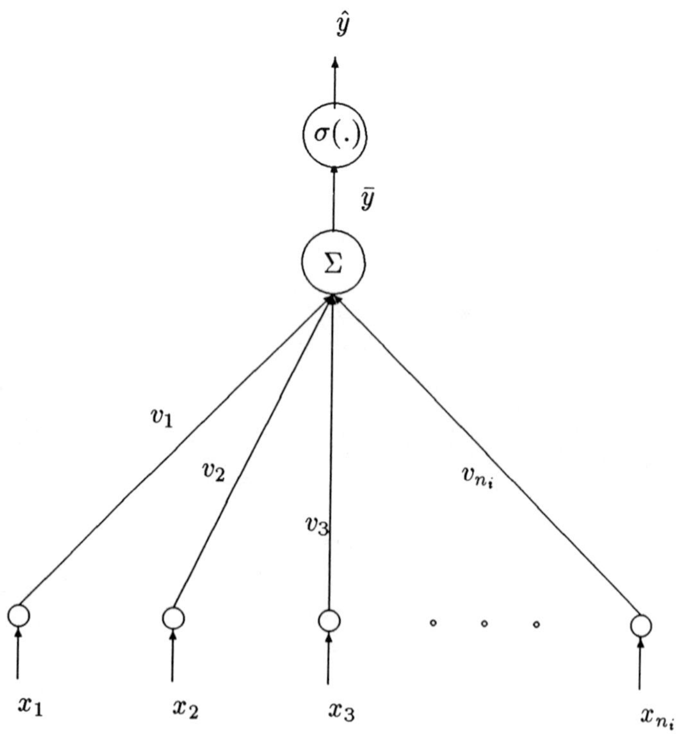

Figure 2.3: An equivalent representation of an artificial neuron.

is exhibited in Figure 2.3. Here the processing of the neuron is broken into two steps, i.e., a weighted sum of the inputs followed by the nonlinearity $\sigma(.)$. The result of the first stage $\bar{y}$, also called the *analog output,* is the input of the nonlinear function $\sigma(.)$, whose output $\hat{y}$ is the *binary output.* If $\sigma(x) = x$, then $\hat{y} = \bar{y}$ and the neuron is a linear neuron.

According to the previous discussion, the output variable $\hat{y}$ is binary. At equilibrium, $\hat{y}$ can be either $+1(1)$, which represents high activity, or $-1(0)$, which represents low activity. The output of each neuron could be determined by a hard-limiting nonlinearity. However, for analytical reasons, the hard-limiting nonlinearities are

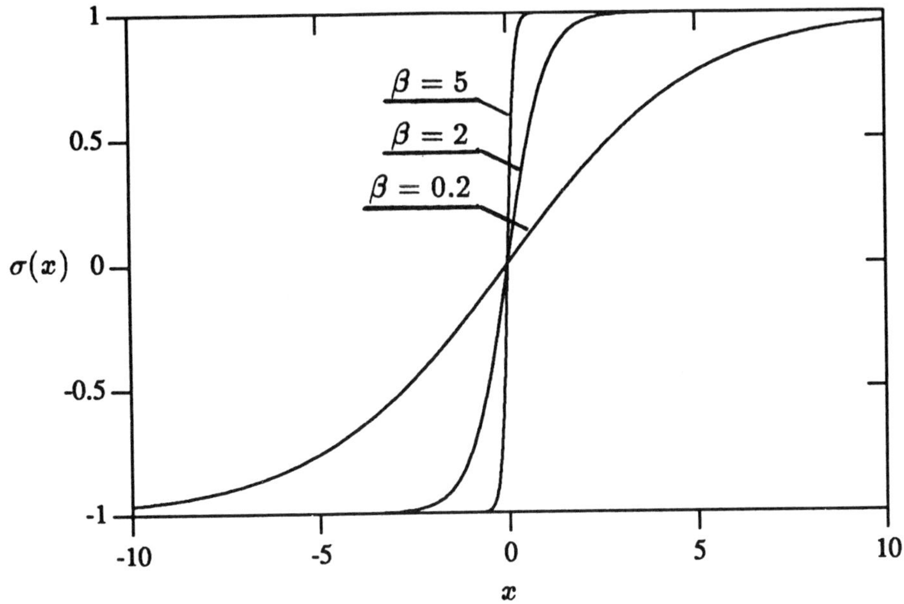

Figure 2.4: $\sigma(x) = \tanh(\beta x)$ for various values of the parameter β.

often approximated by soft-limiters. In the case that the two states of the neurons are $+1$ and -1, $\sigma(.)$ could be approximated by the hyperbolic tangent, that is,

$$\sigma(x) = \xi^{-1}(x) = \tanh(\beta x) \tag{2.2.4}$$

It can easily be verified that the corresponding function $\xi(.)$ which determines the decay of the neuron activity is given by

$$\xi(x) = -\frac{1}{2\beta}\ln(\frac{1-x}{1+x}) \tag{2.2.5}$$

Figures 2.4 and 2.5 show the pair of functions $\sigma(.)$ and $\xi(.)$, defined by (2.2.4) and (2.2.5), respectively, for different values of the parameter β. The form of the function $\xi(.)$, shown in Figure 2.5, reveals the role of the decay term in the time evolution of (2.2.1). If the activity $\hat{y}$ is in the neighborhood of zero, the decay term is negligible and the input term I becomes dominant. The activity $\hat{y}$ increases and approaches $+1$ if $I > 0$. Conversely, if $I < 0$ the activity $\hat{y}$ decreases and approaches -1. Since $\xi(.)$ is a nondecreasing

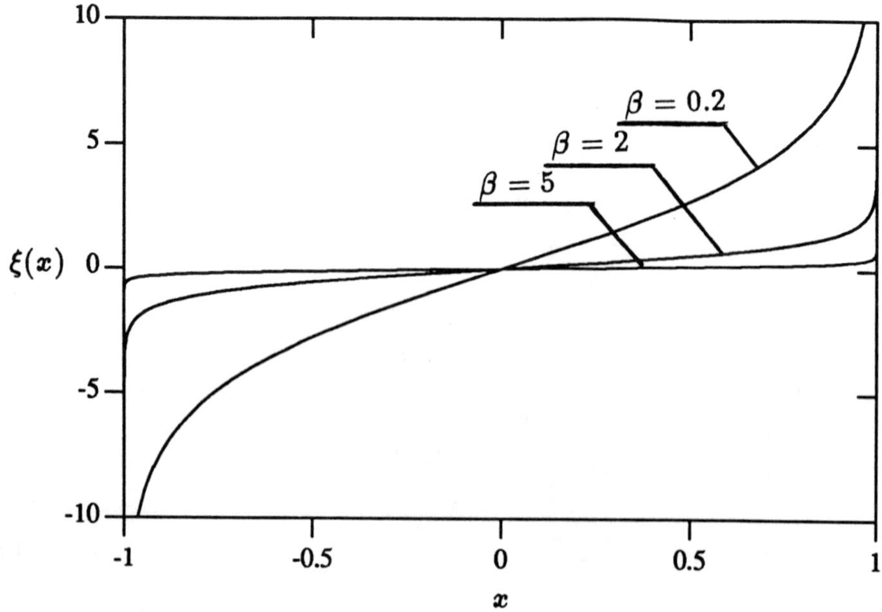

Figure 2.5: $\xi(x) = \sigma^{-1}(x)$ for various values of the parameter β.

function, the effect of the decay term $-\xi(\hat{y})$ becomes increasingly significant as $\hat{y}$ approaches $+1$ or -1. If $I > 0$, $\xi(\hat{y}) > 0$ and the decay term inhibits the convergence of $\hat{y}$ to $+1$. Conversely, if I, $\xi(\hat{y}) < 0$ and the decay term $-\xi(\hat{y})$ inhibits the convergence of $\hat{y}$ to -1. When $\xi(\hat{y}) = I$, (2.2.1) reaches an equilibrium state and $\hat{y}$ stops changing, i.e., $d\hat{y}/dt = 0$.

The assumption that the input term I is a linear function of the input signals leads to the simplest possible artificial neuron described by (2.2.3). An even more general class of artificial neurons can be developed on the basis of the assumption that the input term I is a nonlinear function of the inputs $x_i, i = 1, 2, \ldots, n_i$, that is,

$$I = v_0 + \sum_{j_1=1}^{n_i} v_{j_1} x_{j_1} + \sum_{j_1=1}^{n_i} \sum_{j_2=1}^{n_i} v_{j_1,j_2} x_{j_1} x_{j_2} + \ldots \qquad (2.2.6)$$

Under this assumption, the activity $\hat{y}$ of the neuron at equilibrium

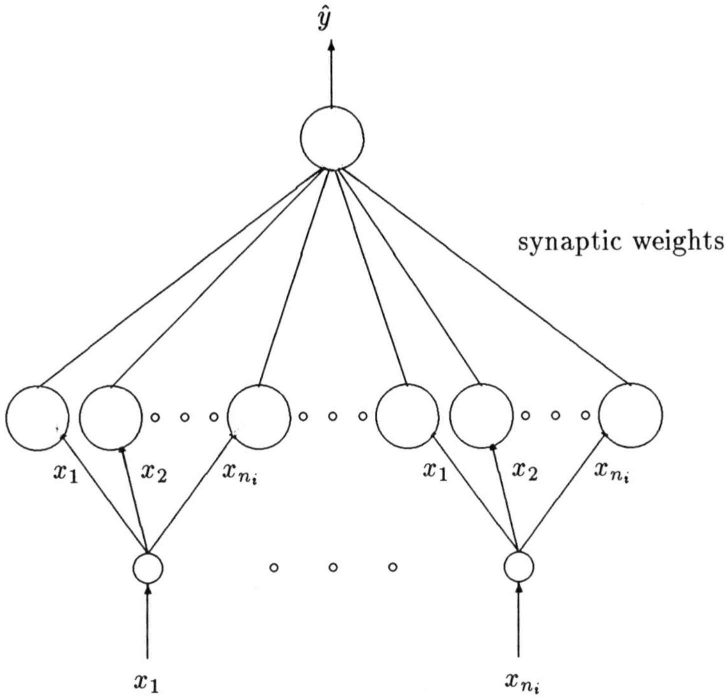

Figure 2.6: A neuron which is used as a building element of a neural network of second order.

is given by

$$\hat{y} = \sigma(v_0 + \sum_{j_1=1}^{n_i} v_{j_1} x_{j_1} + \sum_{j_1=1}^{n_i} \sum_{j_2=1}^{n_i} v_{j_1,j_2} x_{j_1} x_{j_2} + \ldots) \tag{2.2.7}$$

where $\sigma(x) = \xi^{-1}(x)$. Figure 2.6 shows a typical representation of the neuron described by (2.2.7) in the case where the input term I consists of only quadratic terms. The artificial neurons resulting from this assumption would be the basic building elements of neural networks of order higher than one. An interesting interpretation of such neurons results from the observation that the activity $\hat{y}$ of each neuron can be either on (+1) or off (−1). Therefore, a neuron can be seen as a classifier which assigns several sets of input signals to one of the two possible states. This interpretation establishes a relationship between neurons described by (2.2.3) and linear discriminant functions and neurons described by (2.2.7) and nonlinear discriminant functions (Psaltis and Park, 1986). The use of neurons described by (2.2.7) is justified by the existence of sets of input signals which are not linearly separable (Widrow and Lehr, 1990).

2.2.2 Single-layered Neural Networks

The neurons described above can be the basic building elements of certain artificial neural network architectures. Artificial neurons are referred to as *units*, in order to emphasize their participation in the formation of neural network architectures. Consider here the system consisting of n_o units and n_i inputs, illustrated in Figure 2.7. The input of the system is formed by the variables $x_j, j = 1, 2, \ldots, n_i$. Assume that the variables $\hat{y}_i, i = 1, 2, \ldots, n_o$ represent the activity of the n_o units used as building elements of the system. Once again, the dynamic behavior of each of these variables can be described by the following differential equation

$$\frac{d\hat{y}_i}{dt} = -\xi(\hat{y}_i) + I_i \tag{2.2.8}$$

Assuming that there is no interaction between the units, I_i represents the effect of the input signals to the ith unit. Assume, in addition, that I_i is a linear function of the input signals $x_j, j = 1, 2, \ldots, n_i$, that is,

$$I_i = w_{i0} + \sum_{j=1}^{n_i} w_{ij}\, x_j \tag{2.2.9}$$

where w_{ij} are the synaptic weights. Under these assumptions, the output activity of each unit is described by the differential equation

$$\frac{d\hat{y}_i}{dt} = -\xi(\hat{y}_i) + w_{i0} + \sum_{j=1}^{n_i} w_{ij}\, x_j \tag{2.2.10}$$

Assuming that the input signals change slowly or ideally remain constant, the system reaches its equilibrium, which corresponds to

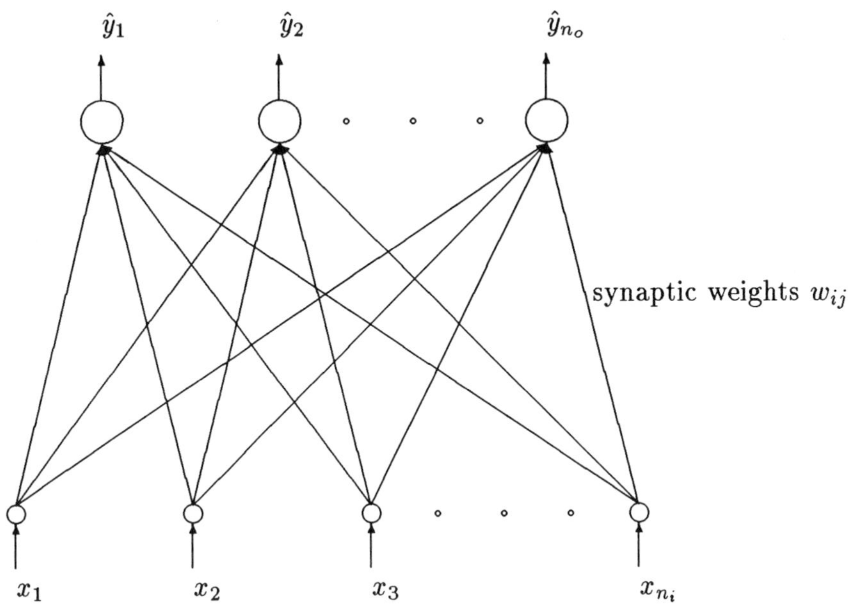

Figure 2.7: A single-layered neural network.

$d\hat{y}_i/dt = 0 \, \forall \, i = 1, 2, \ldots, n_o$. At equilibrium, the system of differential equations (2.1.10) can be written as

$$\hat{y}_i = \sigma(\sum_{j=0}^{n_i} w_{ij} \, x_j) \tag{2.2.11}$$

where $\sigma(x) = \xi^{-1}(x)$ and $x_0 = 1$.

The existence of units described by (2.2.7) suggests an approach for constructing alternative neural network architectures. Consider here a system of n_o units and assume once again that there is no interaction between the units. The dynamic behavior of the variables $\hat{y}_i, i = 1, 2, \ldots, n_o$ which represent the activity of the units can be described by (2.2.8). The input term I_i can be a nonlinear function of the input variables $x_i, i = 1, 2, \ldots, n_i$. Suppose, for simplicity, that I_i is a quadratic function of $x_i, i = 1, 2, \ldots, n_i$, that is,

$$I_i = \sum_{j_1=1}^{n_i} \sum_{j_2=1}^{n_i} w_{i,j_1,j_2}\, x_{j_1}\, x_{j_2} \tag{2.2.12}$$

Under this assumption, the output activity of each unit is described by the differential equation

$$\frac{d\hat{y}_i}{dt} = -\xi(\hat{y}_i) + \sum_{j_1=1}^{n_i} \sum_{j_2=1}^{n_i} w_{i,j_1,j_2}\, x_{j_1}\, x_{j_2} \tag{2.2.13}$$

When the system (2.2.13) reaches its equilibrium, i.e., $d\hat{y}_i/dt = 0$,

$$\hat{y}_i = \sigma\Big(\sum_{j_1=1}^{n_i} \sum_{j_2=1}^{n_i} w_{i,j_1,j_2}\, x_{j_1}\, x_{j_2}\Big) \tag{2.2.14}$$

The system developed on the basis of these assumptions is a neural network of second order, which is shown in Figure 2.8. A similar procedure provides neural networks of order higher than two. However, the definition of high-order neural networks varies in some other approaches (Chen et al., 1986). According to (2.2.14), the output activity of units used in forming a second-order neural network depends only on quadratic terms formed from pairs of the input variables $x_i, i = 1, 2, \ldots, n_i$. According to an alternative definition, a second-order neural network is built of units whose response depends on quadratic as well as linear terms (Chen et al., 1986). Nevertheless, the development of a second-order neural network using units with response given by (2.2.14) facilitates its training and its performance evaluation. There is evidence that the performance of a neural network improves as its order increases (Psaltis et al., 1988). However, the size of the network and time required for its training both increase significantly with order. For example, a second-order neural network with n_o output units which is fed by n_i input variables consists of $n_o \times n_i^2$ synaptic weights compared to $n_o \times n_i$ synaptic weights of the corresponding first-order neural network. Chapter 7 presents an extensive study of the properties, performance, and training of high-order neural networks. Among the most remarkable results is that the training of any neural network of order higher than one can be transformed to the training of a neural network of first order.

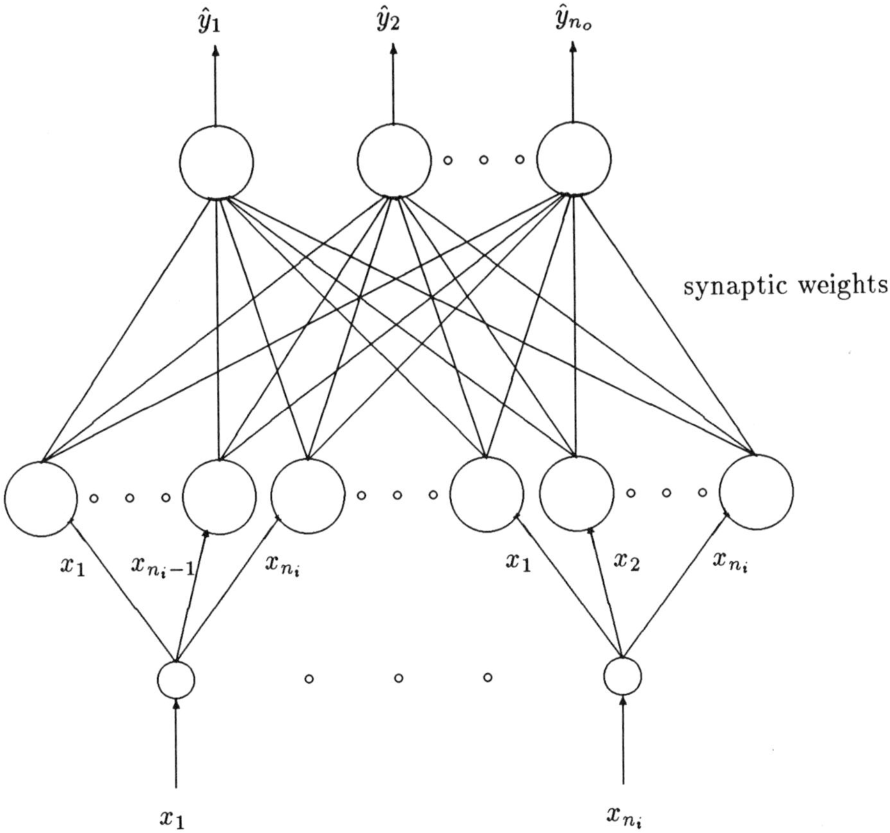

Figure 2.8: A neural network of second order.

2.2.3 Multi-layered Neural Networks

Early research in neural networks focused on simple neural networks with one layer of linear or nonlinear output units. However, it was realized very early that the complexity of real problems could hardly be met by such simple network architectures. Some researchers proposed more complicated architectures in an attempt to overcome the deficiencies of single-layered neural networks (Widrow and Lehr, 1990). Nevertheless, it was the introduction of multi-layered neural networks that gave new momentum to neural network research (Rumelhart et al., 1986). A multi-layered neural network contains

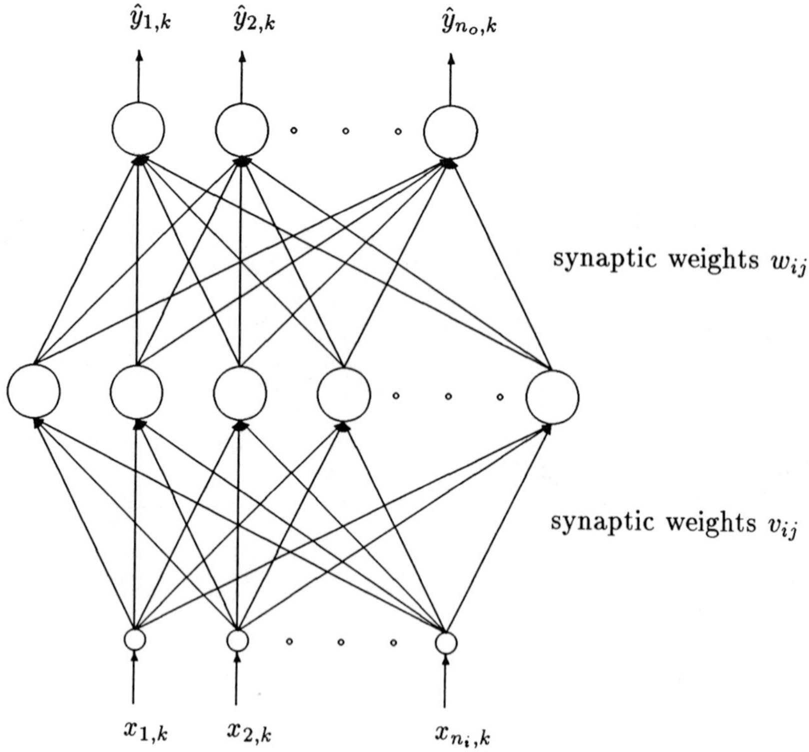

Figure 2.9: A neural network with one layer of hidden units.

one or more layers of *hidden* units between the input and output layers. A hidden unit may be connected to an input, an output unit, or a hidden unit belonging to another layer. Figure 2.9 shows a fully connected multi-layered neural network with one layer of hidden units. The hidden units are connected to the input units through the synaptic weights v_{pq}, which form the matrix $\mathbf{V}$, and to the output units through the synaptic weights w_{pq}, which form the matrix $\mathbf{W}$.

Consider the neural network with one layer of hidden units, shown in Figure 2.9. The input variables $x_i, i = 1, 2, \ldots, n_i$ provide the input of the n_h hidden units. The output activity $\hat{h}_j$ of each of the hidden units is described by the differential equation

$$\frac{d\hat{h}_j}{dt} = -\zeta(\hat{h}_j) + v_{j0} + \sum_{\ell=1}^{n_i} v_{j\ell}\, x_\ell \tag{2.2.15}$$

The hidden layer feeds the output layer of n_o units. The activity $\hat{y}_i$ of each output unit is determined by the differential equation

$$\frac{d\hat{y}_i}{dt} = -\xi(\hat{y}_i) + w_{i0} + \sum_{j=1}^{n_h} w_{ij}\, \hat{h}_j \tag{2.2.16}$$

The activity of each output unit depends on the response of the system of n_h differential equations defined by (2.2.16), in which the inputs are the outputs of the hidden layer. Therefore, it is necessary that the system (2.2.15) reaches its equilibrium faster than the system (2.2.16). In other words, it is required that the units belonging to the hidden layer reach their equilibrium state faster than the units which belong to the output layer. The nonlinearity $\zeta(.)$ determines how fast (2.2.15) reaches its equilibrium. Assuming that $\zeta^{-1}(x) = \tanh(\beta x)$, the shape of the nonlinearity $\zeta(.)$ depends on the parameter β. As β increases, $\zeta(.)$ approaches asymptotically a hard-limiting nonlinearity. If x is in the neighborhood of zero, $\zeta(x)$ can be approximated as $\zeta(x) \approx x/\beta$. If β is large and $\hat{h}_j$ is around zero, $\zeta(\hat{h}_j)$ is close to zero and, therefore, the effect of $-\zeta(\hat{h}_j)$ in the time evolution of (2.2.15) is negligible. In fact, if β is large, $-\zeta(\hat{h}_j)$ starts to inhibit the convergence of $\hat{h}_j$ to one of the two states $+1$ or -1 only if $\hat{h}_j$ is sufficiently close to one of them. On the other hand, if β is small, $-\zeta(\hat{h}_j)$ starts to inhibit the convergence of $\hat{h}_j$ to $+1$ or -1 even if $\hat{h}_j$ is in the neighborhood of zero. This is an indication that the speed of convergence of $\hat{h}_j$ to $+1$ or -1 depends on the shape of $\zeta(.)$. The closer $\zeta(.)$ is to a hard-limiting nonlinearity, the faster is the convergence of each $\hat{h}_j$ to its equilibrium state. This is a verification that the units belonging to the hidden layer of a multi-layered neural network should be highly nonlinear.

Consider here that the neural system described by the differential equations (2.2.15) and (2.2.16) has reached equilibrium. According to (2.2.15), the response of each hidden unit at equilibrium is given by

$$\hat{h}_j = \rho(\sum_{\ell=0}^{n_i} v_{j\ell} x_\ell) \tag{2.2.17}$$

where $x_0 = 1$ and $\rho(x) = \zeta^{-1}(x)$. Similarly, (2.2.16) indicates that the response of each output unit at equilibrium is

$$\hat{y}_i = \sigma(\sum_{j=0}^{n_h} w_{ij}\, \hat{h}_j) \tag{2.2.18}$$

where $\hat{h}_0 = 1$ and $\sigma(x) = \xi^{-1}(x)$. A direct extension of the above discussion provides multi-layered neural networks with more than one layers of hidden units between the input and the output layers. Figure 2.10 shows a multi-layered neural network with L layers of hidden units.

A genuine multi-layered neural network consists of nonlinear hidden units, although the output units could be linear in some applications. This requirement can be justified by considering, as a counterexample, that the hidden units of the network shown in Figure 2.9 are linear. Consider that the input of the network is the vector $\mathbf{x}$. If the hidden units are assumed to be linear, the output of the hidden units form the vector $\hat{\mathbf{h}}^* = [\hat{h}_1\, \hat{h}_2\, \dots\, \hat{h}_{n_h}]$, which is given by

$$\hat{\mathbf{h}} = \mathbf{V}\mathbf{x} \tag{2.2.19}$$

The output of the hidden layer feeds the output of the network through the synaptic weights w_{pq}. The analog output of the network is the vector $\bar{\mathbf{y}}^* = [\bar{y}_1\, \bar{y}_2\, \dots\, \bar{y}_{n_o}]$, which is given by

$$\bar{\mathbf{y}} = \mathbf{W}\hat{\mathbf{h}} = \mathbf{W}(\mathbf{V}\mathbf{x}) = (\mathbf{W}\mathbf{V})\mathbf{x} \tag{2.2.20}$$

This last equation indicates that a multi-layered neural network with linear hidden units is in fact equivalent to a single-layered neural network. The multi-layered neural network considered here is equivalent to a single-layered neural network with matrix of synaptic weights $\mathbf{U} = \mathbf{W}\mathbf{V}$.

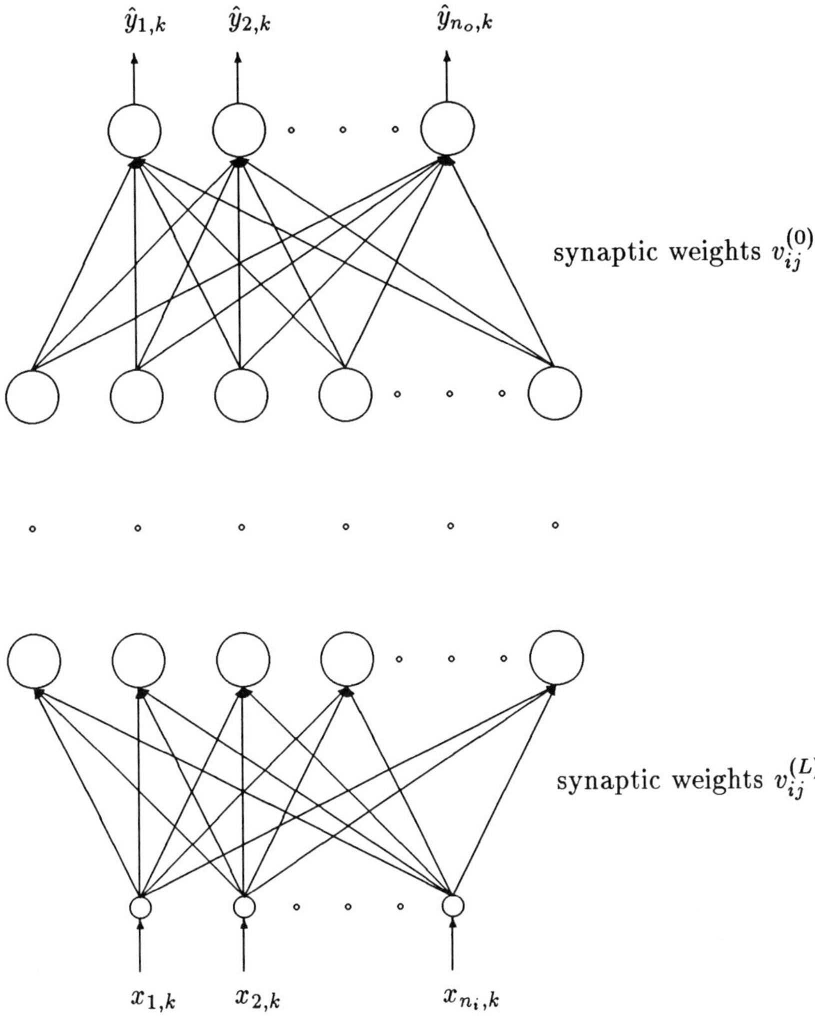

Figure 2.10: A neural network with L layers of hidden units.

2.2.4 Supervised Training of Feed-forward Neural Networks

A feed-forward neural network maps a set of input vectors to a set of output vectors. For a given input $\mathbf{x}$, the response of the network is given by

$$\hat{\mathbf{y}} = T(\mathbf{x}) \tag{2.2.21}$$

where $T(.)$ is, in general, a nonlinear operator which depends on the architecture of the neural network considered. A neural network can be trained to carry out a certain task by appropriately creating a *training set,* that is, a set of output-input vector pairs $(\mathbf{y}_k, \mathbf{x}_k), k = 1, 2, \ldots, m$, also called *associations.* The network is trained by adjusting the synaptic weights in such a way that when the input of the network is $\mathbf{x}_k$ its response is the corresponding output $\mathbf{y}_k$. For a given application, the training set is determined by an *external teacher.* This kind of learning is usually called *supervised learning.*

For a given set of synaptic weights, the response of the network to the input $\mathbf{x}_k$ is

$$\hat{\mathbf{y}}_k = T(\mathbf{x}_k) \tag{2.2.22}$$

The synaptic weights of the network can be determined during the training by minimizing a discrepancy measure $E_k = d(\mathbf{y}_k, \hat{\mathbf{y}}_k)$ between the desired output $\mathbf{y}_k$ and the corresponding actual response $\hat{\mathbf{y}}_k$ for all of the output-input pairs in the training set. This is usually achieved by minimizing

$$E = \sum_{k=1}^{m} E_k = \sum_{k=1}^{m} \sum_{i=1}^{n_o} \phi(y_{i,k} - \hat{y}_{i,k}) \tag{2.2.23}$$

where $\phi(.)$ is a positive definite, convex, continuous and differentiable everywhere function. If a network consists of n_o output units, the discrepancy between the expected output $\mathbf{y}_k$ and the estimate $\hat{\mathbf{y}}_k$ provided by the network is usually measured by the quadratic error

$$E_k = \frac{1}{2} \sum_{i=1}^{n_o} (y_{i,k} - \hat{y}_{i,k})^2 \tag{2.2.24}$$

Feed-forward neural networks have conventionally been trained by minimizing the quadratic error between the expected output and the estimate provided by the network. A feed-forward neural network trained by minimizing the quadratic error criterion approximates the Bayes optimal discriminant function (Ruck et al., 1990). If the output of the network is binary, the network can alternatively be trained by maximizing the relative entropy criterion (Solla et al., 1988; Baum and Wilczek, 1988; Hopfield, 1987). If the binary outputs of the network take the values ± 1, the relative entropy criterion used in (Solla et al., 1988) should be slightly modified. In this case, the network can be trained by maximizing the following function (Hertz et al., 1991)

$$E_k = -\frac{1}{2}\sum_{i=1}^{n_o}\left[(1-y_{i,k})\ln\frac{1-y_{i,k}}{1-\hat{y}_{i,k}} + (1+y_{i,k})\ln\frac{1+y_{i,k}}{1+\hat{y}_{i,k}}\right] \tag{2.2.25}$$

The training of a feed-forward neural network is usually performed in *adaptation cycles.* Each adaptation cycle involves the adaptation of all the synaptic weights of the network with respect to all the associations which form the training set. During each adaptation cycle, the synaptic weights are updated by minimizing $E_k \,\forall\, k = 1, 2, \ldots, m$.

2.2.5 Learning Algorithms for Single-Layered Neural Networks

Consider the single-layered neural network shown in Figure 2.7. Assuming that a particular pattern $\mathbf{x}_k$ is the input to the network, the corresponding analog output of the network consists of the elements $\bar{y}_{i,k} = w_{i0} + \sum_{j=1}^{n_i} w_{ij}\, x_{j,k} \,\forall\, i = 1, 2, \ldots, n_o$. Ideally, the binary output of the network is formed by the elements

$$y^o_{i,k} = \text{sgn}(\bar{y}_{i,k}) = \text{sgn}(\sum_{j=0}^{n_i} w_{ij}\, x_{j,k}) \,\forall\, i = 1, 2, \ldots, n_o \tag{2.2.26}$$

where $x_{0,k} = 1 \,\forall\, k = 1, 2, \ldots, m$ and sgn(.) is the signum function, defined as $\text{sgn}(x) = -1$ if $x < 0$ and $\text{sgn}(x) = +1$ if $x \geq 0$.

The training of single-layered neural networks is based here on the minimization of E, given by (2.2.23), where $y_{i,k}, i = 1, 2, \ldots, n_o$ are the elements of the expected output $\mathbf{y}_k$ and $\hat{y}_{i,k}, i = 1, 2, \ldots, n_o$ the corresponding estimates provided by the network, given by

$$\hat{y}_{i,k} = \sigma(\bar{y}_{i,k}) = \sigma(\sum_{j=0}^{n_i} w_{ij}\, x_{j,k}) \,\forall\, i = 1, 2, \ldots, n_o \tag{2.2.27}$$

where $x_{0,k} = 1 \,\forall\, k = 1, 2, \ldots, m$ and $\sigma(.)$ is a continuous and differentiable everywhere function. If the output of the network is analog, $\sigma(x) = x$. If the output of the network consists of binary elements ± 1, $\sigma(x) = \tanh(x)$.

Learning Based on Least-squares Techniques

Consider here the training of a single-layered neural network with respect to its analog output, consisting of the elements $\bar{y}_{i,k} = \mathbf{x}_k^* \mathbf{w}_i = \mathbf{w}_i^* \mathbf{x}_k$. In the case where $\phi(x) = \frac{1}{2}x^2$, the synaptic weights of the network can be determined on the basis of well-known least-squares techniques. The optimal estimate of the rows $\mathbf{w}_i$ of the matrix of synaptic weights $\mathbf{W}$ can be determined by

$$\frac{\partial E}{\partial \mathbf{w}_i} = -\sum_{k=1}^{m} \mathbf{x}_k\, (y_{i,k} - \mathbf{x}_k^*\, \mathbf{w}_i) = 0 \,\forall\, i = 1, 2, \ldots, n_o \tag{2.2.28}$$

It can easily be verified that $\mathbf{w}_i$ is the solution of the set of linear equations

$$\mathbf{X}_m\, \mathbf{X}_m^*\, \mathbf{w}_i = \mathbf{X}_m\, \mathbf{y}_{i,m} \,\forall\, i = 1, 2, \ldots, n_o \tag{2.2.29}$$

where $\mathbf{X}_m$ is the matrix with column vectors $\mathbf{x}_k, k = 1, 2, \ldots, m$, i.e., $\mathbf{X}_m = [\mathbf{x}_1\, \mathbf{x}_2\, \ldots\, \mathbf{x}_m]$ and $\mathbf{y}_{i,m}^* = [y_{i,1}\, y_{i,2}\, \ldots\, y_{i,m}]$. The combination of the n_o equations defined by (2.2.29) provides

$$\mathbf{W}\, \mathbf{X}_m\, \mathbf{X}_m^* = \mathbf{Y}_m\, \mathbf{X}_m^* \tag{2.2.30}$$

where $\mathbf{Y}_m$ is the matrix with column vectors $\mathbf{y}_k, k = 1, 2, \ldots, m$, that is, $\mathbf{Y}_m = [\mathbf{y}_1\, \mathbf{y}_2\, \ldots\, \mathbf{y}_m]$. The optimal solution of (2.2.30) in the least-squares sense is

$$\mathbf{W} = \mathbf{Y}_m\, \mathbf{X}_m^*\, (\mathbf{X}_m\, \mathbf{X}_m^*)^+ \tag{2.2.31}$$

where $\mathbf{A}^+$ denotes the generalized inverse of any rectangular matrix $\mathbf{A}$ (Rao and Mitra, 1971; Albert, 1972). The same estimate for the matrix of synaptic weights was derived by Kohonen in the early seventies (Kohonen and Ruohonen, 1973; Kohonen, 1974). In fact, Kohonen argued that a linear single-layered neural network can be trained with respect to the associations $(\mathbf{y}_k, \mathbf{x}_k), k = 1, 2, \ldots, m$ by solving the matrix equation

$$\mathbf{Y}_m = \mathbf{W}\,\mathbf{X}_m \tag{2.2.32}$$

The optimal solution of (2.2.32) in the least-squares sense is given by

$$\begin{aligned} \mathbf{W} &= \mathbf{Y}_m\,\mathbf{X}_m^+ \\ &= \mathbf{Y}_m\,(\mathbf{X}_m^*\,\mathbf{X}_m)^+\,\mathbf{X}_m^* = \mathbf{Y}_m\,\mathbf{X}_m^*\,(\mathbf{X}_m\,\mathbf{X}_m^*)^+ \end{aligned} \tag{2.2.33}$$

Chapter 3 presents the development of two recursive least-squares algorithms for the training of single-layered neural networks with respect to their analog output.

Consider now the assumption that the input vectors $\mathbf{x}_k, k = 1, 2, \ldots, m$ are uncorrelated, that is, $\mathbf{X}_m^*\,\mathbf{X}_m = n_i\,\mathbf{I}$. Under this assumption, the optimal estimate of the matrix of synaptic weights becomes

$$\mathbf{W} = \frac{1}{n_i}\,\mathbf{Y}_m\,\mathbf{X}_m^* \tag{2.2.34}$$

The learning scheme described by (2.2.34) is known in the literature as the *outer-product rule* or *correlational learning.* The outer-product rule is among the earliest learning schemes which appeared in the literature (Anderson, 1972; Nakano, 1972; Kohonen, 1972). However, the outer-product rule is not an optimal learning scheme in any sense. In fact, the major advantage of this scheme seems to be its simplicity. The performance of neural networks trained on the basis of this learning scheme is rather poor. Chapter 6 presents an extensive investigation of the relationship between the outer-product rule and optimal least-squares training.

Learning Algorithms Based on the Delta Rule

The Delta Rule, introduced by Widrow et al some decades ago, was one of the contributions which have strongly influenced neural network research (Widrow and Winter, 1988; Widrow et al., 1988; Widrow and Lehr, 1990). Although the Delta Rule is widely known from its application in adaptive filtering (Widrow et al., 1975; Widrow and Stearns, 1985), its simplicity and flexibility made it a particularly attractive tool for the training of neural networks (Widrow and Lehr, 1990). However, the learning algorithms based on the Delta Rule are characterized by slow convergence and, in some situations, can be trapped in local minima.

The training of single-layered neural networks is frequently based on the observation that the minimization of the objective function $E = \sum_{k=1}^{m} E_k$ can be performed by sequentially minimizing $E_k = \frac{1}{2}\sum_{i=1}^{n_o}(y_{i,k} - \hat{y}_{i,k})^2$ for $k = 1, 2, \ldots, m$, using the Delta Rule. It is shown in Appendix A2.1 that the synaptic weights w_{pq} can be updated by

$$\mathbf{w}_{p,k} = \mathbf{w}_{p,k-1} + \alpha\, \varepsilon_{p,k}^{o}\, \mathbf{x}_k \tag{2.2.35}$$

where α is a positive real number, called the *learning rate*, and

$$\varepsilon_{p,k}^{o} = \sigma'(\bar{y}_{p,k})\,(y_{p,k} - \hat{y}_{p,k}) \tag{2.2.36}$$

Consider that the network is trained with respect to its analog output, that is $\hat{y}_{i,k} = \bar{y}_{i,k}$. Under this assumption, the update equation is given by (2.2.35), where $\varepsilon_{i,k}^{o}$ can be obtained from (2.2.36) for $\sigma(x) = x$, that is,

$$\varepsilon_{i,k}^{o} = e_{i,k} = y_{i,k} - \hat{y}_{i,k} \tag{2.2.37}$$

The reduction of this error during the training of the network is revealed by the following analysis (Widrow et al., 1988; Widrow and Lehr, 1990). Assuming that $\hat{y}_{i,k} = \bar{y}_{i,k} = \mathbf{x}_k^{*}\,\mathbf{w}_i$, the difference between the error evaluated at $\mathbf{w}_i = \mathbf{w}_{i,k}$ and the error evaluated at $\mathbf{w}_i = \mathbf{w}_{i,k-1}$ is given by

$$\begin{aligned}\Delta\varepsilon_{i,k}^{o} &= (y_{i,k} - \mathbf{x}_k^{*}\,\mathbf{w}_{i,k}) - (y_{i,k} - \mathbf{x}_k^{*}\,\mathbf{w}_{i,k-1}) \\ &= -\mathbf{x}_k^{*}\,(\mathbf{w}_{i,k} - \mathbf{w}_{i,k-1})\end{aligned} \tag{2.2.38}$$

According to (2.2.35),

$$\mathbf{w}_{i,k} - \mathbf{w}_{i,k-1} = \alpha\, \varepsilon_{i,k}^{o}\, \mathbf{x}_k \tag{2.2.39}$$

The substitution of (2.2.39) in (2.2.38) provides

$$\Delta\varepsilon^o_{i,k} = -\alpha \, ||\mathbf{x}_k||^2 \, \varepsilon^o_{i,k} \tag{2.2.40}$$

This last equation indicates that the change of the error $\Delta\varepsilon^o_{i,k}$ is proportional to the error $\varepsilon^o_{i,k} = y_{i,k} - \hat{y}_{i,k}$. During the training of the network, the error $\varepsilon^o_{i,k}$ decreases after each adaptation cycle. Therefore, the reduction of the error during a certain adaptation cycle is smaller than the error reduction during the previous adaptation cycle. This behavior explains the low rate of convergence achieved by this algorithm, especially after the initial adaptation cycles.

This network could alternatively be trained with respect to the output of the soft-limiting nonlinearities. In this case, the update equation is given by (2.2.35), where $\varepsilon^o_{i,k}$ is obtained from (2.2.36) for $\sigma(x) = \tanh(x)$. It can easily be verified that $\sigma'(x) = 1 - \sigma(x)^2$. Since $\hat{y}_{i,k} = \sigma(\bar{y}_{i,k})$, (2.2.36) amounts to

$$\varepsilon^o_{i,k} = (1 - \hat{y}^2_{i,k})(y_{i,k} - \hat{y}_{i,k}) \tag{2.2.41}$$

If the network is trained with respect to its analog output, the adaptation of the synaptic weights which connect the input units and the ith output unit is determined by the corresponding error $e_{i,k} = y_{i,k} - \hat{y}_{i,k}$. In the case where the network is trained with respect to its binary output, the adaptation of the same synaptic weights is also affected by the term $1 - \hat{y}^2_{i,k} = (1 - \hat{y}_{i,k})(1 + \hat{y}_{i,k})$. This term depends on the particular soft-limiting nonlinearity which determines the function of the output units. The term $1 - \hat{y}^2_{i,k}$ approaches zero when $\hat{y}_{i,k}$ approaches its target $y_{i,k}$, which takes the values +1 or −1. On the other hand, this term attains its maximum value +1 when $\hat{y}_{i,k} = 0$, i.e., $\hat{y}_{i,k}$ is far away from its target. In conclusion, the effect of the error $e_{i,k} = y_{i,k} - \hat{y}_{i,k}$ on the adaptation of the corresponding synaptic weights is not uniform. In fact, the effect of the error $e_{i,k}$ is enhanced when the corresponding output is far away from its target and suppressed otherwise.

2.2.6 Learning Algorithms for Multi-layered Neural Networks

Consider the network with one layer of hidden units shown in Figure 2.9. Assuming that the pattern $\mathbf{x}_k$ is the input to the network, the corresponding outputs of the hidden units are

$$\hat{h}_{j,k} = \rho(\bar{h}_{j,k}) = \rho(\sum_{\ell=0}^{n_i} v_{j\ell}\, x_{\ell,k}) \tag{2.2.42}$$

where $x_{0,k} = 1 \,\forall\, k = 1, 2, \ldots, m$. Let $\hat{\mathbf{h}}_k$ be the $(n_h + 1) \times 1$ vector formed by the outputs of the hidden units (2.2.42) as follows

$$\hat{\mathbf{h}}_k^* = [\hat{h}_{0,k}\, \hat{h}_{1,k}\, \hat{h}_{2,k}\, \ldots\, \hat{h}_{n_h,k}] \tag{2.2.43}$$

where $\hat{h}_{0,k} = 1 \,\forall\, k = 1, 2, \ldots, m$. Clearly, $\hat{\mathbf{h}}_k$ is a function of the corresponding pattern $\mathbf{x}_k$ as well as the synaptic weights of the lower network v_{pq}. The synaptic weights v_{pq} are determined by the learning algorithm during the training process. Therefore, for each pair $(\mathbf{y}_k, \mathbf{x}_k)$ of output-input vectors, the learning algorithm creates a vector $\hat{\mathbf{h}}_k$, which is often called the *internal representation* of the pair $(\mathbf{y}_k, \mathbf{x}_k)$. The external teacher plays no role in the creation of these internal representations. In fact, the creation of the internal representations is the result of the distribution of the existing information over the synaptic weights of the network. The outputs of the network are given by

$$\hat{y}_{i,k} = \sigma(\bar{y}_{i,k}) = \sigma(\sum_{j=0}^{n_h} w_{ij}\, \hat{h}_{j,k}) \,\forall\, i = 1, 2, \ldots, n_o \tag{2.2.44}$$

Consider also the network with L layers of hidden units, shown in Figure 2.10. In this case, the estimate $\hat{\mathbf{y}}_k$ of the expected output pattern $\mathbf{y}_k$ is formed by the elements

$$\hat{y}_{i,k} = \sigma(\sum_{j=0}^{n_1} w_{ij}\, \hat{h}_{j,k}^{(1)}) \,\forall\, i = 1, 2, \ldots, n_o \tag{2.2.45}$$

The outputs of the units which belong to the hidden layers indexed by $r = 1, 2, \ldots, L$ are given by

$$\hat{h}_{j,k}^{(r)} = \rho(\sum_{\ell=0}^{n_{r+1}} v_{j\ell}^{(r)}\, \hat{h}_{\ell,k}^{(r+1)}) \,\forall\, r = 1, 2, \ldots, L \tag{2.2.46}$$

where $\hat{h}_{\ell,k}^{(L+1)} = x_{\ell,k} \, \forall \, \ell = 0, 1, \ldots, n_i$.

A neural network with one or more layers of hidden units can be trained using the well-known Error Back Propagation algorithm (Werbos, 1974; Rumelhart et al., 1986). According to this algorithm, the synaptic weights of the network are updated during the training by propagating back the error between the expected output and the output provided by the network.

The Error Back Propagation Algorithm

This algorithm is derived in Appendix A2.2 by sequentially minimizing the objective function E_k, defined by (2.2.23), for $k = 1, 2, \ldots, m$. The update equation for the synaptic weights w_{pq} is obtained as follows

$$\mathbf{w}_{p,k} = \mathbf{w}_{p,k-1} + \alpha \, \varepsilon_{p,k}^{o} \, \hat{\mathbf{h}}_k \tag{2.2.47}$$

where α is the learning rate and

$$\varepsilon_{p,k}^{o} = -\sigma'(\bar{y}_{p,k}) \frac{\partial \phi(e_{p,k})}{\partial \hat{y}_{p,k}} \tag{2.2.48}$$

It is also shown in Appendix A2.2 that the synaptic weights v_{pq} can be updated through the equation

$$\mathbf{v}_{p,k} = \mathbf{v}_{p,k-1} + \alpha \, \varepsilon_{p,k}^{h} \, \mathbf{x}_k \tag{2.2.49}$$

where

$$\varepsilon_{p,k}^{h} = (1 - \hat{h}_{p,k}^2) \sum_{i=1}^{n_o} \varepsilon_{i,k}^{o} \, w_{ip} \tag{2.2.50}$$

If the output of the network is analog, $\hat{y}_{i,k} = \sigma(\bar{y}_{i,k})$, where $\sigma(x) = x$. In this case, the network is trained by minimizing the quadratic error function (2.2.24) which corresponds to $\phi(e_{i,k}) = \frac{1}{2} e_{i,k}^2$. For this error criterion,

$$\frac{\partial \phi(e_{i,k})}{\partial \hat{y}_{i,k}} = \frac{\partial \phi(e_{i,k})}{\partial e_{i,k}} \frac{\partial e_{i,k}}{\partial \hat{y}_{i,k}} = -(y_{i,k} - \hat{y}_{i,k}) \tag{2.2.51}$$

Since $\sigma'(x) = 1$, (2.2.48) gives

$$\varepsilon_{i,k}^{o} = y_{i,k} - \hat{y}_{i,k} \tag{2.2.52}$$

If the output of the network is binary, $\hat{y}_{i,k} = \sigma(\bar{y}_{i,k})$, where $\sigma(x) = \tanh(x)$. In this case, $\sigma'(x) = 1 - \sigma(x)^2$. If the network is trained by minimizing the quadratic error (2.2.24), $\phi(e_{i,k}) = \frac{1}{2}\, e_{i,k}^2$ and $d\phi(e_{i,k})/d\hat{y}_{i,k} = -(y_{i,k} - \hat{y}_{i,k})$. In this case, (2.2.48) becomes

$$\varepsilon_{i,k}^{o} = (1 - \hat{y}_{i,k}^2)(y_{i,k} - \hat{y}_{i,k}) \tag{2.2.53}$$

If a network with binary output is trained by maximizing the relative entropy criterion,

$$\phi(e_{i,k}) = \frac{1}{2}\left[(1 - y_{i,k}) \ln \frac{1 - y_{i,k}}{1 - \hat{y}_{i,k}} + (1 + y_{i,k}) \ln \frac{1 + y_{i,k}}{1 + \hat{y}_{i,k}}\right] \tag{2.2.54}$$

For this error measure,

$$\frac{\partial \phi(e_{i,k})}{\partial \hat{y}_{i,k}} = \frac{1}{2}\left(\frac{1 - y_{i,k}}{1 - \hat{y}_{i,k}} - \frac{1 + y_{i,k}}{1 + \hat{y}_{i,k}}\right) = -\frac{y_{i,k} - \hat{y}_{i,k}}{1 - \hat{y}_{i,k}^2} \tag{2.2.55}$$

In this case, $\varepsilon_{i,k}^{o}$ is given by

$$\varepsilon_{i,k}^{o} = y_{i,k} - \hat{y}_{i,k} \tag{2.2.56}$$

The Error Back Propagation algorithm can be generalized for the training of neural networks with more than one layer of hidden units, such as the network shown in Figure 2.10. The update equations for the synaptic weights of this network are obtained in Appendix A2.2 as

$$\mathbf{v}_{p,k}^{(r)} = \mathbf{v}_{p,k-1}^{(r)} + \alpha\, \varepsilon_{p,k}^{(r)}\, \hat{\mathbf{h}}_k^{(r+1)}\; \forall r = 0, 1, \ldots, L \tag{2.2.57}$$

where $\hat{\mathbf{h}}_k^{(L+1)} = \mathbf{x}_k$, $\varepsilon_{p,k}^{(0)} = \varepsilon_{p,k}^{o}$ is given above, and

$$\varepsilon_{p,k}^{(r)} = (1 - \hat{h}_{p,k}^{(r)\,2}) \sum_{\ell=1}^{n_{r-1}} \varepsilon_{\ell,k}^{(r-1)}\, v_{\ell p}^{(r-1)}\; \forall r = 1, 2, \ldots, L \tag{2.2.58}$$

The training of a multi-layered neural network begins from a random initial estimate of the synaptic weights. Usually, the initial set of synaptic weights is formed by small real numbers provided by a random number generator. The learning rate α is another parameter which must be chosen by the user. An adequate selection of the learning rate α is crucial for the learning speed that can be achieved by the algorithm. A very small α may slow down the convergence rate of the algorithm. On the other hand, a relatively large α may force the algorithm to oscillate between two points of the parameter space and never reach a minimum. The input vectors can be either binary or analog. It is suggested that the elements of an analog input be normalized. Usually, the elements of the input vectors are normalized between 0 and 1 or between -1 and $+1$. The normalization of the input overcomes numerical problems that could possibly arise otherwise.

Among the first experiments used for the justification of the Error Back Propagation algorithm was the training of multi-layered feed-forward neural networks to realize the Exclusive OR function, which is a well-known linearly nonseparable problem (Hertz et al., 1991). The n to n encoder and the parity problems were also used to exhibit the capabilities of the Error Back Propagation algorithm (Rumelhart et al., 1986). These studies provided valid and convincing arguments against the criticism of the Perceptrons and the resulting suspicion about the viability of neural network research. However, the use of the Error Back Propagation algorithm in more complicated real-world problems revealed some of its disadvantages. One of the major problems revealed by the use of the Error Back Propagation algorithm in large-scale problems is its slow convergence and the time required for training. The possibility that this gradient descent based algorithm may be trapped in local minima is another serious disadvantage, which also relates to the performance of the trained network. The *generalization ability* of feed-forward neural networks trained by the Error Back Propagation algorithm is another issue of great concern in real world applications. Generalization ability is the terminology used in the literature to describe the property of feed-forward neural networks trained on a finite set of input-output associations to provide "reasonable" responses to inputs not contained in the training set. The ability of a feed-forward neural network to generalize also depends on its size and architecture, that is, the number of hidden layers and the number of units per layer. Most of the problems associated with the training of feed-forward neural networks are addressed in the chapters which follow. Chapter 3 presents learning algorithms which achieve higher convergence rates than the Error Back Propagation algorithm. Chapter 4 presents fast learning algorithms based on a generalized criterion for training. Chapter 5 presents learning algorithms capable of determining the architecture of a feed-forward neural network while performing its training.

Alternative Learning Algorithms

The gradient descent based minimization of an error function is not the only possible path towards the development of learning algorithms for feed-forward neural networks. Motivated by different shortcomings of the Error Back Propagation algorithm, several researchers attempted the development of alternative learning algorithms for feed-forward neural networks. A successful alternative to the Error Back Propagation algorithm must simultaneously satisfy several, often conflicting, requirements. These requirements include the reduction of the time and computational requirements required for training, the guarantee that the algorithm will not be trapped in local minima, and the ability of the trained network to achieve better generalization. An alternative to the Error Back Propagation algorithm was based on the combination of gradient descent based minimization and linear programming (Shawe-Taylor and Cohen, 1990). This algorithm avoids a wrong direction in the weight space by solving a linear program at each adaptation cycle. Despite its complexity, this algorithm may also be trapped in local minima. Also, the application of this algorithm to large-scale problems is restricted by the excessive number of arithmetic operations performed during each adaptation cycle. The internal representations are created by the Error Back Propagation algorithm during the training of the network. Krogh et al. introduced a cost function for learning in feed-forward neural networks which is an explicit function of the internal representations in addition to the synaptic weights (Krogh et al., 1990). The learning problem is formulated according to this approach as a training of two single-layered neural networks and a search for internal representations. The Error Back Propagation algorithm can be interpreted as a limiting case of this algorithm.

Krogh et al. indicated that the algorithm based on the minimization of this cost function finds successful solutions more often than the Error Back Propagation algorithm (Krogh et al., 1990). Grossman et al. proposed a learning algorithm based on the choice of internal representations for a network with a single output unit and one layer of binary hidden units (Grossman et al., 1988). Instead of minimizing an error function, this algorithm is based on an efficient search in the space of internal representations. When a correct set of internal representations is found, the synaptic weights are evaluated by a local Perceptron learning rule. This learning scheme is useful in situations where the external teacher has some information about the desired internal representations. Another advantage is that the algorithm may be implemented using only integer-valued weights. However, during the learning phase this algorithm requires the storage of the internal representations of all training patterns, and the extension of this algorithm to the training of networks with more than one output unit is not a trivial task. Engel argued that the rounding of the real-valued synaptic weights provided by the Error Back Propagation algorithm could degrade the performance of the trained network (Engel, 1988). On the basis of this argument, he applied simulated annealing to the problem of training feed-forward neural networks with discrete-valued synaptic weights. Although this algorithm works well for simple tasks such as the Exclusive OR problem, the performance of the algorithm in large-scale problems and the generalization of the networks trained by this algorithm is an open problem (Engel, 1988).

2.2.7 Unsupervised Training of Feed-forward Neural Networks

Feed-forward neural networks are usually trained by an external teacher. In fact, the popularity of the Error Back Propagation algorithm made feed-forward neural networks almost synonymous with supervised learning. In supervised learning the network separates the input parameter space on the basis of some examples or associations presented to it. In unsupervised learning there is no external teacher. A feed-forward neural network may be trained without a teacher according to some learning rule which imposes a certain condition on its output. Feed-forward neural networks trained without a teacher may measure the correlation of the input data, identify certain features, or perform principal component analysis (Hertz et al., 1991). The unsupervised training of feed-forward neural networks was extensively studied during the past decade (Linsker, 1986a; Linsker, 1986b, Linsker, 1986c, Linsker, 1988; Oja, 1982; Oja, 1983; Oja, 1989; Yuille et al., 1989; Sanger, 1989a; Sanger, 1989b; Sanger, 1990; Banzhaf and Haken, 1990). A common ingredient in all these studies is the Hebbian learning rule, which is presented here. The Hebbian learning rule is a biologically inspired scheme, which has strongly influenced unsupervised learning (Hebb, 1949; Anderson, 1983).

The presentation of the Hebbian rule begins with the single unit shown in Figures 2.2 and 2.3. It is initially assumed that the unit is linear. Under this assumption, the output of this unit is given in terms of the input vector $\mathbf{x}^* = [x_1\, x_2\, \dots\, x_{n_i}]$ and the vector of the synaptic weights $\mathbf{v}^* = [v_1\, v_2\, \dots\, v_{n_i}]$ by

$$\hat{y} = \bar{y} = \mathbf{v}^* \mathbf{x} = \mathbf{x}^* \mathbf{v} \tag{2.2.59}$$

According to the Hebbian rule, the change of each synaptic weight v_i is proportional to the product of its input x_i and its output $\hat{y}$. The behavior of the system in time may be studied by introducing the index k which represents discrete time. If the weight vector $\mathbf{v} = \mathbf{v}_{k-1}$ maps the input $\mathbf{x} = \mathbf{x}_k$ into the output $\hat{y} = \hat{y}_k$, the Hebbian rule provides

$$\Delta v_{i,k-1} = v_{i,k} - v_{i,k-1} = \alpha\, \hat{y}_k\, x_{i,k} \tag{2.2.60}$$

The update equation provided by the Hebbian rule may also be obtained by applying the gradient descent method to maximize $\hat{y}^2$ or, equivalently, to minimize $J(\mathbf{v}) = -\frac{1}{2}\hat{y}^2$. Clearly,

$$\Delta \mathbf{v}_{k-1} = -\alpha \frac{\partial J(\mathbf{v})}{\partial \mathbf{v}}\Big|_{\mathbf{v}=\mathbf{v}_{k-1}} = \alpha\, \hat{y}_k\, \mathbf{x}_k \tag{2.2.61}$$

On the average, the change of the vector of synaptic weights $\mathbf{v} = \mathbf{v}_{k-1}$ is given by

$$\begin{aligned} E\{\Delta \mathbf{v}\} &= \alpha\, E\{\mathbf{x}\, \hat{y}\} \\ &= \alpha\, E\{\mathbf{x}\, \mathbf{x}^*\, \mathbf{v}\} = \alpha\, E\{\mathbf{x}\, \mathbf{x}^*\}\, \mathbf{v} = \alpha\, \mathbf{C}\, \mathbf{v} \end{aligned} \tag{2.2.62}$$

Obviously, $E\{\Delta \mathbf{v}\}$ is proportional to the gradient of the energy function

$$\begin{aligned} E\{J\} &= E\{-\frac{1}{2}\hat{y}^2\} \\ &= E\{-\frac{1}{2}\mathbf{v}^*\mathbf{x}\mathbf{x}^*\mathbf{v}\} = -\frac{1}{2}\mathbf{v}^* E\{\mathbf{x}\mathbf{x}^*\}\mathbf{v} = -\frac{1}{2}\mathbf{v}^*\mathbf{C}\mathbf{v} \end{aligned} \tag{2.2.63}$$

The basic problem with the update equation (2.2.60) is that the synaptic weights grow without bound and the learning never stops. In fact, there are only unstable fixed points for the update equation based on the Hebbian rule (Hertz et al., 1991).

Oja overcame the problems associated with the Hebbian rule by considering the normalized version of the update equation (2.2.60), that is,

$$v_{i,k} = \frac{v_{i,k-1} + \alpha\, \hat{y}_k\, x_{i,k}}{[\sum_{i=1}^{n_i}(v_{i,k-1} + \alpha\, \hat{y}_k\, x_{i,k})^2]^{\frac{1}{2}}} \tag{2.2.64}$$

According to the normalized Hebbian rule (2.2.64), $\sum_{i=1}^{n_i} v_{i,k}^2 = 1$. The update equation (2.2.64) controls the growth of the synaptic weights, by forcing the Euclidean norm of the weight vector $\mathbf{v}$ to be equal to unity after each adaptation. Oja suggested that the normalized Hebbian rule (2.2.64) can be approximated by (Oja, 1982)

$$v_{i,k} = v_{i,k-1} + \alpha\, \hat{y}_k\,(x_{i,k} - \bar{y}_k\, v_{i,k-1}) \tag{2.2.65}$$

This update equation is known as the Oja's learning rule. It is shown in Appendix A2.3 that the update equation (2.2.65) can be obtained by expanding the normalized Hebbian rule (2.2.64) around α and using a series of approximations (Oja, 1982). The update rule (2.2.65) is strictly local, in the sense that the change of each synaptic weight v_i depends exclusively on the terms $\hat{y}\, x_i$ and $\hat{y}\, v_i$, which are available at this position. Oja's learning rule is closely related to Hebbian learning. Clearly, the term $\hat{y}\, x_i$ of Oja's update rule realizes Hebbian learning. The term $-\hat{y}^2\, v_i$ is a decay or forgetting term which guarantees the stability of the adaptation of $\mathbf{v}$.

The update equation (2.2.65) can be written in a vector form as

$$\mathbf{v}_k = \mathbf{v}_{k-1} + \alpha\, \hat{y}_k\,(\mathbf{x}_k - \bar{y}_k\, \mathbf{v}_{k-1}) \tag{2.2.66}$$

If the unit is linear, $\hat{y}_k = \bar{y}_k = \mathbf{v}_{k-1}^* \mathbf{x}_k$ and the update equation (2.2.66) can be written as

$$\begin{aligned} \Delta \mathbf{v}_{k-1} &= \alpha \, (\mathbf{x}_k - \mathbf{v}_{k-1} \, \mathbf{v}_{k-1}^* \, \mathbf{x}_k) \, \mathbf{x}_k^* \, \mathbf{v}_{k-1} \\ &= \alpha \, (\mathbf{I} - \mathbf{v}_{k-1} \, \mathbf{v}_{k-1}^*) \, \mathbf{x}_k \, \mathbf{x}_k^* \, \mathbf{v}_{k-1} \end{aligned} \tag{2.2.67}$$

The system under consideration reaches equilibrium when $E\{\Delta \mathbf{v}\} = 0$. At equilibrium, the vector of synaptic weights $\mathbf{v} = \mathbf{v}_{k-1}$ is obtained by

$$E\{\Delta \mathbf{v}\} = \alpha \, (\mathbf{I} - \mathbf{v} \, \mathbf{v}^*) \, E\{\mathbf{x} \, \mathbf{x}^*\} \, \mathbf{v} = 0 \tag{2.2.68}$$

If $\mathbf{C} = E\{\mathbf{x} \, \mathbf{x}^*\}$, (2.2.68) amounts to

$$E\{\Delta \mathbf{v}\} = \alpha \, [\mathbf{C} \, \mathbf{v} - (\mathbf{v}^* \, \mathbf{C} \, \mathbf{v}) \, \mathbf{v}] = 0 \tag{2.2.69}$$

or equivalently,

$$\mathbf{C} \, \mathbf{v} = \lambda \, \mathbf{v} \tag{2.2.70}$$

where

$$\lambda = \mathbf{v}^* \, \mathbf{C} \, \mathbf{v} \tag{2.2.71}$$

If $\mathbf{x}$ is zero-mean, $\mathbf{C}$ is the covariance matrix of the input data. According to (2.2.70), $\mathbf{v}$ is an eigenvector of the covariance matrix $\mathbf{C}$. In fact, $\mathbf{v}$ is the first principal component of $\mathbf{x}$, that is, the eigenvector of $\mathbf{C}$ which corresponds to its maximum eigenvalue (Hertz et al., 1991).

The existence of the Hebbian term in the update equation (2.2.66) indicates that Oja's rule determines the weight vector $\mathbf{v}$ which minimizes $E\{-\frac{1}{2}\hat{y}^2\}$ or, equivalently, maximizes $H = E\{\hat{y}^2\}$. On the other hand, the combination of (2.2.70) and (2.2.71) gives

$$\lambda = \mathbf{v}^* \, \lambda \, \mathbf{v} = \lambda \, \mathbf{v}^* \, \mathbf{v} = \lambda \, ||\mathbf{v}||^2 \tag{2.2.72}$$

This latter equation indicates that, when the system reaches equilibrium, the weight vector $\mathbf{v}$ satisfies the condition $||\mathbf{v}||^2 = 1$. In fact, Oja indicated that the weight vector provided by the update equation (2.2.66) maximizes (Oja, 1991)

$$H(\mathbf{v}) = E\{\hat{y}^2\} \tag{2.2.73}$$

under the constraint that

$$||\mathbf{v}||^2 = 1 \tag{2.2.74}$$

Oja proposed a similar learning rule that can be used for the unsupervised training of a single nonlinear unit (Oja, 1991; Oja et al., 1991). In this case, the output of the single unit shown in Figure 2.2 is given by

$$\hat{y} = \sigma(\bar{y}) = \sigma(\mathbf{v}^* \mathbf{x}) = \sigma(\mathbf{x}^* \mathbf{v}) \tag{2.2.75}$$

If the unit is nonlinear, the update equation (2.2.66) is modified as follows

$$\mathbf{v}_k = \mathbf{v}_{k-1} + \gamma\, \psi(\bar{y}_k)\, (\mathbf{x}_k - \bar{y}_k\, \mathbf{v}_{k-1}) \tag{2.2.76}$$

where $\bar{y}_k = \mathbf{v}_{k-1}^* \mathbf{x}_k$ and $\psi(.)$ is the *learning function*, defined in terms of the output nonlinearity $\sigma(.)$ as follows

$$\psi(z) = \frac{d\sigma^2(z)}{dz} \tag{2.2.77}$$

The form of the learning function (2.2.77) can be obtained by determining the update equation provided by the Hebbian rule in the case where the output unit is nonlinear. This can be achieved by using the gradient descent method to minimize $J(\mathbf{v}) = -\frac{1}{2}\,\hat{y}^2$, where $\hat{y} = \sigma(\bar{y}) = \sigma(\mathbf{v}^* \mathbf{x})$. Clearly,

$$\begin{aligned} \Delta \mathbf{v}_{k-1} &= -\alpha \frac{\partial J(\mathbf{v})}{\partial \mathbf{v}}\Big|_{\mathbf{v}=\mathbf{v}_{k-1}} \\ &= \frac{\alpha}{2} \frac{\partial \sigma^2(\bar{y}_k)}{\partial \mathbf{v}_{k-1}} = \frac{\alpha}{2} \frac{d\sigma^2(\bar{y}_k)}{d(\bar{y}_k)} \frac{\partial(\mathbf{v}_{k-1}^* \mathbf{x}_k)}{\partial \mathbf{v}_{k-1}} = \frac{\alpha}{2}\, \psi(\bar{y}_k)\, \mathbf{x}_k \end{aligned} \tag{2.2.78}$$

where $\psi(.)$ is given by (2.2.77). The minimization of $J(\mathbf{v})$ under the constraint (2.2.74) results in the learning rule (2.2.76).

The definition of the learning function (2.2.77) implies that $\sigma^2(.)$ must be a continuous and differentiable everywhere function. Nevertheless, Oja et al. provided recently a list of possible output nonlinearities that are not necessarily continuous and their corresponding learning functions are not necessarily monotone. The choice of these learning functions was restricted only by the requirement that a learning function must be positive for positive values of the argument and negative for negative values (Oja et al., 1991). Consider here that $\sigma(.)$ is the sigmoid nonlinearity $\sigma(x) = \tanh(x)$, which is extensively used in this chapter. In this case, $\psi(z) = 2\,\sigma(z)(1-\sigma^2(z))$. Since $\hat{y}_k = \sigma(\bar{y}_k)$, $\psi(\bar{y}_k) = 2\,\hat{y}_k\,(1-\hat{y}_k^2)$, and the update equation (2.2.76) becomes

$$\mathbf{v}_k = \mathbf{v}_{k-1} + \gamma\,\hat{y}_k\,(1-\hat{y}_k^2)\,(\mathbf{x}_k - \bar{y}_k\,\mathbf{v}_{k-1}) \tag{2.2.79}$$

The use of the sigmoid nonlinearity tanh(.) generates the term $1-\hat{y}_k^2$, which is familiar from the training of feed-forward neural networks with binary output. Clearly, the term $1-\hat{y}_k^2$ affects the adaptation of the synaptic weights on the basis of the proximity of the output $\hat{y}_k$ to the values ± 1. The update rule (2.2.76) is essentially the generalization of the rule proposed for the linear unit. In fact, the update rule (2.2.66) can be simply obtained from (2.2.76) if $\sigma(x) = x$. In this case, (2.2.77) gives $\psi(\bar{y}_k) = 2\,\bar{y}_k = 2\,\hat{y}_k$. The resulting update equation is identical with (2.2.66), provided that $\alpha = 2\,\gamma$.

The Hebbian learning rule was also the basis for other learning rules reported in the literature (Linsker, 1986; Linsker, 1988; Yuille et al., 1989). Yuille et al. proposed that the update equation for a single linear unit may be obtained by using the gradient descent method to minimize

$$J(\mathbf{v}) = -\frac{1}{2}\,\hat{y}^2 + \frac{1}{4}\,(\sum_{i=1}^{n_i} v_i^2)^2 \tag{2.2.80}$$

The first term of (2.2.80) imposes the requirement that the training maximizes $\hat{y}^2$. The second term of (2.2.80) produces a decay term which guarantees the existence of stable states. From (2.2.80),

$$\Delta\mathbf{v}_{k-1} = -\alpha\,\frac{\partial J(\mathbf{v})}{\partial\mathbf{v}}|_{\mathbf{v}=\mathbf{v}_{k-1}} = \alpha\,(\hat{y}_k\,\mathbf{x}_k - ||\mathbf{v}_{k-1}||^2\,\mathbf{v}_{k-1}) \tag{2.2.81}$$

The update equation (2.2.81) contains a Hebbian term as well as a decay term, which is different than that provided by Oja's rule. Since the decay term of (2.2.81) contains the Euclidean norm $||\mathbf{v}||^2 = \sum_{i=1}^{n_i} v_i^2$, the update of each synaptic weight by (2.2.81) is not local.

On the average, the change of the synaptic weights $\Delta\mathbf{v} = \Delta\mathbf{v}_{k-1}$ is given by

$$\begin{aligned} E\{\Delta\mathbf{v}\} &= \alpha\,(E\{\mathbf{x}\,\mathbf{x}^*\,\mathbf{v}\} - \|\mathbf{v}\|^2\,\mathbf{v}) \\ &= \alpha\,(\mathbf{C}\,\mathbf{v} - \|\mathbf{v}\|^2\,\mathbf{v}) \end{aligned} \tag{2.2.82}$$

According to the above analysis, $E\{\Delta\mathbf{v}\}$ is proportional to the gradient of the energy function

$$\begin{aligned} E\{J\} &= -\frac{1}{2}\,E\{\hat{y}^2\} + \frac{1}{4}\,\|\mathbf{v}\|^4 \\ &= -\frac{1}{2}\,E\{\mathbf{v}^*\,\mathbf{x}\,\mathbf{x}^*\,\mathbf{v}\} + \frac{1}{4}\,\|\mathbf{v}\|^4 \\ &= -\frac{1}{2}\,\mathbf{v}^*\,\mathbf{C}\,\mathbf{v} + \frac{1}{4}\,\|\mathbf{v}\|^4 \end{aligned} \tag{2.2.83}$$

When the system updated by (2.2.81) reaches equilibrium, $E\{\Delta\mathbf{v}\} = 0$. Under this assumption, (2.2.82) gives

$$\mathbf{C}\,\mathbf{v} = \|\mathbf{v}\|^2\,\mathbf{v} \tag{2.2.84}$$

Thus, $\mathbf{v}$ is the eigenvector which corresponds to the maximum eigenvalue of $\mathbf{C}$, given in this case as $\lambda = \|\mathbf{v}\|^2$.

Oja also extended his unsupervised learning rule to single-layered neural networks with $n_o > 1$ output units (Oja, 1989; Oja, 1991; Oja et al., 1991). Consider the single-layered neural network shown in Figure 2.7 and assume that $n_o \leq n_i$, where n_i is the length of the input vectors. This network is trained by maximizing the criterion

$$J(\mathbf{w}_i, i = 1, 2, \ldots, n_o) = E\{\sum_{i=1}^{n_o} \hat{y}_i^2\} = E\{\sum_{i=1}^{n_o} \sigma(\mathbf{w}_i^*\,\mathbf{x})^2\} \tag{2.2.85}$$

under the constraint that the n_o vectors of synaptic weights $\mathbf{w}_i, i = 1, 2, \ldots, n_o$ are orthonormal, that is,

$$\mathbf{w}_i^*\,\mathbf{w}_j = \delta_{ij} \tag{2.2.86}$$

where δ_{ij} is the Kronecker delta, defined as $\delta_{ij} = 1$ if $i = j$ and $\delta_{ij} = 0$ if $i \neq j$. The network trained on the basis of the above conditions performs principal component analysis (Krogh and Hertz, 1990; Oja, 1991).

A single-layered neural network with linear units satisfies the above conditions if the weight matrix $\mathbf{W} = [\mathbf{w}_1\, \mathbf{w}_2 \ldots \mathbf{w}_{n_o}]$ is updated according to the following rule

$$\mathbf{W}_k = \mathbf{W}_{k-1} + \alpha\,(\mathbf{I} - \mathbf{W}_{k-1}\,\mathbf{W}^*_{k-1})\,\mathbf{x}_k\,\mathbf{x}^*_k\,\mathbf{W}_{k-1} \tag{2.2.87}$$

The derivation of this update equation is presented in Appendix A2.3 (Oja, 1992). The update equation (2.2.87) produces a set of orthonormal vectors which span the same subspace spanned by the first n_o principal components (Krogh and Hertz, 1990). The corresponding network is called principal component analysis (PCA) network (Oja, 1991). Since $n_o \leq n_i$, the PCA network compresses the input data. Taking into account that $\hat{\mathbf{y}}_k = \mathbf{W}^*_{k-1}\,\mathbf{x}_k$, (2.2.87) can also be written as

$$\mathbf{W}_k = \mathbf{W}_{k-1} + \alpha\,(\mathbf{x}_k\,\hat{\mathbf{y}}^*_k - \mathbf{W}_{k-1}\,\hat{\mathbf{y}}_k\,\hat{\mathbf{y}}^*_k) \tag{2.2.88}$$

According to (2.2.88), the update equation for each weight vector is given by

$$\begin{aligned} \mathbf{w}_{i,k} &= \mathbf{w}_{i,k-1} + \alpha\,\hat{y}_{i,k}\,(\mathbf{x}_k - \mathbf{W}_{k-1}\,\hat{\mathbf{y}}_k) \\ &= \mathbf{w}_{i,k-1} + \alpha\,\hat{y}_{i,k}\,(\mathbf{x}_k - \sum_{\ell=1}^{n_o} \hat{y}_{\ell,k}\,\mathbf{w}_{\ell,k-1}) \end{aligned} \tag{2.2.89}$$

Sanger recently proposed an alternative learning rule for the unsupervised training of single-layered neural networks formed by linear units (Sanger, 1989). Sanger's rule may be obtained by modifying the update equation (2.2.88) as

$$\mathbf{W}_k = \mathbf{W}_{k-1} + \alpha\,(\mathbf{x}_k\,\hat{\mathbf{y}}^*_k - \mathbf{W}_{k-1}\,U(\hat{\mathbf{y}}_k\,\hat{\mathbf{y}}^*_k)) \tag{2.2.90}$$

where $U(\mathbf{A})$ is the matrix obtained from $\mathbf{A}$ by setting all its lower diagonal elements equal to zero. It can easily be verified that each vector of synaptic weights can be updated according to Sanger's rule by

$$\begin{aligned} \mathbf{w}_{i,k} &= \mathbf{w}_{i,k-1} + \alpha\,\hat{y}_{i,k}\,(\mathbf{x}_k - \sum_{\ell=1}^{i} \hat{y}_{\ell,k}\,\mathbf{w}_{\ell,k-1}) \\ &= \mathbf{w}_{i,k-1} + \alpha\,\hat{y}_{i,k}\,[(\mathbf{x}_k - \sum_{\ell=1}^{i-1} \hat{y}_{\ell,k}\,\mathbf{w}_{\ell,k-1}) - \hat{y}_{i,k}\,\mathbf{w}_{i,k-1}] \end{aligned} \tag{2.2.91}$$

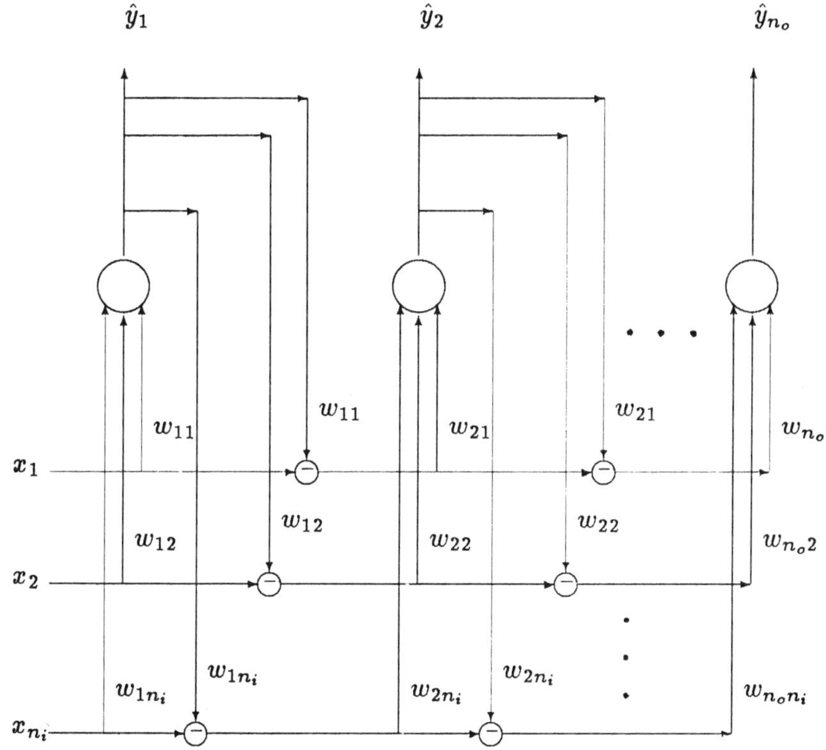

Figure 2.11: A network implementing Sanger's learning rule.

Sanger's rule also converges to a set of orthonormal weight vectors $\mathbf{w}_i, i = 1, 2, \ldots, n_o$. Moreover, there is a simple relationship between the resulting weight vectors $\mathbf{w}_i, i = 1, 2, \ldots, n_o$ and the first n_o principal components. Suppose that $\lambda_i, i = 1, 2, \ldots, n_o$ are the n_o eigenvalues of $\mathbf{C}$ and let $\mathbf{e}_i, i = 1, 2, \ldots, n_o$ be the corresponding normalized eigenvectors. Assume also that $\lambda_1 \geq \lambda_2 \geq \ldots \geq \lambda_{n_o}$. It can be shown that Sanger's rule converges to the weight vectors $\mathbf{w}_i \rightarrow \pm\mathbf{e}_i$. In other words, Sanger's rule orders the weight vectors $\mathbf{w}_i, i = 1, 2, \ldots, n_o$ in such a way that they provide the direction of the first n_o principal components. In this sense, Sanger's rule is directly related to the Karhunen-Loeve transform (Hertz, et al.,

1991). Figure 2.11 shows a network proposed for the implementation of Sanger's learning rule (Sanger, 1989). This implementation is based on the observation that the last term of the summation involved in Sanger's update equation can be separated as indicated by (2.2.91). Oja discusses the relationship between different linear learning rules in a forthcoming publication (Oja, 1992).

Oja's rule may be extended to single-layered neural networks formed by nonlinear units (Oja et al., 1991). The matrix of synaptic weights $\mathbf{W}$ of such a network is updated according to the learning rule

$$\mathbf{W}_k = \mathbf{W}_{k-1} + \gamma\,(\mathbf{I} - \mathbf{W}_{k-1}\,\mathbf{W}^*_{k-1})\,\mathbf{x}_k\,\psi(\mathbf{x}^*_k\,\mathbf{W}_{k-1}) \qquad (2.2.92)$$

where $\psi(.)$ is the learning function, defined in terms of the output nonlinearity $\sigma(.)$ by (2.2.77). Once again, the update equation (2.2.87) for the network with linear units can be interpreted as the special case of (2.2.92) which corresponds to $\sigma(x) = x$ and $\alpha = 2\,\gamma$. If the output units are nonlinear, the orthonormal vectors $\mathbf{w}_i, i = 1, 2, \ldots, n_o$ do not span the principal component analysis space. If $\sigma(x) = \tanh(x)$, the network trained by the update rule (2.2.92) performs *robust* principal component analysis. In this case, $\psi(\bar{y}) = 2\,\hat{y}\,(1 - \hat{y}^2)$, where $\hat{y} = \tanh(\bar{y}) = \tanh(\mathbf{w}^*\mathbf{x})$. As a result, the effect of the input vectors which produce a large analog output $\bar{y}$ is suppressed during the training due to the term $1 - \hat{y}^2$.

2.3 Feed-back Neural Networks

In feed-back neural networks, the input information defines the initial activity state of a feed-back system. After some state transitions the asymptotic equilibrium state is identified as the outcome of the computation. The Hopfield model is the simplest and most widely used feed-back neural architecture (Hopfield, 1982; Hopfield, 1984). The time evolution of the Hopfield network is described by a set of coupled nonlinear differential equations. In the case of symmetric synaptic weights, a Lyapunov function can be assigned to this set of differential equations. The existence of a Lyapunov function guarantees the convergence of the network to some local minimum, regardless of its initial state. Under certain assumptions, the

Hopfield network can function as a content-addressable associative memory. The popularity of the Hopfield model was probably the result of Hopfield's argument that such a network can perform fast optimization. The basic idea behind this argument is that the Lyapunov function is formulated in terms of the objective function to be minimized. The resulting set of differential equations will then have stable states corresponding to local minima of the objective function. The Hopfield neural network is a special case of a more general class of feed-back neural network models, proposed and analyzed by Cohen and Grossberg (Cohen and Grossberg, 1983).

2.3.1 A General Model for Feed-back Neural Networks

Cohen and Grossberg considered the neural network models which can be described by the following system of coupled nonlinear differential equations

$$\frac{dx_i}{dt} = a_i(x_i)\,[b_i(x_i) - \sum_{j=1}^{n} c_{ij}\, d_j(x_j)] \qquad (2.3.1)$$

where the matrix $\mathbf{C} = [c_{ij}]$ is symmetric, i.e., $c_{ij} = c_{ji}$, $a_i(.)$ are positive definite functions, i.e., $a_i(x_i) \geq 0$, and $d_j(.)$ are monotone functions, i.e., $d_j'(x_j) \geq 0$. The system of differential equations defined by (2.3.1) can be used to represent a neural network. According to this interpretation, x_i is the activity level of the ith neuron. In addition, $d_j(x_j)$ is the output of the jth neuron and c_{ij} is a synaptic weight which represents the strength of connection between the ith and jth neurons. The sum $\sum_{j=1}^{n} c_{ij} d_j(x_j)$ represents the net input from the neurons which form the network to the ith neuron. Under the assumption that $a_i(x_i) \geq 0$, the sign of the right hand term of (2.3.1) depends on the relationship between the net input to the ith neuron and the intrinsic function $b_i(x_i)$. In fact, the activity of the ith neuron decreases if and only if the net input exceeds $b_i(x_i)$.

In general, a dynamical system defined by the system of differential equations

$$\frac{dx_i}{dt} = F_i(x_1, x_2, \ldots, x_n) \qquad (2.3.2)$$

is *competitive* if

$$\frac{\partial F_i}{\partial x_k} \leq 0 \, \forall \, k \neq i \tag{2.3.3}$$

It can be shown that for the particular system defined by (2.3.1)

$$\frac{\partial F_i}{\partial x_k} = -c_{ik} \, a_i(x_i) \, d_k'(x_k) \tag{2.3.4}$$

Since it is assumed that $a_i(x_i) \geq 0$ and $d_k'(x_k) \geq 0$, the sign of $\partial F_i / \partial x_k$ depends only on the sign of the synaptic weights c_{ik}. Clearly, this system is competitive if $c_{ik} \geq 0$, i.e., the connections between the neurons are all inhibitory. It is also intuitively reasonable that if the effect of each neuron on every other neuron is inhibitory, then they all compete with one another. In general, there is no guarantee that the synaptic weights are all positive or negative. Therefore, (2.3.1) defines a mixed cooperative-competitive dynamical system.

The stability of a dynamical system is of critical importance if the system represents a neural network. The stability of a dynamical system is guaranteed if there exists a Lyapunov function for this system. A Lyapunov function is a function on the state space which is nonincreasing along the trajectories. Cohen and Grossberg provided a rigorous proof that there exists a global Lyapunov function for the models described by (2.3.1), which is of the form

$$V = -\sum_{i=1}^{n} \int^{x_i} b_i(\xi_i) \, d_i'(\xi_i) \, d\xi_i + \frac{1}{2} \sum_{j=1}^{n} \sum_{k=1}^{n} c_{jk} \, d_j(x_j) \, d_k(x_k) \tag{2.3.5}$$

The crucial point of this proof is that V satisfies the basic property of a Lyapunov function. The time derivative of V can be evaluated by (Cohen and Grossberg, 1983)

$$\frac{dV}{dt} = \sum_{i=1}^{n} \frac{\partial V}{\partial x_i} \frac{dx_i}{dt} \tag{2.3.6}$$

It can easily be verified that since $c_{ij} = c_{ji}$,

$$\frac{\partial V}{\partial x_i} = -d_i'(x_i) \, [b_i(x_i) - \sum_{j=1}^{n} c_{ij} \, d_j(x_j)] \tag{2.3.7}$$

The combination of (2.3.1) and (2.3.7) with (2.3.6) gives

$$\frac{dV}{dt} = -\sum_{i=1}^{n} a_i(x_i)\, d_i'(x_i)[b_i(x_i) - \sum_{j=1}^{n} c_{ij}\, d_j(x_j)]^2 \tag{2.3.8}$$

Since it was assumed that $a_i(x_i) \geq 0$ and $d_i'(x_i) \geq 0$, then $dV/dt \leq 0$. When $dV/dt = 0$, the system reaches its equilibrium.

A special transformation, known as Σ S Exchange, can be used to show that the general model (2.3.1) is a generalization of various neural network models which have been proposed independently over the years (Grossberg, 1988b). More specifically, it was shown that for some specific selection of the functions $a_i(.)$ and $b_j(.)$ the general model is mathematically equivalent to neural network models such as the McCulloch and Pitts model (McCulloch and Pitts, 1943) and the Brain-State-in-a-Box model (Anderson et al., 1977).

2.3.2 The McCulloch-Pitts Model

The dynamic behavior of the McCulloch-Pitts model is based on the following discrete-time equation (McCulloch and Pitts, 1943; Grossberg, 1988b)

$$x_j(t+1) = \text{sgn}(\sum_{k=1}^{n} w_{jk}\, x_k(t) - \theta_j) \tag{2.3.9}$$

where w_{jk} are the synaptic weights, θ_j is a threshold, and sgn(.) is the well-known signum function, defined as $\text{sgn}(x) = +1$ if $x \geq 0$ and $\text{sgn}(x) = -1$ if $x \leq 0$. The effect of the threshold θ_j on the analysis can be avoided by introducing a function, say $s_j(.)$, which is the signum function translated by the corresponding threshold θ_j, that is,

$$s_j(x) = \text{sgn}(x - \theta_j) \tag{2.3.10}$$

After this definition, the dynamical equation (2.3.9) can be written as

$$x_j(t+1) = s_j(\sum_{k=1}^{n} w_{jk}\, x_k(t)) \tag{2.3.11}$$

Consider now the continuous-time version of the dynamical equation (2.3.11), that is,

$$\frac{dx_j}{dt} = -x_j + s_j(\sum_{k=1}^{n} w_{jk}\, x_k(t)) \tag{2.3.12}$$

The model (2.3.12) can be interpreted as a special case of the general model (2.3.1) by introducing the change of variables

$$y_i = \sum_{j=1}^{n} w_{ij}\, x_j \tag{2.3.13}$$

Multiplying (2.3.12) by w_{ij} and summing over j gives

$$\sum_{j=1}^{n} w_{ij} \frac{dx_j}{dt} = -\sum_{j=1}^{n} w_{ij}\, x_j + \sum_{j=1}^{n} w_{ij}\, s_j(y_j) \tag{2.3.14}$$

The substitution of (2.3.13) into (2.3.14) gives

$$\frac{dy_i}{dt} = -y_i + \sum_{j=1}^{n} w_{ij}\, s_j(y_j) \tag{2.3.15}$$

which is clearly a special case of (2.3.1) in which $a_i(x_i) = 1$, $b_i(x_i) = -x_i$, $c_{ij} = -w_{ij}$, and $d_j(.) = s_j(.)$. The definition (2.3.13) was crucial in transforming the dynamical model (2.3.12), involving a nonlinearity of a sum, to the equivalent model (2.3.15), involving the sum of nonlinearities. This was the motivation for introducing this transformation as the Σ S Exchange (Grossberg, 1988b).

2.3.3 The Brain-State-in-a-Box Model

The Brain-State-in-a-Box model is described by the discrete-time equation (Anderson et al., 1977; Grossberg, 1988b)

$$x_i(t+1) = S(x_i(t) + \alpha \sum_{j=1}^{n} v_{ij}\, x_j(t)) \tag{2.3.16}$$

The coefficients v_{ij} are symmetric, i.e., $v_{ij} = v_{ji}$, and the nonlinearity $S(.)$ is defined by

$$S(x) = \begin{cases} -F & \text{if } x \le -F \\ x & \text{if } -F \le x \le +F \\ +F & \text{if } x \ge +F \end{cases} \tag{2.3.17}$$

Clearly, $S(x)$ is a linear function of x until x reaches the saturation limit $\pm F$. It was shown that each x_i approaches $\pm F$ as the system (2.3.16) evolves in time (Anderson et al., 1977). In fact, the name of the model was inspired by the observation that the vector $\mathbf{x} = (x_1, x_2, \ldots, x_n)$ approaches a corner of the "box" $(\pm F, \pm F, \ldots, \pm F)$ as the system (2.3.16) reaches equilibrium.

The continuous-time version of the dynamical equation (2.3.16) is

$$\frac{dx_i}{dt} = -x_i + S(\sum_{j=1}^{n} w_{ij}\, x_j) \qquad (2.3.18)$$

where $w_{ij} = \delta_{ij} + \alpha\, v_{ij}$, and δ_{ij} is the Kronecker delta, defined as $\delta_{ij} = 1$ if $i = j$ and $\delta_{ij} = 0$ if $i \neq j$. Consider once again the time-dependent variables

$$y_i = \sum_{j=1}^{n} w_{ij}\, x_j \qquad (2.3.19)$$

Applying the Σ S Exchange in similar fashion transforms the system (2.3.18) into the following equivalent system

$$\frac{dy_i}{dt} = -y_i + \sum_{j=1}^{n} w_{ij}\, S(y_j) \qquad (2.3.20)$$

Clearly, the Brain-State-in-a-Box model is also a special case of the general feed-back neural network model described by (2.3.1). The only difference is that in this model $S(.)$ is not a differentiable everywhere function. Therefore, the dynamic behavior of this model cannot be studied on the basis of a Lyapunov function of the form (2.3.5), whose evaluation would involve the derivative of $S(.)$.

2.3.4 The Hopfield Model

Grossberg indicated that the Hopfield model may be interpreted as a special case of the general model (2.3.1) (Grossberg, 1988b). This interpretation was based on the assumption that $a_i(.)$ is a constant, that is,

$$a_i(x_i) = \frac{1}{C_i} \qquad (2.3.21)$$

and $b_j(.)$ is a linear function of the form

$$b_i(x_i) = -\frac{1}{R_i}\, x_i + I_i \tag{2.3.22}$$

If, in addition, $c_{ij} = -T_{ij}$ and $d_j(x_j) = g_j(x_j)$ the general model (2.3.1) provides the well-known Hopfield model, described by the following set of differential equations

$$C_i\, \frac{dx_i}{dt} = -\frac{x_i}{R_i} + I_i + \sum_{j=1}^{n} T_{ij}\, y_j \tag{2.3.23}$$

where

$$y_j = g_j(x_j) \tag{2.3.24}$$

It was independently proposed by Hopfield that there exists a global Lyapunov function for the system (2.3.23), given by (Hopfield, 1984)

$$V = \sum_{i=1}^{n} \frac{1}{R_i} \int^{g_i(x_i)} g_i^{-1}(\xi) d\xi - \sum_{i=1}^{n} I_i\, y_i - \frac{1}{2} \sum_{j=1}^{n} \sum_{k=1}^{n} T_{jk}\, y_j\, y_k \tag{2.3.25}$$

Hopfield considered that $g_i(.)$ is a sigmoid function and, therefore, the existence of its inverse is guaranteed (Hopfield, 1984). However, Grossberg argued that the Lyapunov function of the form (2.3.25) imposes the additional constraint of invertibility on the function $g_i(.)$ (Grossberg, 1988b). Moreover, Grossberg indicated that the existence of a Lyapunov function of the system (2.3.23) is simply guaranteed by the interpretation of the system (2.3.23) as a simple case of the general system (2.3.1). In fact, in this particular case (2.3.5) becomes

$$V = \sum_{i=1}^{n} \frac{1}{R_i} \int^{x_i} \xi_i\, g_i'(\xi_i)\, d\xi_i - \sum_{i=1}^{n} I_i\, y_i - \frac{1}{2} \sum_{j=1}^{n} \sum_{k=1}^{n} T_{jk}\, y_j\, y_k \tag{2.3.26}$$

The model (2.3.23), which is known as the Hopfield model, is certainly a simple case of the general model defined by (2.3.1). The Σ S Exchange transformation suggests that there exists an implicit relationship between this model and other neural models in the literature, including the early McCulloch-Pitts model. Nevertheless,

Hopfield provided some interesting interpretations for the dynamical system (2.3.23) and the corresponding Lyapunov function. One reason behind the popularity of the model (2.3.23) was Hopfield's argument that the continuous version of the model (2.3.23) can be implemented by an electrical circuit. Figure 2.12 shows a generic electrical circuit consisting of resistors, capacitors, and amplifiers proposed by Hopfield to represent the model (2.3.23) (Hopfield, 1984). In Figure 2.12, each neuron is represented by an amplifier. The output voltage y_i of the ith amplifier is given in terms of its input x_i by $y_i = g_i(x_i)$. The function $g_i(.)$ which characterizes the input-output relation of the ith amplifier is the sigmoid function shown in Figure 2.4. Since there exist excitatory as well as inhibitory connections between the neurons, Hopfield assigned to each amplifier a normal (+) and an inverted (−) output. One of the two outputs of the jth amplifier is connected to the input of the ith amplifier through the resistor R_{ij}, defined by $R_{ij} = 1/|T_{ij}|$, and represented in Figure 2.12 by a circle. If the synaptic weight T_{ij} is excitatory, i.e., $T_{ij} > 0$, the resistor is connected to the normal output of the jth amplifier. On the other hand, this resistor is connected to the inverted output of the jth amplifier if the synaptic weights T_{ij} is inhibitory, i.e., $T_{ij} < 0$. The circuit shown in Figure 2.12 also includes an externally supplied current I_i for each neuron. Although not shown in Figure 2.12, the ith amplifier also has an input resistor ρ_i and an input capacitor C_i, both connected in parallel to a reference ground. The analysis of the circuit indicates that the time constant τ_i of the ith neuron is given by

$$\tau_i = R_i C_i \tag{2.3.27}$$

where

$$\frac{1}{R_i} = \frac{1}{\rho_i} + \sum_{j=1}^{n} \frac{1}{R_{ij}} \tag{2.3.28}$$

Hopfield suggested that the shape of the sigmoid nonlinearity used in the model (2.3.23) determines the effect of the first term of (2.3.25), which corresponds to the decay of the activity x_i in the model (2.3.23) (Hopfield, 1984). Consider the model (2.3.23), where

$$y_j = g_j(\beta x_j) \tag{2.3.29}$$

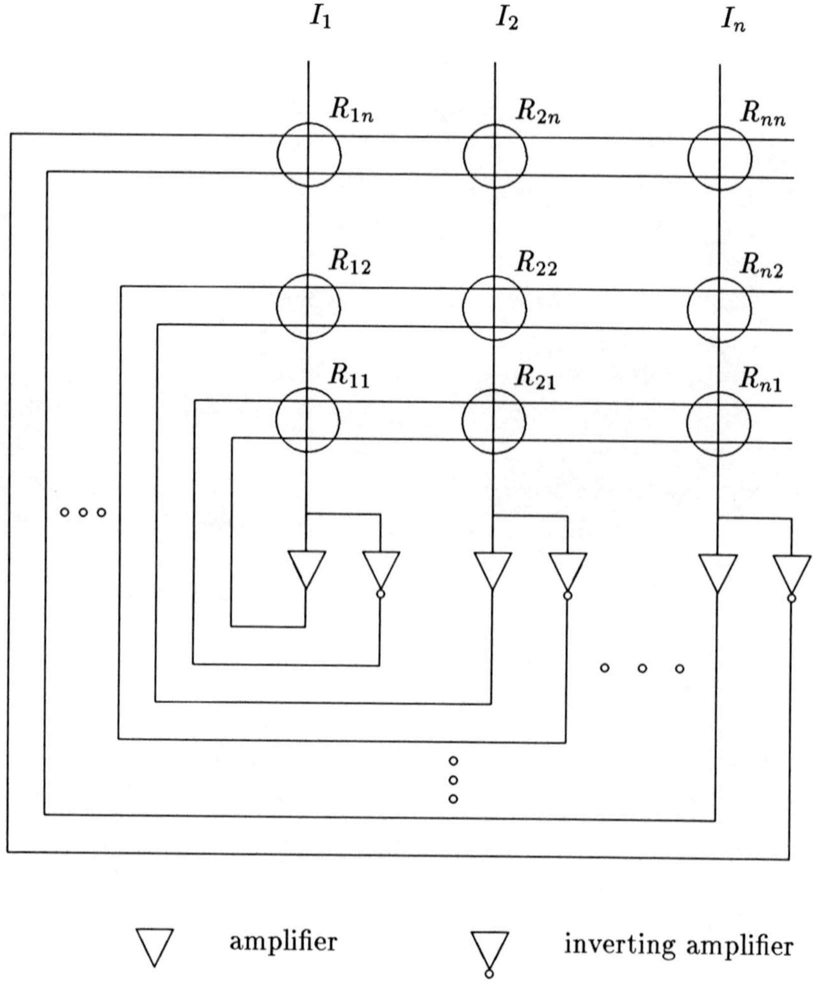

Figure 2.12: The Hopfield model.

and β is a gain which determines the steepness of the sigmoid nonlinearity. Since the inverse function in this case is $x_j = (1/\beta)g_j^{-1}(y_j)$, the Lyapunov function of the system defined by (2.3.23) and (2.3.29) is given by

$$V = \frac{1}{\beta}\sum_{i=1}^{n}\frac{1}{R_i}\int^{y_i} g_i^{-1}(\xi)\,d\xi - \sum_{i=1}^{n} I_i\,y_i - \frac{1}{2}\sum_{j=1}^{n}\sum_{k=1}^{n} T_{jk}\,y_j\,y_k \quad (2.3.30)$$

If $\beta < 1$, the effect of the first term of (2.3.30) increases. As β increases, i.e., the sigmoid nonlinearity asymptotically approaches a hard-limiting nonlinearity, the effect of the first term of (2.3.30) decreases. In the high-gain case, that is, when $\beta \gg 1$, its effect becomes negligible and (2.3.30) can be approximated by

$$V = -\sum_{i=1}^{n} I_i\,y_i - \frac{1}{2}\sum_{j=1}^{n}\sum_{k=1}^{n} T_{jk}\,y_j\,y_k \quad (2.3.31)$$

For a given set of synaptic weights T_{ij} and bias parameters I_i, the system defined by (2.3.23) converges from a certain initial state to *stable states.* In the high-gain case, i.e., when $\beta \gg 1$, these stable states are the local minima of the quantity (2.3.31) and tend to be corners of a hypercube in state space. Based on the existence of stable states for the system (2.3.23), Hopfield argued that the corresponding network can be used to solve complicated optimization problems. This can be achieved by selecting the synaptic weights T_{ij} and the bias parameters I_i in such a way that (2.3.31) represents the objective function to be minimized, and then allowing the corresponding network to evolve in time from some initial state-estimate. The analysis presented so far indicates that the time evolution of the system defined by (2.3.23) and (2.3.29) converges to a local minimum of the objective function represented by (2.3.31). Among the optimization problems reported by Hopfield and reviewed in chapter 9, the most impressive is probably the development of a neural network which provides "satisfactory" solutions to the traveling salesman problem, which is a well-known NP-complete problem (Hopfield and Tank, 1985). The efficient use of neural networks in optimization assumes the implementation of the analog system (2.3.23) by electrical and eventually optical hardware. In fact, the simulation of the time evolution of (2.3.23) on a digital computer using numerical integration techniques results in processing requirements comparable to conventional optimization techniques. The importance of neural networks in optimization applications has been criticized because of the scaling problems associated with the implementation of the analog system (2.3.23). This criticism is only one facet of the ongoing argument regarding the consideration of neural networks as computing devices.

Hopfield also indicated that the network described by (2.3.23) and (2.3.29) can operate as a content addressable memory (CAM). A CAM is a system which stores a set of memory states, that is, binary vectors with elements $+1$ and -1. A CAM retrieves a particular stored memory state when presented with a noisy or incomplete version of this state at its inputs. When the network operates as a CAM, the information is not localized but distributed over the synaptic weights of the network. The network described by the model (2.3.23) can be made to function as a CAM by selecting its synaptic weights in such a way that the stable states of the system are in fact the memory states. In the high-gain case, i.e., when $\beta \gg 1$, the memory states are local minima of the quantity (2.3.31). The input or key-word of the CAM plays the role of the initial state of the network described by (2.3.23). After evolving in time, this network converges to the memory state which is the "closest" to the key-word.

Hopfield suggested that the neural network described by (2.3.23) operates as a CAM which stores the memory states $\mathbf{y}_k$, $k = 1, 2, \ldots, m$ if $I_i = 0$ and

$$T_{ij} = \sum_{k=1}^{m} y_{i,k}\, y_{j,k} \tag{2.3.32}$$

This formation of the synaptic weights was proposed very early in the literature and is known as the outer-product rule or correlational learning (Anderson, 1972; Kohonen, 1972; Nakano, 1972). Hopfield justified this learning scheme by showing that this formation of the synaptic weights guarantees that the memory states correspond to the local minima of (2.3.31) (Hopfield, 1984). In fact, if $I_i = 0$ and the synaptic weights are given by (2.3.32), then for current state vector $\mathbf{y}$, (2.3.31) becomes

$$V = -\frac{1}{2}\sum_{k=1}^{m} c(\mathbf{y}_k, \mathbf{y})^2 \tag{2.3.33}$$

where $\mathbf{y}_k, k = 1, 2, \ldots, m$ are the stored patterns, and $c(\mathbf{y}_k, \mathbf{y})$ is the direction cosine between the vectors $\mathbf{y}_k$ and $\mathbf{y}$, defined by

$$c(\mathbf{y}_k, \mathbf{y}) = \mathbf{y}_k^* \mathbf{y} = \sum_{i=1}^{n} y_{i,k}\, y_i \tag{2.3.34}$$

If the initial state of the network $\mathbf{y}$ is a random vector, the contribution of each term $c(\mathbf{y}_k, \mathbf{y})^2, k = 1, 2, \ldots, m$ is close to zero. But if $\mathbf{y}$ is close to $\mathbf{y}_k$, $c(\mathbf{y}_k, \mathbf{y})$ is approximately equal to one and, therefore, V approaches a local minimum, i.e., $V \approx -\frac{1}{2}\, n^2$.

The validity of the previous analysis depends on the number m of stable states relative to the length n of the stored patterns. Hopfield suggested that this analysis is valid if $m \le 0.15\, n$, while other researchers verified that reliable operation of the Hopfield model as a CAM is guaranteed if $m \le n/(2 \log n)$ (McEliece et al., 1987). Abu-Mostafa and Jacques argued that the latter upper bound is based on the assumption that the synaptic weights T_{ij} are formed by the outer-product rule (Abu-Mostafa and Jacques, 1985). They also indicated that, regardless of the formation of the synaptic weights, the Hopfield network cannot have more than n arbitrary stable states.

The Kanerva's sparse distributed memory model attracted significant attention as an efficient alternative to the Hopfield model (Kanerva, 1988). Kanerva's model may be viewed as a network with three layers, with the middle layer playing the role of the hidden layer. In order for this system to realize an auto-associative network, the output layer can be fed back into the input layer (Keeler, 1988). Chou suggested that the capacity of Kanerva's associative memory grows exponentially with the length of the stored patterns, as opposed to the sub-linear growth in the capacity of the Hopfield model (Chou, 1988). Keeler indicated that the advantage of Kanerva's associative memory is that the number of stored patterns is independent of the size of the patterns (Keeler, 1988). However, Keeler contradicted Chou's result regarding the capacity of this memory model by suggesting that the efficiency of Kanerva's

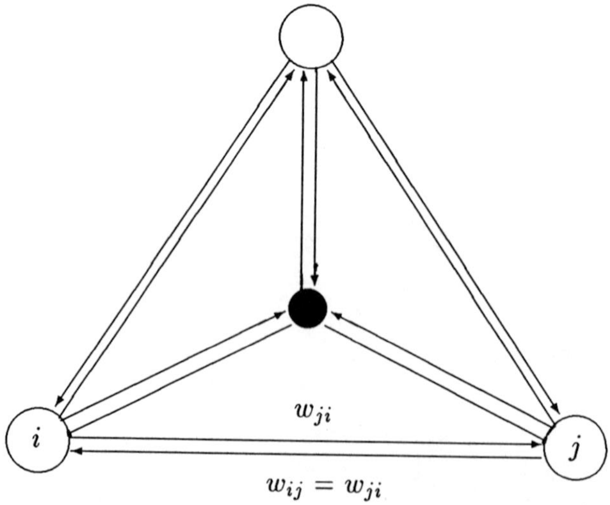

Figure 2.13: A Boltzmann machine consisting of three visible units and one hidden unit.

associative memory in terms of information stored per bit is the same as for Hopfield-type models.

2.3.5 The Boltzmann Machine

The neural network which realizes a Boltzmann machine is composed of stochastic basic building elements or units (Ackley et al., 1985; Hinton and Sejnowski, 1986). The network consists of both *visible* and *hidden* units, connected by synaptic weights w_{ij} which are symmetric, i.e., $w_{ij} = w_{ji}$. Figure 2.13 shows a fully connected Boltzmann machine consisting of three visible units and one hidden unit. The architecture of the network which realizes a Boltzmann machine is close to that of the Hopfield model. The important difference is the existence of hidden units, i.e., units which are not connected to the outside environment. In fact, the existence of hidden units reveals a similarity between the Boltzmann machine and multi-layered, feed-forward neural networks. Consider a network which realizes a Boltzmann machine and let $\bar{s}_i$ be the total effect of the surrounding units on the ith unit, which is given by

$$\bar{s}_i = \sum_j w_{ij}\, s_j \tag{2.3.35}$$

The probability that the output of the ith unit takes the value $+1$ is given by

$$\text{Prob}(s_i = +1) = \frac{1}{1 + \exp(-\frac{2}{T}\,\bar{s}_i)} \tag{2.3.36}$$

On the other hand, the probability that the output of the ith unit is -1 is

$$\text{Prob}(s_i = -1) = 1 - \text{Prob}(s_i = +1) = \frac{1}{1 + \exp(+\frac{2}{T}\,\bar{s}_i)} \tag{2.3.37}$$

The main idea behind the training of the Boltzmann machine is based upon statistical mechanics and, in particular, the Boltzmann-Gibbs distribution which is briefly described here (Hertz et al., 1991). Consider a system with a finite set of states. Suppose that E_α is the total energy of the system at the state α and let $T > 0$ be the absolute temperature of the system. In general, the system may fluctuate in some definite direction. The system is at thermal equilibrium if it fluctuates around constant average values. When the system is at thermal equilibrium, the probability of occurrence of the state α is provided by the Boltzmann-Gibbs distribution as

$$P_\alpha = \frac{1}{Z}\exp(-E_\alpha / k_B T) \tag{2.3.38}$$

where k_B is the Boltzmann's constant and Z is the *partition function,* defined by

$$Z = \sum_\alpha \exp(-E_\alpha / k_B T) \tag{2.3.39}$$

Assume that $s_i^{\alpha\beta}$ denotes the actual value of s_i when the visible units are in state α and the hidden units are in state β. The energy of the network when the visible and hidden units are in states α

and β, respectively, was defined as (Ackley et al., 1985; Hinton and Sejnowski, 1986)

$$E_{\alpha\beta} = -\sum_{j}\sum_{i<j} w_{ij}\, s_i^{\alpha\beta}\, s_j^{\alpha\beta} \tag{2.3.40}$$

Since the outputs of each unit are ± 1, and the synaptic weights are symmetric, i.e., $w_{ij} = w_{ji}$, the energy $E_{\alpha\beta}$ can also be written as

$$E_{\alpha\beta} = -\frac{1}{2}\sum_{i}\sum_{j} w_{ij}\, s_i^{\alpha\beta}\, s_j^{\alpha\beta} \tag{2.3.41}$$

The probability $P'(V_\alpha H_\beta)$ that the visible units are in state α and the hidden units are in state β is provided by the Boltzmann-Gibbs distribution as

$$P'(V_\alpha H_\beta) = \frac{1}{Z}\exp(-\frac{1}{T}E_{\alpha\beta}) \tag{2.3.42}$$

where

$$Z = \sum_{\alpha}\sum_{\beta}\exp(-\frac{1}{T}E_{\alpha\beta}) \tag{2.3.43}$$

The partition function Z is a measure of the number of states that are readily accessible at temperature T (Bounds, 1986). If the temperature T is very high, Z approaches the total number of states in the system. Assuming that the network realizing a Boltzmann machine consists of n_v visible units and n_h hidden units, the total number of states is $2^{n_v} \times 2^{n_h} = 2^{(n_v+n_h)}$. The actual probability that the visible units are in state α is given by

$$P'(V_\alpha) = \sum_{\beta} P'(V_\alpha H_\beta) = \sum_{\beta}\frac{1}{Z}\exp(-\frac{1}{T}E_{\alpha\beta}) \tag{2.3.44}$$

Let $P(V_\alpha)$ be the desired probability that the visible units are in state α. The training of the Boltzmann machine is based on the minimization of a discrepancy measure between the actual probability $P'(V_\alpha)$ that the visible units are in state α and the corresponding probability $P(V_\alpha)$ over the states α. It was proposed that the discrepancy measure is the relative entropy of the desired probabilities with respect to the actual ones, defined by (Ackley et al., 1985; Hinton and Sejnowski, 1986)

$$G = \sum_{\alpha} P(V_\alpha)\ln\frac{P(V_\alpha)}{P'(V_\alpha)} \tag{2.3.45}$$

The minimization of G can be performed using the gradient descent method, which is based on the following update equation (Ackley et al., 1985; Hinton and Sejnowski, 1986)

$$\Delta w_{ij} = -\eta \frac{\partial G}{\partial w_{ij}} \tag{2.3.46}$$

The evaluation of $\partial G / \partial w_{ij}$ is presented in Appendix A2.4 (Ackley et al., 1985; Hinton and Sejnowski, 1986). The combination of (2.3.46) and (A2.4.11) provides the following update equation for the synaptic weights

$$\Delta w_{ij} = \eta \frac{1}{T} \left[\sum_{\alpha} \sum_{\beta} P(V_\alpha H_\beta)\, s_i^{\alpha\beta}\, s_j^{\alpha\beta} - \sum_{\alpha} \sum_{\beta} P'(V_\alpha H_\beta)\, s_i^{\alpha\beta}\, s_j^{\alpha\beta} \right] \tag{2.3.47}$$

or

$$\Delta w_{ij} = \eta \frac{1}{T} \left[p_{ij} - p'_{ij} \right] \tag{2.3.48}$$

where

$$p'_{ij} = \sum_{\alpha} \sum_{\beta} P'(V_\alpha H_\beta)\, s_i^{\alpha\beta}\, s_j^{\alpha\beta} \tag{2.3.49}$$

corresponds to unlearning in a free-running system, and

$$\begin{aligned} p_{ij} &= \sum_{\alpha} \sum_{\beta} P(V_\alpha H_\beta)\, s_i^{\alpha\beta}\, s_j^{\alpha\beta} \\ &= \sum_{\alpha} P(V_\alpha) \sum_{\beta} P(H_\beta / V_\alpha)\, s_i^{\alpha\beta}\, s_j^{\alpha\beta} \end{aligned} \tag{2.3.50}$$

In (2.3.50), $\sum_\beta P(H_\beta / V_\alpha)\, s_i^{\alpha\beta}\, s_j^{\alpha\beta}$ is the "average" value of the product $s_i\, s_j$, given that the visible units are clamped in state α. The probabilities of occurrence of each state α provide the weights in the evaluation of p_{ij} by the weighted summation over all the states α, as indicated in (2.3.50). As a result, p_{ij} is the "desired" average value of $s_i s_j$.

The role of the temperature T is crucial in the training of a Boltzmann machine. If the temperature T is very low, learning may be difficult because there are only few accessible states. If the temperature T is very high, the convergence of the learning algorithm may be poor due to the existence of many available states. Computer simulations indicate that there exists a window of temperature at which learning in a Boltzmann machine is possible (Bounds, 1986). An alternative to the a priori selection of a fixed temperature is the use of a simulated annealing procedure. According to this procedure, the temperature T is gradually reduced from a high value to the working value. The gradual reduction of the temperature during the training significantly reduces the time required by the Boltzmann machine to reach equilibrium (Hertz et al., 1991).

The major disadvantage of Boltzmann machines is the excessive computational effort required for their training. Peterson and Anderson proposed a deterministic version of Boltzmann machines in an attempt to speed-up the learning process (Peterson and Anderson, 1987). They suggested that the time required for learning may be reduced 10 to 30 times by using a mean field annealing process. Sejnowski proposed third order Boltzmann machines, where triplets of units interact through symmetric conjunctive interactions (Sejnowski, 1986). He also proposed the corresponding learning algorithm, which can also be extended to higher-order connections. Sejnowski speculated that the rate of learning for internal representations in a high-order Boltzmann machine should be much faster than that of a second-order Boltzmann machine based on pairwise interactions.

Ackley et al. trained Boltzmann machines to perform simple tasks, such as the realization of an n to n encoder (Ackley et al., 1985). Lippmann presented applications of Boltzmann machines in speech processing and recognition (Lippmann, 1989). Kohonen et al. presented an evaluation of the performance of Boltzmann machines in a pattern recognition problem (Kohonen et al., 1988). They found that a Boltzmann machine was more efficient than a feed-forward neural network trained by the Error Back Propagation algorithm.

2.3.6 Recurrent Back Propagation

The popularity of the Error Back Propagation algorithm motivated the application of the gradient descent method to the training of feed-back or recurrent neural networks. Pineda proposed that a recurrent neural network can be trained to realize an input to output mapping by a gradient descent based learning algorithm, known as Recurrent Back Propagation (Pineda, 1987). Similar algorithms were almost simultaneously proposed by other researchers (Almeida, 1987; Rohwer and Forrest, 1987). Although the derivation of these algorithms is based on the gradient descent method, there are slight differences in the network architectures used and the details of the algorithms. For example, the algorithm proposed by Rohwer and Forrest involves a matrix inversion (Rohwer and Forrest, 1987). In the other two approaches this matrix inversion is indirectly performed by an auxiliary network whose dynamic behavior is closely related to the learning equations provided by gradient descent (Almeida, 1987; Pineda, 1987). The presentation which follows is based on Pineda's approach (Pineda, 1987; Pineda, 1988a; Pineda, 1988b; Pineda, 1989).

The dynamic behavior of a recurrent neural network is described by the set of coupled nonlinear differential equations

$$\tau \frac{dz_i}{dt} = -z_i + \rho(u_i) + I_i \tag{2.3.51}$$

where z_i represents the activity of the ith unit, I_i represents external bias, $\rho(.)$ is a sigmoid nonlinearity, and u_i is defined in terms of the synaptic weights w_{ij} connecting the ith unit with other units in the network as follows

$$u_i = \sum_j w_{ij} z_j \tag{2.3.52}$$

When the system (2.3.51) reaches equilibrium, i.e., $dz_i/dt = 0$,

$$z_i^* = \rho(u_i^*) + I_i = \rho(\sum_j w_{ij} z_j^*) + I_i \tag{2.3.53}$$

The units of the network which belong to a certain subset A are the input units, while the units belonging to a subset Ω are the output units. The hidden units are members of neither A nor Ω. A unit may be simultaneously an input and an output unit. In other words, the intersection $A \cap \Omega$ of the sets A and Ω is not necessarily an empty set. The external bias I_i is related to the input $x_i, i \in A$ of the network by

$$I_i = \begin{cases} x_i & \text{if } i \in A \\ 0 & \text{otherwise} \end{cases} \tag{2.3.54}$$

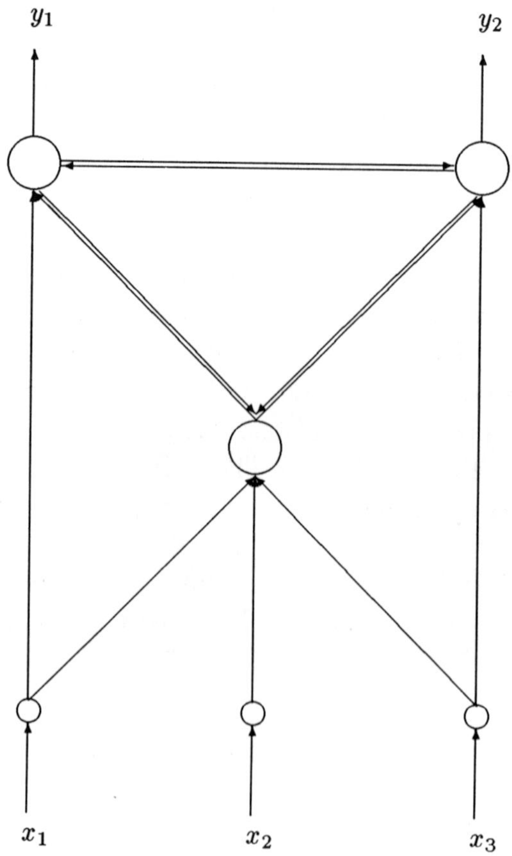

Figure 2.14: A recurrent neural network.

This last equation indicates that only the input units are affected by an external bias. Figure 2.14 shows a recurrent network with three external inputs, one hidden unit and two output units.

The derivation of this algorithm provided a simple alternative to early claims that a recurrent neural network may be trained by unfolding it to an equivalent feed-forward neural network. This claim is based on the argument that, for a finite time, for every recurrent neural network there exists an equivalent feed-forward neural network. It should be emphasized here that the derivation of the algorithm which follows is based on the assumption that there exists at least one stable fixed point. A fixed point corresponds to the equilibrium state of the set of differential equations defined by (2.3.51).

The training of a recurrent neural network guarantees that the output units of the network converge to some desired output values $y_k, k \in \Omega$ by updating its synaptic weights. The synaptic weights of the network are updated by minimizing the quadratic error function evaluated at equilibrium as

$$E = \frac{1}{2} \sum_k E_k^2 \tag{2.3.55}$$

where

$$E_k = \begin{cases} y_k - z_k^* & \text{if } k \in \Omega \\ 0 & \text{otherwise} \end{cases} \tag{2.3.56}$$

Appendix A2.5 presents the derivation of the Recurrent Back Propagation algorithm, based on the minimization of the total error E by the gradient descent method (Pineda, 1987). This derivation reveals that the adaptation of the synaptic weights also depends on the evolution in time of an auxiliary network, whose dynamic behavior is described by

$$\frac{dv_i}{dt} = -v_i + \sum_p v_p \, \rho'(u_p) \, w_{pi} + E_i \tag{2.3.57}$$

This differential equation involves a sum of nonlinearities, as opposed to the equation (2.3.51) which involves the nonlinearity of a sum. The external bias is provided in this case by the errors E_i between the states of the output units and the expected outputs, defined by (2.3.56). Clearly, E_i is nonzero only if the ith unit is an output unit.

The algorithm derived in Appendix A2.5 can be summarized as follows:

- The stable states z_i^* of the units are obtained by allowing the system (2.3.51) to evolve in time and reach equilibrium.

- The stable states v_i^* of the dynamical system described by (2.3.57) are obtained by allowing this system to reach equilibrium, given the output errors E_i resulting from the equilibrium state of (2.3.51).

- The synaptic weights of the network are updated by

$$\Delta w_{pq} = \alpha\, \rho'(u_p^*)\, v_p^*\, z_q^* \tag{2.3.58}$$

Simard et al. showed that the Recurrent Back Propagation algorithm does not necessarily converge to a stable fixed point (Simard et al., 1989). In addition, they indicated that this algorithm is sensitive to increases in the learning rate α. The performance of the networks trained by the Recurrent Back Propagation algorithm deteriorates when the learning task increases in complexity. More specifically, they indicated that as the complexity of the learning task increases, there is a higher possibility that the gradient descent based algorithm may be trapped in local minima. Simard et al. suggested that this particular problem may be partially solved by combining gradient descent with simulated annealing. Pearlmutter argued that the Recurrent Back Propagation algorithm proposed by Pineda deals only with the limiting behavior of this class of networks and suppresses all other temporal behavior (Pearlmutter, 1989). Another criticism regards the basic assumption upon which the derivation of the Recurrent Back Propagation algorithm is based, that is, that there exists a simple fixed point. As an alternative, Pearlmutter proposed a procedure for the evaluation of $\partial E/\partial w_{ij}$, where E is an arbitrary functional of the temporal trajectory of the states of a continuous recurrent network and w_{ij} are the synaptic weights of the network. Under certain assumptions, the algorithm based on this approach provides the Recurrent Back Propagation algorithm.

The Recurrent Back Propagation algorithm was used to train a network to calculate stereo disparity in random-dot stereograms (Qian and Sejnowski, 1988b). Giles et al. showed that a second-order recurrent network learns to infer regular grammars from positive and negative strings of training samples (Giles, et al., 1990; Giles et al., 1991). Kamimura explored the possibility of acquisition of syntactic competence with recurrent neural networks, by training a fully recurrent neural network to discriminate between well-formed and ill-formed sentences (Kamimura, 1991).

2.4 Self-organizing Neural Networks

Humans have the unique ability to use their past experiences in order to adapt to unpredictable changes of their environment. In the context of learning systems, such adaptation with no involvement of an external teacher is called *self-organization.* This section presents the basics of two competitive self-organizing neural architectures.

2.4.1 The Kohonen's Feature Map

In a self-organizing neural network, neighboring cells compete in their activation by means of mutual lateral interactions, and develop adaptively into specific detectors of different signal patterns. This category of learning is called *competitive, unsupervised,* or *self-organizing.* The self-organizing feature map, proposed by Kohonen, is probably the most representative neural system in this category (Kohonen, 1990). The self-organizing feature map is a sheet-like neural network, the cells of which become specifically tuned to various input signal patterns or classes of patterns through an unsupervised learning process. In this network, only one cell or local group of cells at a time provides the active response to the current input. The location of the responses tend to become ordered as if some meaningful coordinate system for different input features were being created over the network. The spatial location or coordinates of a cell in the network then correspond to a particular domain of input signal patterns. Each cell or local group of cells acts like a separate decoder for the same input. It is thus the presence or absence of an active response at that location, and not so much the exact input-output transformation or the magnitude of the response, that provides an interpretation of the input information. Self-organizing feature maps have been used for tasks similar to those to which other more traditional neural networks have been applied, such as pattern recognition, robotics, and process control (Kohonen, 1989).

A two-dimensional map consists of cells which form a hexagonal or rectangular grid. Figure 2.15 shows a two-dimensional feature map whose cells form a rectangular grid. The map receives as input the vector $\mathbf{x}^* = [x_1, x_2, \ldots, x_n]$, which is connected in parallel to

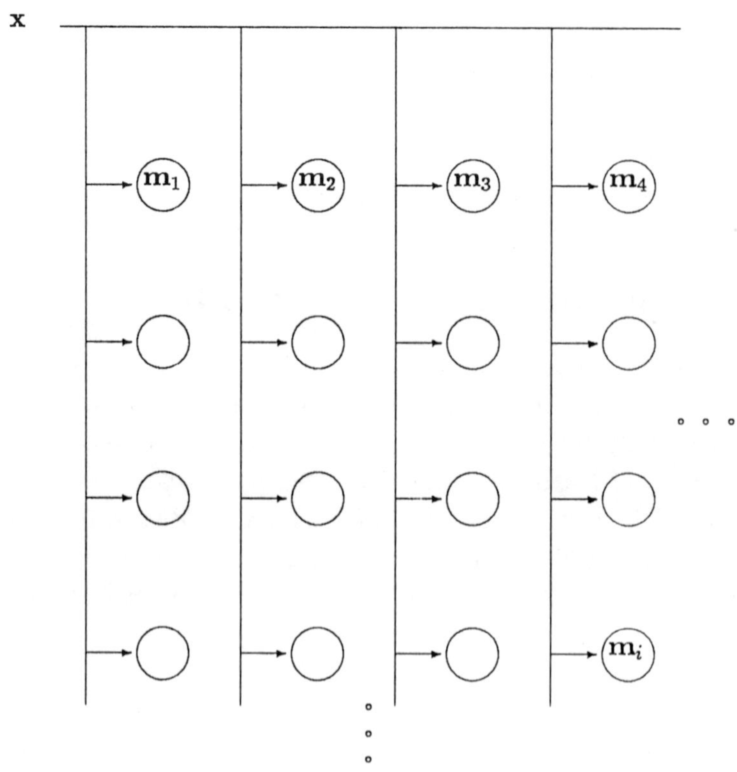

Figure 2.15: A two-dimensional self-organizing map.

all the cells. A weight vector is assigned to each one of the cells. Let $\mathbf{m}_i^* = [m_{i1}, m_{i2}, \ldots, m_{in}]$ be the vector which corresponds to the ith cell of the map. The weights m_{ij} which form each weight vector $\mathbf{m}_i$ are the adaptive parameters of the system. Each weight vector may be interpreted as an "image" that will be compared to the input $\mathbf{x}^* = [x_1, x_2, \ldots, x_n]$. The similarity between the input vector $\mathbf{x}$ and each of the weight vectors is usually measured by their Euclidean distance. Another possible measure of similarity between the input vector and each weight vector $\mathbf{m}_i$ is their inner product $< \mathbf{x}, \mathbf{m}_i >= \mathbf{x}^* \mathbf{m}_i$.

Kohonen proposed an adaptive process which determines the

weight vector of each cell in such a way that every cell becomes sensitive to a particular subset of input signals in a regular order. This process is called *ordering* of the map, since it results in a spatial ordering of the output vectors. During the ordering process, the cells of the map are not updated independently but as topologically related subsets. The selection of the subset of cells to be updated at each learning step requires the definition of a *center cell.* During each learning step, the center cell is that whose weight vector $\mathbf{m}_c$ is the closest to the input vector $\mathbf{x}$ in the Euclidean distance sense., i.e.,

$$||\mathbf{x} - \mathbf{m}_c|| = \min_i ||\mathbf{x} - \mathbf{m}_i|| \tag{2.4.1}$$

The subset of cells whose weight vectors are updated at a given learning step is determined by defining a neighborhood $N_c(t)$ around the center cell (Kohonen, 1989; Kohonen, 1990). Kohonen also suggested that the quality of the ordering improves if the neighborhood N_c changes with time, i.e., $N_c = N_c(t)$. He also suggested that $N_c = N_c(t)$ must be very wide when the ordering process begins and shrink monotonically with time. Figure 2.16 shows an example of a topological neighborhood in three different instances of the ordering process. A wide neighborhood in the initial steps guarantees a rough global ordering of the weight vectors, while the shrinking of the neighborhood N_c during the ordering improves the spatial resolution of the map.

The weight vectors of the cells can be updated according to the following rule

$$\mathbf{m}_i(t+1) = \begin{cases} \mathbf{m}_i(t) + \alpha(t)\,[\mathbf{x}(t) - \mathbf{m}_i(t)] & \text{if } i \in N_c(t) \\ \mathbf{m}_i(t) & \text{if } i \notin N_c(t) \end{cases} \tag{2.4.2}$$

where $\alpha(t)$ is the "adaptation gain" which satisfies $0 < \alpha(t) < 1$. During the initial steps, $\alpha(t)$ should be constant and close to unity. After the initial steps, $\alpha(t)$ should decrease monotonically with time. It was indicated that $\alpha(t)$ could be a linear or an exponential function of t (Kohonen, 1990). For example, Kohonen suggested that $\alpha(t)$ may be evaluated at every step as $\alpha(t) = 0.9\,(1 - t/1000)$.

The update rule (2.4.2) requires the selection of the neighborhood $N_c(t)$ at every step. This can be avoided by updating the

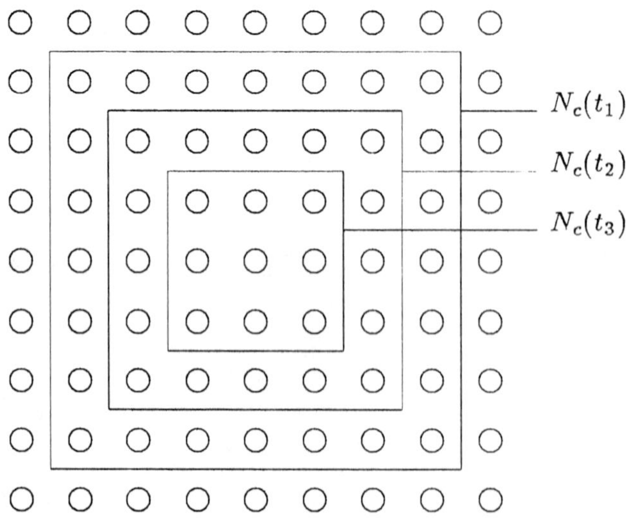

Figure 2.16: A topological neighborhood ($t_1 < t_2 < t_3$).

weight vectors according to the following generalized rule

$$\mathbf{m}_i(t+1) = \mathbf{m}_i(t) + h_{ci}(t)\left[\mathbf{x}(t) - \mathbf{m}_i(t)\right] \tag{2.4.3}$$

where $h_{ci} = h_{ci}(t)$ is a scalar "kernel" function. Obviously, the update equation (2.4.2) is a special case of the generalized update equation (2.4.3) resulting after the assumption that

$$h_{ci}(t) = \begin{cases} \alpha(t) & \text{if } i \in N_c(t) \\ 0 & \text{if } i \notin N_c(t) \end{cases} \tag{2.4.4}$$

Kohonen argued that h_{ci} could be a function of the relative distance between the center cell and the ith cell (Kohonen, 1990). Assuming that $\mathbf{r}_c$ and $\mathbf{r}_i$ are the coordinates of the center and the ith cells respectively, h_{ci} could be the following bell curved function (Kohonen, 1990)

$$h_{ci} = h_{ci}(t) = h_o(t)\exp(-||\mathbf{r}_i - \mathbf{r}_c||^2/\sigma(t)^2) \tag{2.4.5}$$

where $h_o = h_o(t)$ and $\sigma = \sigma(t)$ are both decreasing functions of time. The definition of $h_{ci} = h_{ci}(t)$ guarantees that only the weight vectors of cells which are adjacent to the center cell are substantially updated.

The activation of a cell in an ordered map by a specific input vector relates to the replacement of a vector by its closest codebook vector in classical *vector quantization* (Linde et al., 1980; Gray, 1984; Gersho and Gray, 1992). There is a similarity between the ordering of a feature map and the codebook design in vector quantization. In fact, Kohonen presented some extensions of the above algorithm which are strongly affected by statistical decision theory and the early work in vector quantization (Kohonen, 1990).

The ordering of feature maps is achieved through a simple update rule. However, the mathematical study of the self-organization process is a rather difficult task. Kohonen presented some elements of such a mathematical analysis (Kohonen, 1989). The self-organization process has also been exhibited through simple computer simulations (Kohonen, 1989; Kohonen, 1990). The performance of an ordered map depends strongly on the number of learning steps during the ordering process. In fact, the larger the number of steps, the better the performance of the map. As a "rule of thumb", Kohonen suggested that the number of steps must be 500 times the number of network units.

Self-organizing feature maps have been successfully used in speech recognition (Kohonen, 1989; Kohonen, 1990), image compression (Nasrabadi and Feng, 1988), control of robot arms and other robotics applications (Graf and LaLonde, 1988; Martinez et al., 1990; Ritter et al., 1989), automatic synthesis of digital systems (Hemani and Postula, 1990), optimization problems (Angeniol, et al., 1988), object recognition (Lampinen and Oja, 1990a), etc.

Learning Vector Quantization

The self-organizing feature map approximates the input signals by localized weight vectors. If the feature map is used as a classifier, the cells must be grouped into subsets which correspond to discrete classes. In this case, the ordering of the map by the update rules described above is not enough. After the original unsupervised ordering, the map is trained to function as a classifier using a training set of input vectors whose classification is already known. Kohonen proposed learning algorithms for the fine tuning of the map which follows its original ordering, which are known as *learning vector quantization* (Kohonen et al., 1988; Kohonen, 1990).

Suppose that several codebook vectors are assigned by an algorithm to a certain class. Each of these vectors is labeled with a symbol which reveals its inclusion in this class. If an input vector $\mathbf{x}$ is closer to a codebook vector $\mathbf{m}_i$, it is classified into the particular class which includes the codebook vector $\mathbf{m}_i$. The resulting weight vectors are assigned to a certain class by using a training set of input vectors, i.e., input vectors whose classification is already known. The input vectors belonging to this training set are presented to the map. Each cell $\mathbf{m}_i$ is assigned to a certain class if it is the closest to the majority of the input vectors which belong to this class.

The identity of each codebook vector within a certain class is not particularly important, since several codebook vectors may represent the same class. This is an indication that if a map is trained to function as a classifier the main objective is to determine near-optimal boundaries between the classes. Kohonen proposed a learning procedure for the refinement of the original classification, which is called learning vector quantization 1 (LVQ1). The main objective of this learning strategy is to create more accurate borders between the various classes by forcing the codebook vectors away from the decision boundaries. An input vector $\mathbf{x}$ is assigned to a certain class if its closest codebook vector belongs to this class. Suppose that the codebook vector $\mathbf{m}_c$ is the closest to the input vector $\mathbf{x}$, resulting in the assignment of $\mathbf{x}$ to the correspongding class c. Assuming that a training input vector $\mathbf{x}$ is presented to the map, each codebook vector $\mathbf{m}_i$ is updated according to the following rule:

If $i = c$ and $\mathbf{x}$ is classified correctly:

$$\mathbf{m}_i(t+1) = \mathbf{m}_i(t) + \alpha(t)\,[\mathbf{x}(t) - \mathbf{m}_i] \qquad (2.4.6)$$

If $i = c$ and $\mathbf{x}$ is classified incorrectly:

$$\mathbf{m}_i(t+1) = \mathbf{m}_i(t) - \alpha(t)\,[\mathbf{x}(t) - \mathbf{m}_i] \qquad (2.4.7)$$

If $i \neq c$:

$$\mathbf{m}_i(t+1) = \mathbf{m}_i(t) \tag{2.4.8}$$

Once again, $\alpha(t)$ is a scalar gain between 0 and 1 which decreases monotonically in time. In this case, the training process starts with a small value of $\alpha(t)$ and decreases to zero after a sufficiently large number of steps.

The learning process determined by the LVQ1 algorithm is supervised, since it depends on the existence of a training set of input vectors with known classification. Kohonen attempted to improve the LVQ1 algorithm on the basis of arguments strongly influenced by classical Bayesian decision theory (Duda and Hart, 1973). The Learning Vector Quantization 2 (LVQ2) and the Learning Vector Quantization 3 (LVQ3) are two alternative algorithms proposed for the refinement of feature maps functioning as classifiers (Kohonen, 1990).

2.4.2 Adaptive Resonance Theory

The development of autonomous systems learning in a changing environment involves the solution of the *stability-plasticity dilemma* (Carpenter and Grossberg, 1988). Such a system must remain plastic, or adaptive, to significant events and simultaneously remain stable to insignificant events. In addition, such a system must preserve previously acquired knowledge while learning new things in such a way that previous memories are not washed away when new information is presented to the system.

Grossberg developed adaptive resonance theory (ART) as a solution to the stability-plasticity dilemma (Grossberg, 1987c). A key idea in ART is that learned top-down expectations prevent the removal of previously acquired memories by focusing attention on bottom-up information. As a result, new information may consistently be stored into the system.

Carpenter and Grossberg proposed several self-organizing neural architectures for the implementation of the theory (Carpenter and Grossberg, 1988). In general, these architectures create stable recognition codes in real time in response to arbitrary sequences of input patterns. The ART 1 was the original system, developed to

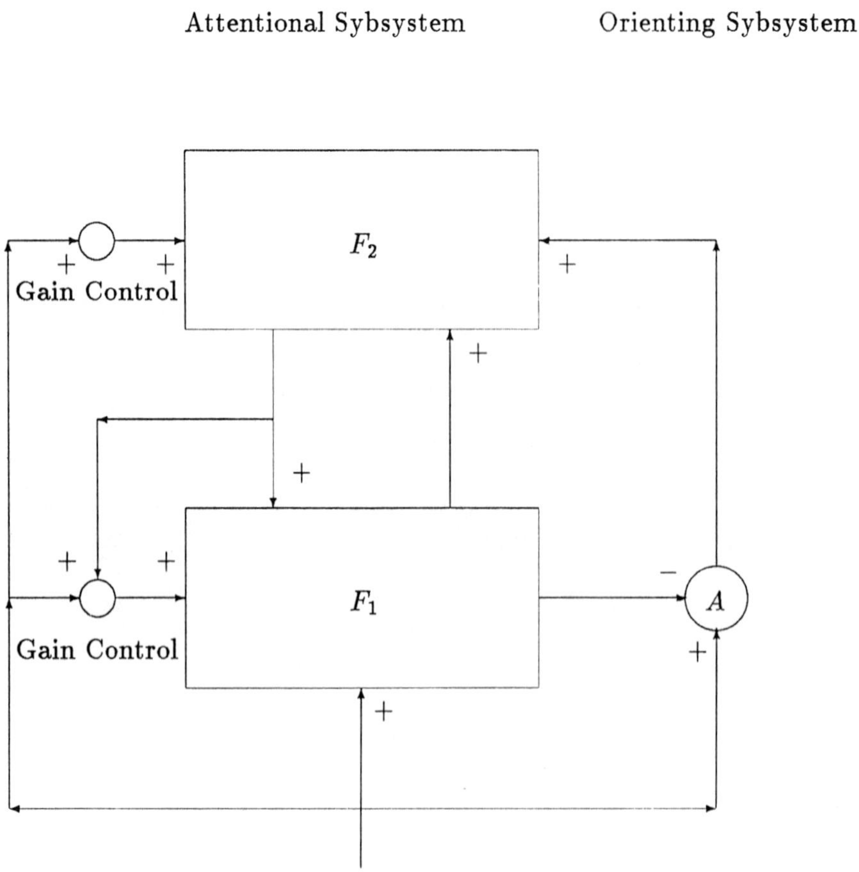

Figure 2.17: The ART 1 neural architecture.

accept only binary input patterns (Carpenter and Grossberg, 1987a; Carpenter and Grossberg, 1988; Moore, 1989).

Figure 2.17 shows the general structure of the ART 1 neural architecture. The ART 1 architecture consists of an attentional subsystem and an orienting subsystem. F_1 and F_2 are two successive stages of the attentional subsystem which encode patterns in a short-term memory (STM). The signals in the two pathways between F_1 and F_2 are multiplied by long-term memory (LTM) traces. The STM and LTM processes are modulated by the gain control modules and the orienting subsystem A. The orienting subsystem A generates an output signal, called *reset wave*, in the case where there exists a sufficiently large mismatch between the bottom-up pattern and the top-down expectation pattern. The generation of the reset wave by the orienting subsystem A is based on the 2/3 Rule, which works as follows: When the system is in the bottom-up mode, the orienting subsystem A receives an inhibitory as well as an excitatory signal. The orienting subsystem does not emit a reset wave, provided that the system is designed in such a way that the effect of the inhibitory signal is larger than that of the excitatory signal. The F_1 subsystem distinguishes between a bottom-up signal and a top-down expectation with the help of the gain control module. Another gain control module affects F_2 in such a way that a reset wave is generated by the orienting subsystem A when a significantly large mismatch between bottom-up signal and top-down expectation signal occurs. This is achieved by reducing the effect of the inhibitory signal in the orienting subsystem A and, therefore, forcing A to emit a reset wave. In fact, the emission of a reset wave is controlled by the parameter ρ, which is called *vigilance parameter*. A high vigilance parameter results in a system that is sensitive to small differences between the bottom-up signal and the top-down expectation signal. Conversely, low vigilance results in a system which groups input patterns on the basis of a coarse measure of similarity.

Some of the limitations of the ART 1 were solved in the ART 2, which was developed for analog input patterns (Carpenter and Grossberg, 1987b). The ART 3 was proposed recently as an improvement of the previous architectures (Carpenter and Grossberg, 1989). In the proposed ART architectures, adaptive pattern recognition is a special case of a more general process which involves hypothesis discovery, testing, search, classification, and learning (Carpenter and Grossberg, 1988).

2.5 Discussion

This chapter presented a review of the most commonly used neural network architectures and learning schemes. The selection of the neural network models and learning schemes presented in this chapter was based on a combined criterion comprising theoretical impact, popularity, and usefulness in practical applications. Certainly, there exist in the literature alternative neural network models as well as variations of the neural network models and learning schemes presented in this chapter. In fact, the development of more efficient neural network architectures is a focus of current research, and there is a considerable research effort directed toward the development of more efficient learning schemes and the implementation of neural network systems. The implementation of neural network systems is an important issue which, however, is beyond the scope of this book. The following chapters present some interesting new developments in the training of feed-forward neural networks. Furthermore, this book provides a sense of the efficiency of the existing neural network architectures and learning schemes as well as their potential use in real-world applications.

Appendix A2.1

The Delta Rule learning algorithm is derived by minimizing the objective function E_k, defined by (2.2.24), using the gradient descent method. The update equation for the synaptic weights w_{pq} of the network is provided by the gradient descent method as

$$\mathbf{w}_{p,k} - \mathbf{w}_{p,k-1} = -\alpha \frac{\partial E_k}{\partial \mathbf{w}_p} = \alpha \sum_{i=1}^{n_o} (y_{i,k} - \hat{y}_{i,k}) \frac{\partial \hat{y}_{i,k}}{\partial \mathbf{w}_p} \tag{A2.1.1}$$

It can easily be verified that when $\hat{y}_{i,k} = \sigma(\bar{y}_{i,k})$,

$$\frac{\partial \hat{y}_{i,k}}{\partial \mathbf{w}_p} = \frac{\partial \sigma(\bar{y}_{i,k})}{\partial \bar{y}_{i,k}} \frac{\partial \bar{y}_{i,k}}{\partial \mathbf{w}_p} = \sigma'(\bar{y}_{i,k})\, \mathbf{x}_k\, \delta_{ip} \tag{A2.1.2}$$

where δ_{ip} is the Kronecker delta, defined as $\delta_{ip} = 1$ if $i = p$ and $\delta_{ip} = 0$ if $i \neq p$. The combination of (A2.1.2) and (A2.1.1) provides the update equation for the synaptic weights w_{pq} as follows

$$\mathbf{w}_{p,k} = \mathbf{w}_{p,k-1} + \alpha\, \varepsilon^o_{p,k}\, \mathbf{x}_k \tag{A2.1.3}$$

where

$$\varepsilon^o_{p,k} = \sigma'(\bar{y}_{p,k})\, (y_{p,k} - \hat{y}_{p,k}) \tag{A2.1.4}$$

Appendix A2.2

The derivation of the Error Back Propagation algorithm requires the evaluation of the gradient $\partial E_k / \partial \mathbf{w}_p$. According to the definition of E_k by (2.2.23),

$$\mathbf{w}_{p,k} - \mathbf{w}_{p,k-1} = -\alpha \frac{\partial E_k}{\partial \mathbf{w}_p} = -\alpha \sum_{i=1}^{n_o} \frac{\partial \phi(e_{i,k})}{\partial \hat{y}_{i,k}} \frac{\partial \hat{y}_{i,k}}{\partial \mathbf{w}_p} \tag{A2.2.1}$$

From the definition of $\hat{y}_{i,k}$ by (2.2.44),

$$\frac{\partial \hat{y}_{i,k}}{\partial \mathbf{w}_p} = \frac{\partial \sigma(\bar{y}_{i,k})}{\partial \bar{y}_{i,k}} \frac{\partial \bar{y}_{i,k}}{\partial \mathbf{w}_p} = \sigma'(\bar{y}_{i,k})\, \hat{\mathbf{h}}_k\, \delta_{ip} \tag{A2.2.2}$$

The update equation for the synaptic weights w_{pq} is obtained by combining (A2.2.1) and (A2.2.2) as follows

$$\mathbf{w}_{p,k} - \mathbf{w}_{p,k-1} = \alpha\, \varepsilon_{p,k}^{o}\, \hat{\mathbf{h}}_k \tag{A2.2.3}$$

where

$$\varepsilon_{p,k}^{o} = -\sigma'(\bar{y}_{p,k}) \frac{\partial \phi(e_{p,k})}{\partial \hat{y}_{p,k}} \tag{A2.2.4}$$

The update equation for the synaptic weights v_{pq} can be similarly obtained. From the definition of E_k,

$$\mathbf{v}_{p,k} - \mathbf{v}_{p,k-1} = -\alpha \frac{\partial E_k}{\partial \mathbf{v}_p} = -\alpha \sum_{i=1}^{n_o} \frac{\partial \phi(e_{i,k})}{\partial \hat{y}_{i,k}} \frac{\partial \hat{y}_{i,k}}{\partial \mathbf{v}_p} \tag{A2.2.5}$$

Taking into account the definition of $\hat{y}_{i,k}$ by (2.2.44),

$$\frac{\partial \hat{y}_{i,k}}{\partial \mathbf{v}_p} = \frac{\partial \sigma(\bar{y}_{i,k})}{\partial \bar{y}_{i,k}} \frac{\partial \bar{y}_{i,k}}{\partial \mathbf{v}_p} = \sigma'(\bar{y}_{i,k}) \sum_{j=1}^{n_h} w_{ij} \frac{\partial \hat{h}_{j,k}}{\partial \mathbf{v}_p} \tag{A2.2.6}$$

Finally, the definition of $\hat{h}_{j,k}$ by (2.2.42) gives

$$\frac{\partial \hat{h}_{j,k}}{\partial \mathbf{v}_p} = \frac{\partial \rho(\bar{h}_{j,k})}{\partial \bar{h}_{j,k}} \frac{\partial \bar{h}_{j,k}}{\partial \mathbf{v}_p} = \rho'(\bar{h}_{j,k})\, \mathbf{x}_k\, \delta_{jp} \tag{A2.2.7}$$

The combination of (A2.2.6) and (A2.2.7) with (A2.2.5) provides the following update equation for the synaptic weights v_{pq}

$$\mathbf{v}_{p,k} - \mathbf{v}_{p,k-1} = \alpha\, \varepsilon_{p,k}^{h}\, \mathbf{x}_k \tag{A2.2.8}$$

where

$$\varepsilon_{p,k}^{h} = (1 - \hat{h}_{p,k}^{2}) \sum_{i=1}^{n_o} \varepsilon_{i,k}^{o} \, w_{ip} \tag{A2.2.9}$$

and $\varepsilon_{i,k}^{o}$ is given by (A2.2.4).

The direct generalization of the above analysis provides the Error Back Propagation algorithm for neural networks with more than one layer of hidden units, such as the network shown in Figure 2.10. The estimates provided by this network are given by (2.2.45), which is identical to (2.2.44) for $\hat{\mathbf{h}}_k^{(1)} = \hat{\mathbf{h}}_k$. As a result, the update equation for the synaptic weights $v_{pq}^{(0)}$ can be obtained from (A2.2.3) for $\mathbf{v}_{p,k}^{(0)} = \mathbf{w}_{p,k}$ $\hat{\mathbf{h}}_k = \hat{\mathbf{h}}_k^{(1)}$ as follows

$$\mathbf{v}_{p,k}^{(0)} = \mathbf{v}_{p,k-1}^{(0)} + \alpha \, \varepsilon_{p,k}^{(0)} \, \hat{\mathbf{h}}_k^{(1)} \tag{A2.2.10}$$

where $\varepsilon_{p,k}^{(0)} = \varepsilon_{p,k}^{o}$ is given by (A2.2.4). The synaptic weights $v_{pq}^{(r)}, r = 1, 2, \ldots, L$ can be updated by minimizing the objective function E_k, defined by (2.2.23), through the equation

$$\mathbf{v}_{p,k}^{(r)} - \mathbf{v}_{p,k-1}^{(r)} = -\alpha \frac{\partial E_k}{\partial \mathbf{v}_p^{(r)}} = \alpha \sum_{i=1}^{n_o} \frac{\partial \phi(y_{i,k})}{\partial \hat{y}_{i,k}} \frac{\partial \hat{y}_{i,k}}{\partial \mathbf{v}_p^{(r)}} \tag{A2.2.11}$$

In the case where $r = 1$,

$$\frac{\partial \hat{y}_{i,k}}{\partial \mathbf{v}_p^{(1)}} = \sigma'(\bar{y}_{i,k}) \sum_{j=1}^{n_1} w_{ij} \frac{\partial \hat{h}_{j,k}^{(1)}}{\partial \mathbf{v}_p^{(1)}} \tag{A2.2.12}$$

where

$$\frac{\partial \hat{h}_{j,k}^{(1)}}{\partial \mathbf{v}_p^{(1)}} = (1 - \hat{h}_{j,k}^{(1)\,2}) \, \hat{\mathbf{h}}_k^{(2)} \, \delta_{jp} \tag{A2.2.13}$$

For $r = 1$, the combination of (A2.2.12) and (A2.2.13) with (A2.2.11) gives

$$\mathbf{v}_{p,k}^{(1)} = \mathbf{v}_{p,k-1}^{(1)} + \alpha \, \varepsilon_{p,k}^{(1)} \, \hat{\mathbf{h}}_k^{(2)} \tag{A2.2.14}$$

where

$$\varepsilon_{p,k}^{(1)} = (1 - \hat{h}_{p,k}^{(1)\,2}) \sum_{i=1}^{n_o} \varepsilon_{i,k}^{(0)} \, w_{ip} \tag{A2.2.15}$$

The direct generalization of the above results provides the following update equation for the synaptic weights $v_{pq}^{(r)}, r = 0, 1, \ldots, L$

$$\mathbf{v}_{p,k}^{(r)} = \mathbf{v}_{p,k-1}^{(r)} + \alpha\, \varepsilon_{p,k}^{(r)}\, \hat{\mathbf{h}}_k^{(r+1)} \tag{A2.2.16}$$

where $\hat{\mathbf{h}}_k^{(L+1)} = \mathbf{x}_k$, and

$$\varepsilon_{p,k}^{(r)} = (1 - \hat{h}_{p,k}^{(r)\,2}) \sum_{\ell=1}^{n_{r-1}} \varepsilon_{\ell,k}^{(r-1)}\, v_{\ell p}^{(r-1)}\, \forall r = 1, 2, \ldots, L \tag{A2.2.17}$$

Appendix A2.3

This Appendix presents the derivation of Oja's rule for a single linear unit. The normalized version of the Hebbian rule (2.2.64) can be written as

$$v_{i,k} = \frac{\bar{v}_{i,k}}{\omega(\bar{v}_{i,k}, i = 1, 2, \ldots, n_i)} \tag{A2.3.1}$$

where

$$\bar{v}_{i,k} = v_{i,k-1} + \alpha\, \hat{y}_k\, x_{i,k} \tag{A2.3.2}$$

and

$$\omega(\bar{v}_{i,k}, i = 1, 2, \ldots, n_i) = (\sum_{i=1}^{n_i} \bar{v}_{i,k}^2)^{\frac{1}{2}} \tag{A2.3.3}$$

The Taylor expansion of the function $\omega(.)$ around $\alpha = 0$ gives

$$\begin{aligned} \omega(\bar{v}_{i,k}, i = 1, 2, \ldots, n_i) &= \omega(v_{i,k-1} + \alpha\, \hat{y}_k\, x_{i,k}, i = 1, 2, \ldots, n_i) \\ &= \omega(v_{i,k-1}, i = 1, 2, \ldots, n_i) \\ &+ \alpha \frac{\partial \omega(\bar{v}_{i,k}, i = 1, 2, \ldots, n_i)}{\partial \alpha}|_{\alpha=0} + \ldots \end{aligned} \tag{A2.3.4}$$

For very small α, the terms which contain second or higher powers of α can be ignored. Since $\omega(v_{i,k-1}, i = 1, 2, \ldots, n_i) = (\sum_{i=1}^{n_i} v_{i,k-1}^2)^{\frac{1}{2}}$ $= 1$, (A2.3.4) can be written as

$$\omega(\bar{v}_{i,k}, i = 1, 2, \ldots, n_i) = 1 + \alpha \frac{\partial \omega(\bar{v}_{i,k}, i = 1, 2, \ldots, n_i)}{\partial \alpha}|_{\alpha=0} + \ldots \tag{A2.3.5}$$

Assuming that α is sufficiently small such that $\alpha \frac{\partial \omega(.)}{\partial \alpha}|_{\alpha=0}$ is absolutely less than unity, (A2.3.5) gives

$$\frac{1}{\omega(\bar{v}_{i,k}, i = 1, 2, \ldots, n_i)} = 1 - \alpha \frac{\partial \omega(\bar{v}_{i,k}, i = 1, 2, \ldots, n_i)}{\partial \alpha}|_{\alpha=0} + \ldots \tag{A2.3.6}$$

Ignoring once again the terms involving second or higher powers of α, the combination of (A2.3.1) and (A2.3.6) results in

$$\begin{aligned} v_{i,k} &= \bar{v}_{i,k} - \alpha\, \bar{v}_{i,k} \frac{\partial \omega(.)}{\partial \alpha}|_{\alpha=0} \\ &= v_{i,k-1} + \alpha\, \hat{y}_k\, x_{i,k} - (\alpha\, v_{i,k-1} - \alpha^2\, \hat{y}_k\, x_{i,k}) \frac{\partial \omega(.)}{\partial \alpha}|_{\alpha=0} \end{aligned} \tag{A2.3.7}$$

If α is sufficiently small, (A2.3.7) can be approximated by

$$v_{i,k} = v_{i,k-1} + \alpha\, \hat{y}_k\, x_{i,k} - \alpha\, v_{i,k-1} \frac{\partial \omega(.)}{\partial \alpha}|_{\alpha=0} \tag{A2.3.8}$$

According to the definition of $\omega(.)$ by (A2.3.3),

$$\begin{aligned} \frac{\partial \omega(\bar{v}_{i,k}, i = 1, 2, \ldots, n_i)}{\partial \alpha} &= [\sum_{i=1}^{n_i} (v_{i,k-1} + \alpha\, \hat{y}_k\, x_{i,k})^2]^{-\frac{1}{2}} \\ &\quad \sum_{i=1}^{n_i} (v_{i,k-1} + \alpha\, \hat{y}_k\, x_{i,k})\, \hat{y}_k\, x_{i,k} \end{aligned} \tag{A2.3.9}$$

When $\partial \omega(.)/\partial \alpha$ is evaluated at $\alpha = 0$,

$$\frac{\partial \omega(\bar{v}_{i,k}, i = 1, 2, \ldots, n_i)}{\partial \alpha}|_{\alpha=0} = [\sum_{i=1}^{n_i} v_{i,k-1}^2]^{-\frac{1}{2}}\, \hat{y}_k \sum_{i=1}^{n_i} v_{i,k-1}\, x_{i,k} \tag{A2.3.10}$$

Since $\sum_{i=1}^{n_i} v_{i,k-1}^2 = 1$ and $\hat{y}_k = \bar{y}_k = \sum_{i=1}^{n_i} v_{i,k-1}\, x_{i,k}$, the latter equation amounts to

$$\frac{\partial \omega(\bar{v}_{i,k}, i = 1, 2, \ldots, n_i)}{\partial \alpha}|_{\alpha=0} = \hat{y}_k^2 = \hat{y}_k\, \bar{y}_k \tag{A2.3.11}$$

The combination of (A2.3.8) and (A2.3.11) gives

$$v_{i,k} = v_{i,k-1} + \alpha\, \hat{y}_k\, (x_{i,k} - \bar{y}_k\, v_{i,k-1}) \tag{A2.3.12}$$

which is Oja's learning rule.

The derivation of the update equation for a single-layered neural network with linear units is essentially the generalization of the derivation presented above for a single linear unit. In this case, principal component analysis of the input can be achieved by updating the matrix of synaptic weights $\mathbf{W} = [\mathbf{w}_1 \mathbf{w}_2 \ldots \mathbf{w}_{n_o}]$ as follows (Oja, 1992)

$$\mathbf{W}_k = \bar{\mathbf{W}}_k \, \mathbf{S}_k^{-1} \tag{A2.3.13}$$

where

$$\bar{\mathbf{W}}_k = \mathbf{W}_{k-1} + \alpha \, \mathbf{x}_k \, \mathbf{x}_k^* \, \mathbf{W}_{k-1} \tag{A2.3.14}$$

and $\mathbf{S}_k$ is a matrix depending on $\bar{\mathbf{W}}_k$, which orthonormalizes the columns of $\bar{\mathbf{W}}_k$. Since $\hat{\mathbf{y}}_k = \mathbf{W}_{k-1}^* \, \mathbf{x}_k$, (A2.3.14) is the update equation provided by the Hebbian rule and (A2.3.13) is its normalized version.

It is assumed here that the columns of $\mathbf{W}_{k-1}$ are orthonormal, in compliance with the original requirement. If α is sufficiently small, the columns of $\bar{\mathbf{W}}_k$ are linearly independent and the matrix $\bar{\mathbf{W}}_k^* \bar{\mathbf{W}}_k$ is nonsingular and positive definite. The matrix $\mathbf{S}_k$ is defined in terms of $\bar{\mathbf{W}}_k$ as

$$\mathbf{S}_k = (\bar{\mathbf{W}}_k^* \, \bar{\mathbf{W}}_k)^{\frac{1}{2}} \tag{A2.3.15}$$

The combination of (A2.3.13) and (A2.3.15) gives

$$\begin{aligned} \mathbf{W}_k^* \, \mathbf{W}_k &= \mathbf{S}_k^{-1} \, \bar{\mathbf{W}}_k^* \, \bar{\mathbf{W}}_k \, \mathbf{S}_k^{-1} \\ &= (\bar{\mathbf{W}}_k^* \, \bar{\mathbf{W}}_k)^{-\frac{1}{2}} \, (\bar{\mathbf{W}}_k^* \, \bar{\mathbf{W}}_k)^{\frac{1}{2}} \, (\bar{\mathbf{W}}_k^* \, \bar{\mathbf{W}}_k)^{\frac{1}{2}} \, (\bar{\mathbf{W}}_k^* \, \bar{\mathbf{W}}_k)^{-\frac{1}{2}} \\ &= \mathbf{I} \end{aligned} \tag{A2.3.16}$$

Therefore, the definition of $\mathbf{S}_k$ by (A2.3.15) guarantees that the column vectors of $\mathbf{W}_k$ are orthonormal.

Combining (A2.3.14) and the definition of $\mathbf{S}_k$ by (A2.3.15),

$$\begin{aligned} \mathbf{S}_k^{-1} &= (\bar{\mathbf{W}}_k^* \, \bar{\mathbf{W}}_k)^{-\frac{1}{2}} \\ &= [(\mathbf{W}_{k-1} + \alpha \, \mathbf{x}_k \, \mathbf{x}_k^* \, \mathbf{W}_{k-1})^* \, (\mathbf{W}_{k-1} + \alpha \, \mathbf{x}_k \, \mathbf{x}_k^* \, \mathbf{W}_{k-1})]^{-\frac{1}{2}} \\ &= [\mathbf{I} + 2 \, \alpha \, \mathbf{W}_{k-1}^* \, \mathbf{x}_k \, \mathbf{x}_k^* \, \mathbf{W}_{k-1} + \ldots]^{-\frac{1}{2}} \end{aligned} \tag{A2.3.17}$$

where the dots denote terms including powers of α higher than one. If α is sufficiently small, (A2.3.17) can be approximated by

$$\mathbf{S}_k^{-1} = \mathbf{I} - \alpha \, \mathbf{W}_{k-1}^* \, \mathbf{x}_k \, \mathbf{x}_k^* \, \mathbf{W}_{k-1} + \ldots \tag{A2.3.18}$$

Ignoring the terms involving powers of α higher than one, the combination of (A2.3.14) and (A2.3.18) with (A2.3.13) gives

$$\begin{aligned}\mathbf{W}_k &= (\mathbf{W}_{k-1} + \alpha\,\mathbf{x}_k\,\mathbf{x}_k^*\,\mathbf{W}_{k-1})\,(\mathbf{I} - \alpha\,\mathbf{W}_{k-1}^*\,\mathbf{x}_k\,\mathbf{x}_k^*\,\mathbf{W}_{k-1}) \\ &= \mathbf{W}_{k-1} + \alpha\,\mathbf{x}_k\,\mathbf{x}_k^*\,\mathbf{W}_{k-1} - \alpha\,\mathbf{W}_{k-1}\,\mathbf{W}_{k-1}^*\,\mathbf{x}_k\,\mathbf{x}_k^*\,\mathbf{W}_{k-1} \\ &- \alpha^2\,\mathbf{x}_k\,\mathbf{x}_k^*\,\mathbf{W}_{k-1}\,\mathbf{W}_{k-1}^*\,\mathbf{x}_k\,\mathbf{x}_k^*\,\mathbf{W}_{k-1} \qquad \text{(A2.3.19)}\end{aligned}$$

Ignoring once again the term which includes α^2,

$$\mathbf{W}_k = \mathbf{W}_{k-1} + \alpha\,(\mathbf{I} - \mathbf{W}_{k-1}\,\mathbf{W}_{k-1}^*)\,\mathbf{x}_k\,\mathbf{x}_k^*\,\mathbf{W}_{k-1} \qquad \text{(A2.3.20)}$$

which is the update equation for this network.

Appendix A2.4

This Appendix presents the derivation of a learning algorithm for Boltzmann machines. Assume that $P'(V_\alpha H_\beta)$ is the actual probability that the state of the visible and hidden units is V_α and H_β, respectively, and let $P(V_\alpha H_\beta)$ be the corresponding desired probability. The actual and desired probabilities that the visible units are in state α are denoted by $P'(V_\alpha)$ and $P(V_\alpha)$, respectively. The actual probability $P'(V_\alpha)$ is given by

$$P'(V_\alpha) = \sum_\beta P'(V_\alpha H_\beta) = \frac{1}{Z}\sum_\beta \exp(-\frac{1}{T}E_{\alpha\beta}) \qquad \text{(A2.4.1)}$$

where

$$Z = \sum_\alpha \sum_\beta \exp(-\frac{1}{T}E_{\alpha\beta}) \qquad \text{(A2.4.2)}$$

From the definition of the discrepancy measure G in (2.3.45),

$$\frac{\partial G}{\partial w_{ij}} = \sum_\alpha P(V_\alpha)\frac{\partial}{\partial w_{ij}}\ln\frac{P(V_\alpha)}{P'(V_\alpha)} = -\sum_\alpha \frac{P(V_\alpha)}{P'(V_\alpha)}\frac{\partial P'(V_\alpha)}{\partial w_{ij}} \qquad \text{(A2.4.3)}$$

From the definition of $P'(V_\alpha)$ by (A2.4.1),

$$\frac{\partial P'(V_\alpha)}{\partial w_{ij}} = \frac{\partial}{\partial w_{ij}}\left\{\frac{1}{Z}\sum_\beta \exp(-\frac{1}{T}E_{\alpha\beta})\right\}$$

$$= \frac{1}{Z}\sum_{\beta}\frac{\partial}{\partial w_{ij}}\left\{\exp(-\frac{1}{T}E_{\alpha\beta})\right\}$$
$$- \frac{1}{Z^2}\sum_{\beta}\exp(-\frac{1}{T}E_{\alpha\beta})\sum_{\lambda}\sum_{\mu}\frac{\partial}{\partial w_{ij}}\left\{\exp(-\frac{1}{T}E_{\lambda\mu})\right\} \tag{A2.4.4}$$

According to the definition of the energy $E_{\alpha\beta}$ by (2.3.41),

$$\frac{\partial}{\partial w_{ij}}\left\{\exp(-\frac{1}{T}E_{\alpha\beta})\right\} = -\frac{1}{T}\exp(-\frac{1}{T}E_{\alpha\beta})\, s_i^{\alpha\beta}\, s_j^{\alpha\beta} \tag{A2.4.5}$$

The combination of (A2.4.4) and (A2.4.5) gives

$$\frac{\partial P'(V_\alpha)}{\partial w_{ij}} = -\frac{1}{T}\sum_{\beta}\left[\frac{1}{Z}\sum_{\beta}\exp(-\frac{1}{T}E_{\alpha\beta})\right] s_i^{\alpha\beta} s_j^{\alpha\beta}$$
$$- \frac{1}{T}\left[\frac{1}{Z}\sum_{\beta}\exp(-\frac{1}{T}E_{\alpha\beta})\right]\sum_{\lambda,\mu}\left[\frac{1}{Z}\exp(-\frac{1}{T}E_{\lambda\mu})\right] s_i^{\lambda\mu} s_j^{\lambda\mu}$$
$$= -\frac{1}{T}\left\{\sum_{\beta}P'(V_\alpha H_\beta)s_i^{\alpha\beta}s_j^{\alpha\beta} - P'(V_\alpha)\sum_{\lambda,\mu}P'(V_\lambda H_\mu)s_i^{\lambda\mu}s_j^{\lambda\mu}\right\} \tag{A2.4.6}$$

Substituting (A2.4.6) into (A2.4.3) gives

$$\frac{\partial G}{\partial w_{ij}} = \frac{1}{T}\sum_{\alpha}\sum_{\beta}\frac{P(V_\alpha)}{P'(V_\alpha)}P'(V_\alpha H_\beta)\, s_i^{\alpha\beta}\, s_j^{\alpha\beta}$$
$$- \frac{1}{T}\sum_{\alpha}P(V_\alpha)\sum_{\lambda}\sum_{\mu}P'(V_\lambda H_\mu)\, s_i^{\lambda\mu}\, s_j^{\lambda\mu} \tag{A2.4.7}$$

Clearly, $P(V_\alpha H_\beta)$ can be written in terms of the conditional probability $P(H_\beta/V_\alpha)$ as

$$P(V_\alpha H_\beta) = P(H_\beta/V_\alpha)\, P(V_\alpha) \tag{A2.4.8}$$

Similarly,

$$P'(V_\alpha H_\beta) = P'(H_\beta/V_\alpha)\, P'(V_\alpha) \tag{A2.4.9}$$

At equilibrium, the probability of occurrence of a hidden state given a visible state must be the same if the state of the visible units is clamped or reached by the free-running network. Since $P(H_\beta/V_\alpha) = P'(H_\beta/V_\alpha)$, the combination of (A2.4.8) and (A2.4.9) gives

$$\frac{P(V_\alpha)}{P'(V_\alpha)}\, P'(V_\alpha H_\beta) = P(V_\alpha H_\beta) \tag{A2.4.10}$$

Since $\sum_\alpha P(V_\alpha) = 1$, the combination of (A2.4.7) and (A2.4.10) gives

$$\frac{\partial G}{\partial w_{ij}} = \frac{1}{T}\left\{\sum_\alpha\sum_\beta P(V_\alpha H_\beta)\, s_i^{\alpha\beta}\, s_j^{\alpha\beta} - \sum_\lambda\sum_\mu P'(V_\lambda H_\mu)\, s_i^{\lambda\mu}\, s_j^{\lambda\mu}\right\} \tag{A2.4.11}$$

Appendix A2.5

This Appendix presents the derivation of the Recurrent Back Propagation algorithm, which is based on the minimization of the error function E defined by (2.3.55). According to the gradient descent method,

$$\Delta w_{pq} = -\alpha\, \frac{\partial E}{\partial w_{pq}} = \alpha \sum_k E_k\, \frac{\partial z_k^*}{\partial w_{pq}} \tag{A2.5.1}$$

Taking into account (2.3.53),

$$\frac{\partial z_i^*}{\partial w_{pq}} = \frac{\partial \rho(u_i^*)}{\partial u_i^*}\frac{\partial u_i^*}{\partial w_{pq}} = \rho'(u_i^*)\, \frac{\partial u_i^*}{\partial w_{pq}} \tag{A2.5.2}$$

According to the definition of u_i^* by (2.3.52), $\partial u_i^* / \partial w_{pq}$ can be evaluated in the case where $i = p$ and $i \neq p$ as follows

$$\frac{\partial u_i^*}{\partial w_{pq}} = \sum_j \left(\frac{\partial w_{ij}}{\partial w_{pq}}\, z_j^* + w_{ij}\, \frac{\partial z_j^*}{\partial w_{pq}}\right) = \delta_{ip}\, z_q^* + \sum_j w_{ij}\, \frac{\partial z_j^*}{\partial w_{pq}} \tag{A2.5.3}$$

The combination of (A2.5.2) and (A2.5.3) gives

$$\frac{\partial z_i^*}{\partial w_{pq}} = \rho'(u_i^*)\left(\delta_{ip}\, z_q^* + \sum_j w_{ij}\, \frac{\partial z_j^*}{\partial w_{pq}}\right) \tag{A2.5.4}$$

A simple manipulation of (A2.5.4) gives

$$\sum_j (\mathbf{L})_{ij} \frac{\partial z_j^*}{\partial w_{pq}} = \delta_{ip}\, \rho'(u_i^*)\, z_q^* \tag{A2.5.5}$$

where $\mathbf{L}$ is the matrix with elements

$$(\mathbf{L})_{ij} = \delta_{ij} - \rho'(u_i^*)\, w_{ij} \tag{A2.5.6}$$

Taking into account (A2.5.5), it can easily be verified that $\partial z_k^* / \partial w_{pq}$ is given in terms of the inverse matrix $\mathbf{L}^{-1}$ as

$$\frac{\partial z_k^*}{\partial w_{pq}} = (\mathbf{L}^{-1})_{kp}\, \rho'(u_p^*)\, z_q^* \tag{A2.5.7}$$

The combination of (A2.5.1) and (A2.5.7) gives

$$\Delta w_{pq} = \alpha\, \rho'(u_p^*)\, v_p^*\, z_q^* \tag{A2.5.8}$$

where

$$v_p^* = \sum_k E_k\, (\mathbf{L}^{-1})_{kp} \tag{A2.5.9}$$

The inversion of the matrix $\mathbf{L}$ may be avoided by observing that

$$\begin{aligned} \sum_p v_p^*\, (\mathbf{L})_{pi} &= \sum_p \sum_k E_k\, (\mathbf{L}^{-1})_{kp}\, (\mathbf{L})_{pi} \\ &= \sum_k E_k \sum_p (\mathbf{L}^{-1})_{kp} (\mathbf{L})_{pi} = E_i \end{aligned} \tag{A2.5.10}$$

Substituting $(\mathbf{L})_{pi}$ from (A2.5.6) into (A2.5.10),

$$\sum_p v_p^*\, [\delta_{pi} - \rho'(u_p)\, w_{pi}] = E_i \tag{A2.5.11}$$

or equivalently,

$$v_i^* - \sum_p v_p^*\, \rho'(u_p)\, w_{pi} = E_i \tag{A2.5.12}$$

This last equation corresponds to the equilibrium state of the following dynamical equation

$$\frac{dv_i}{dt} = -v_i + \sum_p v_p\, \rho'(u_p)\, w_{pi} + E_i \tag{A2.5.13}$$

Chapter 3

ELEANNE: Efficient LEarning Algorithms for Neural NEtworks

3.1 Introduction

A large portion of the research done in the recent decades was motivated by the exciting challenge of developing "thinking machines," i.e., machines capable of reliably performing human like operations. Neural networks is a rapidly expanding research field which has recently attracted the attention of engineers and scientists as a realistic alternative for the development of speech and image recognition systems as well as trainable control devices. This expectation is substantiated in principle by recent results in this research area, which have also motivated a new appreciation for the pioneering work on neural networks which appeared some decades ago. However, the transition from the dream to the reality requires an extensive amount of work on the possible applications of neural networks and, even more importantly, the development of efficient learning algorithms for the training of neural networks.

A learning algorithm is judged on the basis of certain rather conflicting requirements, such as *simplicity, flexibility,* and *efficiency.* The simplicity of a learning algorithm relates to the effort required for the reproduction and application of the algorithm by an ordinary programmer. The flexibility of a learning algorithm relates to the analytical and programming effort required to use the algorithm for training neural networks with different architectures. The efficiency of a learning algorithm is evaluated with respect to the computational and time requirements for training a neural network as well as the performance of the resulting network.

Feed-forward neural networks are frequently trained by first-order learning algorithms, which were derived on the basis of the gradient descent method. Although the impact of these algorithms on the neural network research and applications was undoubtedly enormous, it is widely recognized that they suffer from slow convergence. The development of alternative learning algorithms for multi-layered neural networks was attempted by modifying the optimization strategy and/or employing adaptation rules other than the gradient descent.

Parker presented an interesting step towards the development of second-order learning algorithms for neural networks (Parker, 1987). In an attempt to develop second-order learning algorithms for feed-forward neural networks, Watrous considered various methods for approximating the second-order derivatives of the objective function iteratively (Watrous, 1987). Becker and le Cun attempted to improve the convergence of back propagation learning by using an approximation of the Newton method (Becker and le Cun, 1989). More specifically, they suggested that the inversion of the Hessian matrix involved in the Newton method can be simplified by ignoring its off-diagonal elements. The validity of the approximation of the Hessian matrix by its diagonal elements was only experimentally tested (Becker and le Cun, 1989). The diagonal elements of the Hessian matrix were evaluated numerically by using the finite-element method that involves forward differencing of the gradient. From an implementation point of view, the lack of closed-form update equations is a factor affecting the comparison of the resulting algorithm and the well-defined Error Back Propagation algorithm. In an attempt to develop second-order learning algorithms for feed-forward neural networks, Kollias and Anastassiou proposed an analytically complicated and computationally demanding algorithm as an efficient alternative to the Error Back Propagation (Kollias and

Anastassiou, 1989).

Singhal and Wu suggested that the training of a multi-layered neural network can be interpreted as an identification problem for a nonlinear dynamic system which can be solved by using the extended Kalman algorithm (Singhal and Wu, 1989). This algorithm usually converges faster than the Error Back Propagation algorithm, but it is computationally complex (Singhal and Wu, 1989). Puskorius and Feldkamp proposed a learning algorithm for feed-forward neural networks based upon a decoupled extended Kalman filter (Puskorius and Feldkamp, 1991). The derivation of this algorithm is based upon a natural simplification of the global extended Kalman algorithm used by Singhal and Wu by ignoring the interdependence of mutually exclusive groups of synaptic weights.

The previous discussion indicates that the development of efficient learning algorithms for feed-forward neural networks involves essentially the tradeoff between computational complexity and convergence. This chapter presents the development of *well-defined*, *computationally simple*, and *easily applicable* learning algorithms with better convergence properties than the Error Back Propagation algorithm (Karayiannis, 1991a).

This chapter is organized as follows: Section 3.2 presents the derivation of two recursive least-squares learning algorithms for the training of single-layered neural networks with analog output. Section 3.3 presents the development of second-order learning algorithms for single-layered neural networks with binary output. Section 3.4 presents the development of learning algorithms for multi-layered neural networks, based on the algorithms proposed for single-layered networks. The computational complexity of the proposed algorithms is studied in section 3.5. Section 3.6 presents an experimental verification of the efficiency of the proposed learning algorithms along with a comparison to learning algorithms already existing in the literature. Section 3.7 contains concluding remarks.

3.2 Recursive Least-squares Algorithms

Consider the single-layered neural network shown in Figure 2.7 trained with respect to the associations $(\mathbf{y}_k, \mathbf{x}_k), k = 1, 2, \ldots, m$, where $\mathbf{x}_k$ are the input patterns and $\mathbf{y}_k$ the corresponding output patterns. Using vector notation, $\mathbf{x}_k^* = [x_{1,k}\, x_{2,k} \ldots x_{n_i,k}]$, where n_i is the number of inputs, and $\mathbf{y}_k^* = [y_{1,k}\, y_{2,k} \ldots y_{n_o,k}]$, where n_o is the number of output units. The training of this network can be performed with respect to its analog output consisting of the elements $\bar{y}_{i,k}, i = 1, 2, \ldots, n_o$ by employing least-squares techniques. The matrix of synaptic weights resulting from such a formulation is the unique solution of the well-known normal equation. The uniqueness of the matrix of synaptic weights implies that the crucial problem in the training of neural networks using least-squares is the development of computationally efficient least-squares learning algorithms. This section presents the development of two alternative recursive least-squares learning algorithms for single-layered neural networks (Karayiannis, 1991a; Karayiannis and Venetsanopoulos, 1990b).

3.2.1 ELEANNE 1

According to the formulation of the training of single-layered neural networks proposed by Kohonen and Ruohonen, the output patterns $\mathbf{y}_k, k = 1, 2, \ldots, m$ can be related to the input patterns $\mathbf{x}_k, k = 1, 2, \ldots, m$ through the equation (Kohonen and Ruohonen, 1973)

$$\mathbf{y}_k = \mathbf{W}_m\, \mathbf{x}_k\; \forall\, k = 1, 2, \ldots, m \qquad (3.2.1)$$

where $\mathbf{W}_m$ is the matrix of synaptic weights which corresponds to the m associations $(\mathbf{y}_k, \mathbf{x}_k), k = 1, 2, \ldots, m$. If $\mathbf{X}_m(\mathbf{Y}_m)$ is the matrix whose columns are $\mathbf{x}_k(\mathbf{y}_k), k = 1, 2, \ldots, m$, the combination of the m equations given by (3.2.1) amounts to

$$\mathbf{Y}_m = \mathbf{W}_m\, \mathbf{X}_m \qquad (3.2.2)$$

The optimal solution of the matrix equation (3.2.2) in the least-squares sense is given by (Kohonen and Ruohonen, 1973)

$$\mathbf{W}_m = \mathbf{Y}_m\, \mathbf{X}_m^{+} \qquad (3.2.3)$$

where $\mathbf{X}_m^+$ is the generalized inverse of $\mathbf{X}_m$ (Rao and Mitra, 1971; Albert, 1972). The derivation of the closed form expression (3.2.3) for the matrix of synaptic weights $\mathbf{W}_m$ is of little practical value, at least when the evaluation of $\mathbf{W}_m$ is concerned. In addition to the computational burden associated with the evaluation of the matrix of synaptic weights using (3.2.3), or the equivalent formula $\mathbf{W}_m = \mathbf{Y}_m\,(\mathbf{X}_m^*\,\mathbf{X}_m)^+\,\mathbf{X}_m^*$, the training of the network with respect to a new association implies a complete re-evaluation based on the new matrices $\mathbf{Y}_{m+1}$ and $\mathbf{X}_{m+1}$. In practice, the evaluation of the matrix of synaptic weights resulting from this formulation requires a flexible algorithm that is not based on matrix inversion. The recursive evaluation of the optimal matrix of synaptic weights was originally attempted in the early seventies (Kohonen, 1974). A simple and easily implementable recursive algorithm is presented here (Karayiannis, 1991a; Karayiannis and Venetsanopoulos, 1990b).

Let $\mathcal{C}(\mathbf{X}_{m-1})$ be the column space of $\mathbf{X}_{m-1}$, i.e., the linear space spanned by the columns of $\mathbf{X}_{m-1}$. Consider here the update equation which provides $\mathbf{W}_m$ in terms of $\mathbf{W}_{m-1}$ in the nontrivial case where $\mathbf{x}_m \notin \mathcal{C}(\mathbf{X}_{m-1})$. This update equation can easily be obtained by observing that $\mathbf{X}_m(\mathbf{Y}_m)$ can be written as a partitioned matrix of the form

$$\mathbf{X}_m = [\mathbf{X}_{m-1}\;\mathbf{x}_m]\;(\mathbf{Y}_m = [\mathbf{Y}_{m-1}\;\mathbf{y}_m]) \tag{3.2.4}$$

The update equation can be obtained by using the Greville formula for the evaluation of the generalized inverse of a partitioned matrix of the form (3.2.4). According to this formula (Rao and Mitra, 1971),

$$\mathbf{X}_m^+ = [\mathbf{X}_{m-1}\;\mathbf{x}_m]^+ = \begin{bmatrix} \mathbf{X}_{m-1}^+ - \mathbf{X}_{m-1}^+\,\mathbf{x}_m\,\mathbf{z}_m^* \\ \mathbf{z}_m^* \end{bmatrix} \tag{3.2.5}$$

where in the case that $\mathbf{x}_m \notin \mathcal{C}(\mathbf{X}_{m-1})$,

$$\mathbf{z}_m = \frac{(\mathbf{I} - \mathbf{X}_{m-1}\,\mathbf{X}_{m-1}^+)\,\mathbf{x}_m}{||(\mathbf{I} - \mathbf{X}_{m-1}\,\mathbf{X}_{m-1}^+)\,\mathbf{x}_m||^2} \tag{3.2.6}$$

It can easily be seen that $\mathbf{U}_{m-1} = \mathbf{X}_{m-1}\,\mathbf{X}_{m-1}^+$ is the optimal solution in the least-squares sense of the equation $\mathbf{X}_{m-1} = \mathbf{U}_{m-1}\,\mathbf{X}_{m-1}$.

In other words, $\mathbf{U}_{m-1}$ is the optimal matrix of synaptic weights of the neural network trained with respect to the associations $(\mathbf{x}_k, \mathbf{x}_k)$, $k = 1, 2, \ldots, m-1$. The combination of (3.2.4) and (3.2.5) with (3.2.3) gives

$$\begin{aligned} \mathbf{W}_m &= \mathbf{Y}_{m-1}\,\mathbf{X}_{m-1}^{+} + (\mathbf{y}_m - \mathbf{Y}_{m-1}\,\mathbf{X}_{m-1}^{+}\,\mathbf{x}_m)\,\mathbf{z}_m^{*} \\ &= \mathbf{W}_{m-1} + (\mathbf{y}_m - \mathbf{W}_{m-1}\,\mathbf{x}_m)\,\mathbf{z}_m^{*} \end{aligned} \tag{3.2.7}$$

The *auto-associative update error* is defined here as

$$\mathbf{a}_m = \mathbf{x}_m - \mathbf{U}_{m-1}\,\mathbf{x}_m = (\mathbf{I} - \mathbf{U}_{m-1})\,\mathbf{x}_m \tag{3.2.8}$$

Similarly, the *hetero-associative update error* is defined by

$$\mathbf{e}_m = \mathbf{y}_m - \mathbf{W}_{m-1}\,\mathbf{x}_m \tag{3.2.9}$$

According to its definition in (3.2.6), $\mathbf{z}_m$ is the normalized auto-associative update error, that is, $\mathbf{z}_m = \mathbf{a}_m/||\mathbf{a}_m||^2$. In addition, the update equation (3.2.7) can be written in terms of the update errors defined by (3.2.8) and (3.2.9) as follows

$$\mathbf{W}_m = \mathbf{W}_{m-1} + \delta_m\,\mathbf{e}_m\,\mathbf{a}_m^{*} \tag{3.2.10}$$

where $\delta_m^{-1} = ||\mathbf{a}_m||^2$. The importance of the recursive algorithm derived above is that there is no need for the evaluation of the generalized inverse appearing in (3.2.3), or even the conventional inverse. Instead, according to this algorithm, the matrix of synaptic weights $\mathbf{W}_m$ which corresponds to m associations can be simply evaluated by updating the matrix $\mathbf{W}_{m-1}$, which corresponds to $m-1$ associations. In addition, the definition of the update errors by (3.2.8) and (3.2.9) indicates that the evaluation of $\mathbf{W}_m$ also involves the matrix of synaptic weights $\mathbf{U}_{m-1}$, which corresponds to the $m-1$ associations $(\mathbf{x}_k, \mathbf{x}_k)$, $k = 1, 2, \ldots, m-1$. The equation which allows the recursive evaluation of this matrix can be obtained as a special case of (3.2.10) as follows

$$\mathbf{U}_m = \mathbf{U}_{m-1} + \delta_m\,\mathbf{a}_m\,\mathbf{a}_m^{*} \tag{3.2.11}$$

where $\mathbf{a}_m$ is the update error given by (3.2.8). The recursive algorithm derived for the optimal training of any single-layered neural network with respect to the associations $(\mathbf{y}_k, \mathbf{x}_k), k = 1, 2, \ldots, m$ is essentially the generalization of an algorithm proposed for the optimal least-squares training of neural networks in the special case where $\mathbf{y}_k = \mathbf{x}_k \, \forall \, k = 1, 2, \ldots, m$ (Cheung et al., 1987).

Consider now the case where $\mathbf{x}_m \in \mathcal{C}(\mathbf{X}_{m-1})$, that is, $\mathbf{x}_m$ is a linear combination of the column vectors $\mathbf{x}_k, k = 1, 2, \ldots, m-1$ of $\mathbf{X}_{m-1}$. It is shown in Appendix A3.1 that under this assumption

$$\mathbf{W}_m = \mathbf{W}_{m-1} \tag{3.2.12}$$

According to (3.2.12), if the new input pattern $\mathbf{x}_m$ is a linear combination of the input patterns belonging to the already stored associations, the new association $(\mathbf{y}_m, \mathbf{x}_m)$ causes no effect on the matrix of synaptic weights.

The recursive evaluation of $\mathbf{W}_m$ can be initialized by the equation $\mathbf{W}_1 = \mathbf{y}_1 \mathbf{x}_1^+$. If $\mathbf{x}_1$ is a nonzero column vector with elements $+/-1$, $\mathbf{W}_1 = \mathbf{y}_1 \mathbf{x}_1^+ = \mathbf{y}_1 (\mathbf{x}_1^* \mathbf{x}_1)^{-1} \mathbf{x}_1^* = (1/n_i) \mathbf{y}_1 \mathbf{x}_1^*$. It can similarly be shown that the recursive evaluation of $\mathbf{U}_m$ can be initialized by the equation $\mathbf{U}_1 = (1/n_i) \mathbf{x}_1 \mathbf{x}_1^*$. The resulting algorithm, called ELEANNE 1, is summarized in the flowchart presented in Table 3.1.

The auto-associative update error $\mathbf{a}_m$ defined by (3.2.8) contains whatever is "maximally new" in $\mathbf{x}_m$ with respect to the matrix of synaptic weights $\mathbf{U}_{m-1}$. This update error was also called a *novelty with respect to the stored patterns* (Kohonen et al., 1981). Accordingly, the system which extracts the novelty from the input data was called a *novelty filter.* The novelty filter has the property that it enhances any unfamiliar part of a pattern passing through it (Kohonen et al., 1981). The definition of $\mathbf{a}_m$ by (3.2.8) indicates that the transfer operator of the novelty filter is

$$\mathbf{N}_m = \mathbf{I} - \mathbf{U}_m \tag{3.2.13}$$

An additional useful result of the analysis presented above is an algorithm for the recursive evaluation of the transfer operator for the novelty filter, defined by (3.2.13). The combination of (3.2.11) and (3.2.13) amounts to

$$\mathbf{N}_m = \mathbf{I} - \mathbf{U}_m = (\mathbf{I} - \mathbf{U}_{m-1}) - \delta_m \, \mathbf{a}_m \, \mathbf{a}_m^* = \mathbf{N}_{m-1} - \delta_m \, \mathbf{a}_m \, \mathbf{a}_m^* \tag{3.2.14}$$

Start
$\mathbf{W} = 0$
$\mathbf{U} = 0$
1 $E = 0$
$k = 0$(k: number of association)
2 $k \leftarrow k + 1$
$\mathbf{x} = \mathbf{x}_k$
$\mathbf{y} = \mathbf{y}_k$
$\mathbf{a} = \mathbf{x} - \mathbf{U}\,\mathbf{x}$
$\mathbf{e} = \mathbf{y} - \mathbf{W}\,\mathbf{y}$
if: $||\mathbf{a}||^2 = 0$; then: go to 2
$\delta^{-1} = ||\mathbf{a}||^2$
$\mathbf{U} \leftarrow \mathbf{U} + \delta\,\mathbf{a}\,\mathbf{a}^*$
$\mathbf{W} \leftarrow \mathbf{W} + \delta\,\mathbf{e}\,\mathbf{a}^*$
$E \leftarrow E + \frac{1}{2}\,||\mathbf{e}||^2$
if $k < m$; then go to 2
if: $E > E_o$; then: go to 1
Stop

Table 3.1: Flowchart of ELEANNE 1.

The recursive evaluation of the transfer operator of the novelty filter can be initialized by $\mathbf{N}_1 = \mathbf{I} - (1/n_i)\,\mathbf{x}_1\,\mathbf{x}_1^*$.

3.2.2 ELEANNE 2

The remainder of this section presents an alternative recursive least-squares algorithm for the training of single-layered neural networks (Karayiannis, 1991a; Karayiannis and Venetsanopoulos, 1990b). This algorithm is based on an alternative least-squares formulation of the training problem. It is reasonable to expect the resulting learning algorithm to be subject to the same performance limits which characterize the learning algorithm presented above. However, the derivation which follows serves as an introduction to the development of even more sophisticated learning algorithms.

Consider once again the single-layered neural network shown in

Figure 2.7, trained with respect to the set of associations $(\mathbf{y}_k, \mathbf{x}_k)$, $k = 1, 2, \ldots, m$. In addition, let the matrix of synaptic weights be $\mathbf{W} = [\mathbf{w}_1\, \mathbf{w}_2\, \ldots\, \mathbf{w}_{n_o}]^*$, where $\mathbf{w}_i^*, i = 1, 2, \ldots, n_o$ are its rows, each one defined as $\mathbf{w}_i^* = [w_{i1}\, w_{i2}\, \ldots\, w_{in_i}]$. Assuming that the input to the network is $\mathbf{x}_k$, the analog output of the network consists of the elements $\bar{y}_{i,k} = \mathbf{x}_k^*\, \mathbf{w}_i = \sum_{j=1}^{n_i} w_{ij} x_{j,k}, i = 1, 2, \ldots, n_o$. Consider that the ith row of the matrix of synaptic weights is estimated by requiring that $\bar{y}_{i,k} = y_{i,k} \,\forall\, k = 1, 2, \ldots, m$, or equivalently,

$$\mathbf{x}_k^*\, \mathbf{w}_i = y_{i,k} \,\forall\, k = 1, 2, \ldots, m \tag{3.2.15}$$

The combination of the m equations defined by (3.2.15) amounts to

$$\mathbf{X}_m^*\, \mathbf{w}_i = \mathbf{y}_{i,m} \tag{3.2.16}$$

where $\mathbf{X}_m$ is the matrix whose columns are $\mathbf{x}_k, k = 1, 2, \ldots, m$, that is, $\mathbf{X}_m = [\mathbf{x}_1\, \mathbf{x}_2\, \ldots\, \mathbf{x}_m]$, and $\mathbf{y}_{i,m}^* = [y_{i,1}\, y_{i,2}\, \cdots\, y_{i,m}]$.

The optimal solution of (3.2.16) in the least-squares sense is the solution $\mathbf{w}_i = \mathbf{w}_{i,m}$ of the normal equation $(\mathbf{X}_m\, \mathbf{X}_m^*)\, \mathbf{w}_{i,m} = \mathbf{X}_m\, \mathbf{y}_{i,m}$. It follows easily that $\mathbf{w}_i = \mathbf{w}_{i,m}$ minimizes the objective function $E_{i,m} = \frac{1}{2} \sum_{k=1}^{m} (y_{i,k} - \mathbf{x}_k^*\, \mathbf{w}_i)^2$. Assuming that the matrix $\mathbf{X}_m\, \mathbf{X}_m^*$ is nonsingular,

$$\mathbf{w}_{i,m} = (\mathbf{X}_m\, \mathbf{X}_m^*)^{-1}\, \mathbf{X}_m\, \mathbf{y}_{i,m} \tag{3.2.17}$$

The optimal estimate which corresponds to the $m-1$ associations $(\mathbf{y}_k, \mathbf{x}_k), k = 1, 2, \ldots, m-1$ is $\mathbf{w}_{i,m-1} = (\mathbf{X}_{m-1}\, \mathbf{X}_{m-1}^*)^{-1}\, \mathbf{X}_{m-1}\, \mathbf{y}_{i,m-1}$. Since $\mathbf{X}_m$ can be written as a partitioned matrix as indicated by (3.2.4),

$$\mathbf{X}_m\, \mathbf{X}_m^* = \mathbf{X}_{m-1}\, \mathbf{X}_{m-1}^* + \mathbf{x}_m\, \mathbf{x}_m^* \tag{3.2.18}$$

The above formulation suggests a recursive evaluation of $\mathbf{w}_i$, provided that there exists a simple formula relating $(\mathbf{X}_m\, \mathbf{X}_m^*)^{-1}$ and $(\mathbf{X}_{m-1}\, \mathbf{X}_{m-1}^*)^{-1}$. This relationship is provided by the well-known matrix inversion lemma (Noble and Daniel, 1977). According to this lemma, assuming that $\mathbf{A}$ is a nonsingular $n \times n$ matrix, $\mathbf{b}$ is an $n \times 1$ vector, and $1 + \mathbf{b}^* \mathbf{A}^{-1} \mathbf{b} \neq 0$,

$$(\mathbf{A} + \mathbf{b}\, \mathbf{b}^*)^{-1} = \mathbf{A}^{-1} - (1 + \mathbf{b}^*\, \mathbf{A}^{-1}\, \mathbf{b})^{-1}\, \mathbf{A}^{-1}\, \mathbf{b}\, \mathbf{b}^*\, \mathbf{A}^{-1} \tag{3.2.19}$$

To simplify the notation, let $\mathbf{P}_k = (\mathbf{X}_k \mathbf{X}_k^*)^{-1}$. According to this definition, (3.2.18) can be written as follows

$$\mathbf{P}_m^{-1} = \mathbf{P}_{m-1}^{-1} + \mathbf{x}_m \mathbf{x}_m^* \tag{3.2.20}$$

Applying the matrix inversion lemma to (3.2.20) gives

$$\mathbf{P}_m = \mathbf{P}_{m-1} - (1 + \mathbf{x}_m^* \mathbf{P}_{m-1} \mathbf{x}_m)^{-1} \mathbf{P}_{m-1} \mathbf{x}_m \mathbf{x}_m^* \mathbf{P}_{m-1} \tag{3.2.21}$$

It is shown in Appendix A3.2 that the matrix of synaptic weights $\mathbf{W}$ can be updated with respect to the new association $(\mathbf{y}_m, \mathbf{x}_m)$ through the equation

$$\mathbf{W}_m = \mathbf{W}_{m-1} + (1 + \mathbf{x}_m^* \mathbf{P}_{m-1} \mathbf{x}_m)^{-1} (\mathbf{y}_m - \mathbf{W}_{m-1} \mathbf{x}_m) \mathbf{x}_m^* \mathbf{P}_{m-1} \tag{3.2.22}$$

where $\mathbf{P}_m$ can be evaluated recursively by (3.2.21).

The evaluation of the matrix of synaptic weights can be initialized by the equation $\mathbf{W}_1 = \mathbf{y}_1 \mathbf{x}_1^* \mathbf{P}_1$. However, the recursive evaluation of $\mathbf{P}_m$ cannot b~ initialized by $\mathbf{P}_1 = (\mathbf{x}_1 \mathbf{x}_1^*)^{-1}$ (Karayiannis, 1991a). Another problem with this form of the algorithm is the assumption that $\mathbf{X}_m \mathbf{X}_m^*$ is a nonsingular matrix. It can easily be verified that $\mathbf{X}_m \mathbf{X}_m^* = \sum_{k=1}^m \mathbf{x}_k \mathbf{x}_k^*$ is a symmetric, positive semidefinite, but not necessarily positive definite matrix. Therefore, there is no guarantee that the matrix $\mathbf{X}_m \mathbf{X}_m^*$ is nonsingular. As a result, there exists the possibility of instability associated with the recursive evaluation of $\mathbf{P}_m$.

These problems can be overcome by evaluating the estimate resulting from (3.2.17) after replacing $\mathbf{X}_m \mathbf{X}_m^*$ by $\mathbf{X}_m \mathbf{X}_m^* + \eta \mathbf{I}$, where η is a positive real variable approaching zero and $\mathbf{I}$ is the identity matrix. The resulting equation is the following

$$\mathbf{w}_{i,m}(\eta) = (\mathbf{X}_m \mathbf{X}_m^* + \eta \mathbf{I})^{-1} \mathbf{X}_m \mathbf{y}_{i,m} \tag{3.2.23}$$

Clearly, $\mathbf{X}_m \mathbf{X}_m^* + \eta \mathbf{I}$ is a positive definite, and therefore nonsingular, matrix, even if the matrix $\mathbf{X}_m \mathbf{X}_m^*$ is not positive definite. Using this approach, $\mathbf{P}_1$ is given by

$$\mathbf{P}_1 = (\eta \mathbf{I} + \mathbf{x}_1 \mathbf{x}_1^*)^{-1} \tag{3.2.24}$$

Under this assumption, the recursive evaluation of $\mathbf{P}_m$ can be initialized by (3.2.21), with $m = 1$ and $\mathbf{P}_0 = (1/\eta) \mathbf{I}$.

Start
 Initialize $\mathbf{W}$
 1 $E = 0$
 $\mathbf{P} = (1/\eta)\,\mathbf{I}$
 $k = 0$
 2 $k \leftarrow k + 1$
 $\mathbf{x} = \mathbf{x}_k$
 $\mathbf{y} = \mathbf{y}_k$
 $\delta = (1 + \mathbf{x}^*\,\mathbf{P}\,\mathbf{x})^{-1}$
 $\mathbf{e} = \mathbf{y} - \mathbf{W}\,\mathbf{y}$
 $\mathbf{W} \leftarrow \mathbf{W} + \delta\,\mathbf{e}\,\mathbf{x}^*\,\mathbf{P}$
 $\mathbf{P} \leftarrow \mathbf{P} - \delta\,\mathbf{P}\,\mathbf{x}^*\,\mathbf{x}\,\mathbf{P}$
 $\mathbf{e} = \mathbf{y} - \mathbf{W}\,\mathbf{y}$
 $E \leftarrow E + \frac{1}{2}||\mathbf{e}||^2$
 if $k < m$; then go to 2
 if: $E > E_o$; then: go to 1
Stop

Table 3.2: Flowchart of ELEANNE 2.

The combination of the n_o equations given by (3.2.23) amounts to $\mathbf{W}_m(\eta) = \mathbf{Y}_m\,\mathbf{X}_m^*\,(\mathbf{X}_m\,\mathbf{X}_m^* + \eta\,\mathbf{I})^{-1}$. According to a definition of the generalized inverse, $\mathbf{X}_m^+ = \lim_{\eta\to 0}\mathbf{X}_m^*\,(\mathbf{X}_m\,\mathbf{X}_m^* + \eta\,\mathbf{I})^{-1}$. Clearly,

$$\lim_{\eta\to 0}\mathbf{W}_m(\eta) = \mathbf{W}_m = \mathbf{Y}_m\,\mathbf{X}_m^+ \tag{3.2.25}$$

This is exactly the optimal least-squares estimate for the matrix of synaptic weights, given by (3.2.3). Therefore, this analysis provides a full justification for the proposed modification of the original algorithm.

In principle, when $\mathbf{X}_m\,\mathbf{X}_m^*$ is nonsingular, the least-squares solution is obtained after a single pass through the training associations. In practice, even the choice of a positive η very close to zero may lead to an unstable recursion for $\mathbf{P}_m$. A better choice for η, which also guarantees the stability of the recursive evaluation of $\mathbf{P}_m$, is $\eta = 1$. The price paid for this choice is that multiple adaptation cycles are required for the weights to converge to the optimal estimate, where an *adaptation cycle* involves the adaptation of the synaptic weights of the network with respect to all m associations. It has been found that the evaluation of the optimal matrix of synaptic weights in this case typically requires fewer than 10 adaptation cycles. The number of adaptation cycles can be further reduced by choosing a suitable estimate for $\mathbf{W}_0$ in the first adaptation cycle. Such an estimate could be obtained using the outer-product rule. This algorithm, called ELEANNE 2, is summarized in the flowchart presented in Table 3.2.

3.3 Efficient Learning Algorithms for Single-layered Neural Networks

Consider the single-layered neural network shown in Figure 2.7. Assuming that the input of the network is formed by $x_{j,k}, j = 1, 2, \ldots, n_i$, the corresponding analog output of the network is formed by the elements $\bar{y}_{i,k} = w_{i0} + \sum_{j=1}^{n_i} w_{ij} x_{j,k} \; \forall \, i = 1, 2, \ldots, n_o$. Ideally, the binary output of the network is formed by passing the analog output through hard-limiting nonlinearities as follows

$$y_{i,k}^{o} = \mathrm{sgn}(\bar{y}_{i,k}) = \mathrm{sgn}(\mathbf{x}_k^* \, \mathbf{w}_i) = \mathrm{sgn}(\sum_{j=0}^{n_i} w_{ij} x_{j,k}) \; \forall \, i = 1, 2, \ldots, n_o \tag{3.3.1}$$

where $\mathbf{x}_k^* = [x_{0,k} \, x_{1,k} \, x_{2,k} \, \ldots \, x_{n_i,k}]$, $x_{0,k} = 1 \, \forall \, k = 1, 2, \ldots, m$, and sgn(.) is the signum function, defined as $\mathrm{sgn}(x) = -1$ if $x < 0$ and $\mathrm{sgn}(x) = +1$ if $x \geq 0$.

As described in chapter 2, the training of neural networks is frequently based on the minimization of the objective function

$$E = \sum_{k=1}^{m} E_k = \frac{1}{2} \sum_{k=1}^{m} \sum_{i=1}^{n_o} (y_{i,k} - \hat{y}_{i,k})^2 \tag{3.3.2}$$

where $y_{i,k}, i = 1, 2, \ldots, n_o$ are the elements of the expected output $\mathbf{y}_k$ and $\hat{y}_{i,k}, i = 1, 2, \ldots, n_o$ the corresponding estimates provided by the network (Rumelhart et al., 1986). During the training of the network, its output is formed by

$$\hat{y}_{i,k} = \sigma(\bar{y}_{i,k}) = \sigma(\mathbf{x}_k^* \mathbf{w}_i) = \sigma(\sum_{j=0}^{n_i} w_{ij} x_{j,k}) \, , \; i = 1, 2, \ldots, n_o \quad (3.3.3)$$

where $\sigma(.)$ is a continuous, differentiable everywhere function. The minimization of the objective function (3.3.2) can produce two different sets of synaptic weights, depending on the particular choice of the estimates $\hat{y}_{i,k}, i = 1, 2, \ldots, n_o$. Most of the learning algorithms for neural networks developed so far, including those presented in the last section, were based on the assumption that the estimates $\hat{y}_{i,k}, i = 1, 2, \ldots, n_o$ correspond to the analog output of the network, that is, $\sigma(x) = x$. If the output of the network is binary, its response is determined by the sign and not by the magnitude of the analog output components. Training a network with binary output is subject to analytical problems, due to the hard-limiting nonlinearity sgn(.). These problems can be avoided by approximating the hard-limiters by soft-limiting nonlinearities. Such a smooth approximation of the hard-limiter was extensively used in the literature (Rumelhart et al., 1986). It was suggested that when the states of the network are $\{0, 1\}$, a reasonable choice for $\sigma(.)$ is the well-known logistic function, that is, $\sigma(x) = 1/(1+e^{-x})$. However, it has been argued that the symmetric output states $\{-1, +1\}$ are preferable to the nonsymmetric ones $\{0, 1\}$ (Widrow et al., 1988). In this chapter, the two states of the network are chosen to be $\{-1, +1\}$. In this case, a good choice for $\sigma(.)$ is the hyperbolic tangent, that is, $\sigma(x) = \tanh(x)$.

The gradient descent method is widely used in performing the minimization of (3.3.2). The simplicity of this adaptation rule seems to be the basic reason behind its popularity, since the resulting learning algorithms suffer from slow convergence. The minimization of (3.3.2) could alternatively be attempted using an iterative algorithm with better convergence properties, such as a Newton-Raphson type of algorithm. However, the direct application of such an algorithm implies the evaluation of the corresponding Hessian matrix. Since E_k is a function of all the synaptic weights, the evaluation of the second-order partial derivatives of E_k with respect to the unknown parameters will result in a huge Hessian matrix.

An alternative optimization strategy can be developed by observing that $E = \sum_{i=1}^{n_o} E_i$, where

$$E_i = E_{i,m} = \frac{1}{2} \sum_{k=1}^{m} e_{i,k}^2 = \frac{1}{2} \sum_{k=1}^{m} (y_{i,k} - \hat{y}_{i,k})^2 \tag{3.3.4}$$

In addition, it can be observed that

$$E_{i,m} = E_{i,m-1} + \frac{1}{2}(y_{i,m} - \hat{y}_{i,m})^2 \tag{3.3.5}$$

The function $E_{i,m-1}$ depends only on the associations $(\mathbf{y}_k, \mathbf{x}_k), k = 1, 2, \ldots, m-1$. Therefore, the network can be trained with respect to the associations $(\mathbf{y}_k, \mathbf{x}_k), k = 1, 2, \ldots, m-1$ by minimizing $E_{i,m-1}$. On the other hand, the synaptic weights of the network can be updated with respect to the new association $(\mathbf{y}_m, \mathbf{x}_m)$ by minimizing $E_{i,m} = E_{i,m-1} + \frac{1}{2}\, e_{i,m}^2$. These observations suggest that the adaptation of each row $\mathbf{w}_i$ of the matrix of synaptic weights be performed by sequentially considering the associations $(\mathbf{y}_k, \mathbf{x}_k), k = 1, 2, \ldots, m$. Consider, for example, the adaptation of the ith row of the matrix of synaptic weights $\mathbf{W}$. The elements of $\mathbf{w}_i$ are updated with respect to the first association $(\mathbf{y}_1, \mathbf{x}_1)$ by minimizing $E_{i,1} = \frac{1}{2}\, e_{i,1}^2$. The same elements are then updated with respect to the second association $(\mathbf{y}_2, \mathbf{x}_2)$ by minimizing $E_{i,2} = E_{i,1} + \frac{1}{2}\, e_{i,2}^2$. In general, assume that $\mathbf{w}_i$ has been updated with respect to the associations $(\mathbf{y}_k, \mathbf{x}_k), k = 1, 2, \ldots, m-1$ by minimizing $E_{i,m-1}$, resulting in the estimate $\mathbf{w}_i = \mathbf{w}_{i,m-1}$. The ith row of the matrix of synaptic weights $\mathbf{W}$ is then updated with respect to the association $(\mathbf{y}_m, \mathbf{x}_m)$ by minimizing $E_{i,m}$. This can be achieved through the following update equation (Karayiannis, 1991a; Karayiannis and Venetsanopoulos, 1992c)

$$\mathbf{w}_{i,m} = \mathbf{w}_{i,m-1} - \alpha\, \mathbf{H}_{i,m}^{-1}|_{\mathbf{w}_i = \mathbf{w}_{i,m-1}} \frac{\partial E_{i,m}}{\partial \mathbf{w}_i}|_{\mathbf{w}_i = \mathbf{w}_{i,m-1}} \tag{3.3.6}$$

where $\partial E_{i,m}/\partial \mathbf{w}_i$ is the gradient of $E_{i,m}$ with respect to $\mathbf{w}_i$, $\mathbf{H}_{i,m}$ the Hessian matrix, and α is a positive real number, called here the *learning rate*.

According to the definition of $E_{i,m}$, its gradient with respect to $\mathbf{w}_i$ is given by

$$\frac{\partial E_{i,m}}{\partial \mathbf{w}_i} = \left[\frac{\partial E_{i,m}}{\partial w_{i0}} \frac{\partial E_{i,m}}{\partial w_{i1}} \frac{\partial E_{i,m}}{\partial w_{i2}} \cdots \frac{\partial E_{i,m}}{\partial w_{in_i}}\right]^* = \sum_{k=1}^{m} e_{i,k}\mathbf{b}_{i,k} \quad (3.3.7)$$

where $\mathbf{b}_{i,k}$ is the $(n_i+1)\times 1$ vector defined by

$$\mathbf{b}_{i,k}^* = \left[\frac{\partial e_{i,k}}{\partial w_{i0}} \frac{\partial e_{i,k}}{\partial w_{i1}} \frac{\partial e_{i,k}}{\partial w_{i2}} \cdots \frac{\partial e_{i,k}}{\partial w_{in_i}}\right] \quad (3.3.8)$$

The particular form of the objective function $E_{i,m}$ suggests a suitable approximation of the gradient $\partial E_{i,m}/\partial \mathbf{w}_i$ which further reduces the arithmetic operations required for the adaptation of $\mathbf{w}_i$. Taking into account (3.3.5),

$$\frac{\partial E_{i,m}}{\partial \mathbf{w}_i}\Big|_{\mathbf{w}_i=\mathbf{w}_{i,m-1}} = \frac{\partial E_{i,m-1}}{\partial \mathbf{w}_i}\Big|_{\mathbf{w}_i=\mathbf{w}_{i,m-1}} + e_{i,m}\,\mathbf{b}_{i,m} \quad (3.3.9)$$

where $e_{i,m}$ and $\mathbf{b}_{i,m}$ are both evaluated at $\mathbf{w}_i = \mathbf{w}_{i,m-1}$. Assuming that $\mathbf{w}_i = \mathbf{w}_{i,m-1}$ minimizes $E_{i,m-1}$,

$$\frac{\partial E_{i,m-1}}{\partial \mathbf{w}_i}\Big|_{\mathbf{w}_i=\mathbf{w}_{i,m-1}} = 0 \quad (3.3.10)$$

Under this assumption, the gradient of $E_{i,m}$ with respect to $\mathbf{w}_i$ can be evaluated at $\mathbf{w}_i = \mathbf{w}_{i,m-1}$ as follows

$$\frac{\partial E_{i,m}}{\partial \mathbf{w}_i}\Big|_{\mathbf{w}_i=\mathbf{w}_{i,m-1}} = e_{i,m}\,\mathbf{b}_{i,m} \quad (3.3.11)$$

Assume now that $\mathbf{w}_i = \mathbf{w}_{i,m-1}$ is not the exact minimum of $E_{i,m-1}$. In this case, (3.3.10) is not valid. However, the current estimate $\mathbf{w}_i = \mathbf{w}_{i,m-1}$ is the result of a step towards the minimization of $E_{i,m-1}$. Therefore, the major contribution in the evaluation of the gradient of $E_{i,m}$ at $\mathbf{w}_i = \mathbf{w}_{i,m-1}$ comes from the term $e_{i,m}\,\mathbf{b}_{i,m}$. Thus, even if $\mathbf{w}_i = \mathbf{w}_{i,m-1}$ is not the minimum of $E_{i,m-1}$, (3.3.11) provides a satisfactory approximation for the gradient $\partial E_{i,m}/\partial \mathbf{w}_i$ evaluated at $\mathbf{w}_i = \mathbf{w}_{i,m-1}$. The closer the estimate $\mathbf{w}_i = \mathbf{w}_{i,m-1}$ is to the minimum, the better the approximation of the gradient by (3.3.11). Using the gradient approximation (3.3.11), the update equation for each row of the matrix of synaptic weights (3.3.6) becomes

$$\mathbf{w}_{i,m} = \mathbf{w}_{i,m-1} - \alpha\, \mathbf{H}_{i,m}^{-1}|_{\mathbf{w}_i=\mathbf{w}_{i,m-1}}\, \mathbf{b}_{i,m}\, e_{i,m} \tag{3.3.12}$$

Consider here the evaluation of the Hessian matrix for this objective function. From the definition of $E_{i,m}$, the elements of the Hessian matrix $\mathbf{H}_{i,m}$ are given by

$$(\mathbf{H}_{i,m})_{pq} = \frac{\partial^2 E_{i,m}}{\partial w_{ip} \partial w_{iq}} = \sum_{k=1}^{m} \frac{\partial e_{i,k}}{\partial w_{ip}} \frac{\partial e_{i,k}}{\partial w_{iq}} + \sum_{k=1}^{m} e_{i,k} \frac{\partial^2 e_{i,k}}{\partial w_{ip} \partial w_{iq}} \tag{3.3.13}$$

In general, there is no guarantee that the Hessian matrix (3.3.13) is positive definite and, therefore, nonsingular. Also, (3.3.13) does not suggest a way to evaluate the inverse of the Hessian matrix recursively. These analytical problems indicate that the derivation of the learning algorithms should be performed separately in the case where the network is trained with respect to its analog or binary output.

Consider a single-layered neural network trained with respect to its analog output. According to the above analysis, each row $\mathbf{w}_i^*, i = 1, 2, \ldots, n_o$ of the matrix of synaptic weights can be determined by minimizing the objective function (3.3.4), where $\hat{y}_{i,k}, i = 1, 2, \ldots, n_o$ are given by (3.3.3) with $\sigma(x) = x$. The resulting optimization problem can be handled using well-known least-squares techniques. An example is the treatment of this problem presented in section 3.2, which resulted in ELEANNE 2. This problem is approached here using the new optimization strategy presented in this section.

If $\hat{y}_{i,k} = \bar{y}_{i,k}$, (3.3.8) becomes $\mathbf{b}_{i,k} = -\mathbf{x}_k \,\forall\, i = 1, 2, \ldots, n_o$. Since $\partial^2 e_{i,k} / \partial w_{ip} \partial w_{iq} = 0$, the second sum in (3.3.13) vanishes and the definition of $\mathbf{b}_{i,k}$ in (3.3.8) indicates that the Hessian matrix is of the form

$$\mathbf{H}_{i,m} = \sum_{k=1}^{m} \mathbf{b}_{i,k}\, \mathbf{b}_{i,k}^* = \sum_{k=1}^{m} \mathbf{x}_k\, \mathbf{x}_k^* = \mathbf{X}_m\, \mathbf{X}_m^* = \mathbf{H}_m \;\forall\, i = 1, 2, \ldots, n_o \tag{3.3.14}$$

This form of the Hessian matrix fully justifies the formulation of the optimization problem as well as the optimization strategy proposed in this section. Clearly, the Hessian matrix given by (3.3.14) is positive semi-definite. As in section 3.2, the stability of the resulting algorithm can be guaranteed by replacing (3.3.14) by a suitable positive definite, and therefore nonsingular, matrix. As defined in (3.3.14), the resulting Hessian matrix can be written as

$$\mathbf{H}_m^{-1} = (\mathbf{H}_{m-1} + \mathbf{x}_m\, \mathbf{x}_m^*)^{-1} \tag{3.3.15}$$

Assuming that $\mathbf{P}_{m-1} = \mathbf{H}_{m-1}^{-1}$ is available, $\mathbf{P}_m = \mathbf{H}_m^{-1}$ can be evaluated by using the matrix inversion lemma presented in section 3.2. Clearly, from (3.3.15),

$$\mathbf{P}_m = \mathbf{P}_{m-1} - (1 + \mathbf{x}_m^*\, \mathbf{P}_{m-1}\, \mathbf{x}_m)^{-1}\, \mathbf{P}_{m-1}\, \mathbf{x}_m\, \mathbf{x}_m^*\, \mathbf{P}_{m-1} \tag{3.3.16}$$

The update equation for each row of the matrix of synaptic weights can be obtained from (3.3.12) by substituting $\mathbf{H}_{i,m}^{-1} = \mathbf{P}_{i,m} = \mathbf{P}_m$ and $\mathbf{b}_{i,m} = \mathbf{b}_m = -\mathbf{x}_m$, as follows

$$\mathbf{w}_{i,m} = \mathbf{w}_{i,m-1} + \alpha\, \mathbf{P}_m\, \mathbf{x}_m\, e_{i,m} \tag{3.3.17}$$

while $\mathbf{P}_m$ is updated through the recursion given by (3.3.16).

At this point, it would be interesting to investigate the relationship between the update equations provided by this method and the ELEANNE 2, presented in section 3.2. Substituting (3.3.16) into (3.3.17), $\Delta\mathbf{w}_{i,m} = \mathbf{w}_{i,m} - \mathbf{w}_{i,m-1}$ is obtained as

$$\begin{aligned}\Delta\mathbf{w}_{i,m} &= \alpha[\mathbf{I} - (1 + \mathbf{x}_m^*\mathbf{P}_{m-1}\mathbf{x}_m)^{-1}\mathbf{P}_{m-1}\mathbf{x}_m\mathbf{x}_m^*]\mathbf{P}_{m-1}\mathbf{x}_m e_{i,m} \\ &= \alpha[1 - \mathbf{x}_m^*\mathbf{P}_{m-1}\mathbf{x}_m(1 + \mathbf{x}_m^*\mathbf{P}_{m-1}\mathbf{x}_m)^{-1}]\mathbf{P}_{m-1}\mathbf{x}_m e_{i,m}\end{aligned} \tag{3.3.18}$$

Finally, the update equation (3.3.17) is equivalent to

$$\mathbf{w}_{i,m} = \mathbf{w}_{i,m-1} + \alpha\, (1 + \mathbf{x}_m^*\, \mathbf{P}_{m-1}\, \mathbf{x}_m)^{-1}\, \mathbf{P}_{m-1}\, \mathbf{x}_m\, e_{i,m} \tag{3.3.19}$$

For $\alpha = 1$, the update equation (3.3.18) provided by this optimization strategy is identical to the one resulting in section 3.2 by treating the same optimization problem using least-squares techniques. This is not surprising, since the Hessian matrix is exactly evaluated by (3.3.14) in the case where $\hat{y}_{i,k} = \bar{y}_{i,k}$. The optimal least-squares estimate for $\mathbf{w}_i = \mathbf{w}_{i,m-1}$ which corresponds to $m-1$ associations minimizes

$$E_{i,m-1} = \frac{1}{2} \sum_{k=1}^{m-1} (y_{i,k} - \hat{y}_{i,k})^2 \qquad (3.3.20)$$

It is also known that the minimum of (3.3.19) is the particular $\mathbf{w}_i = \mathbf{w}_{i,m-1}$ which satisfies $(\partial E_{i,m-1}/\partial \mathbf{w}_i)|_{\mathbf{w}_i = \mathbf{w}_{i,m-1}} = 0$. Therefore, the estimate of the gradient provided by (3.3.11) coincides with its exact value.

3.3.1 ELEANNE 3

Consider a single-layered neural network trained with respect to its binary output. Assuming that $\mathbf{x}_k$ is the input of the network, the binary output of the network consists of the components given by (3.3.3) with $\sigma(x) = \tanh(x)$. According to the above analysis, each row $\mathbf{w}_i^*, i = 1, 2, \ldots, n_o$ of the matrix of synaptic weights is estimated by minimizing the objective function (3.3.4), where $\hat{y}_{i,k}, i = 1, 2, \ldots, n_o$ are given by (3.3.3) with $\sigma(x) = \tanh(x)$.

The derivation of the learning algorithm resulting from the formulation of the training problem considered here requires the evaluation of the vector $\mathbf{b}_{i,m}$ and the Hessian matrix $\mathbf{H}_{i,m}$ when $\hat{y}_{i,k} = \tanh(\bar{y}_{i,k})$. It is shown in Appendix A3.3 that

$$\mathbf{b}_{i,k} = -(1 - \hat{y}_{i,k}^2)\, \mathbf{x}_k \qquad (3.3.21)$$

It is also shown in Appendix A3.3 that the corresponding Hessian matrix is not guaranteed to be invertible. However, the Hessian matrix can be approximated in this case by

$$\mathbf{H}_{i,m} \approx \sum_{k=1}^{m} \mathbf{b}_{i,k}\, \mathbf{b}_{i,k}^* \qquad (3.3.22)$$

The resulting approximation of the Hessian matrix satisfies the conditions which are necessary for the derivation of a recursive, stable algorithm. Clearly, $\mathbf{H}_{i,m}$ is a positive semi-definite matrix. In addition, the inverse of $\mathbf{H}_{i,m}$ can be evaluated recursively. From the definition of $\mathbf{H}_{i,m}$ by (3.3.22),

$$\mathbf{H}_{i,m}^{-1} = (\mathbf{H}_{i,m-1} + \mathbf{b}_{i,m}\, \mathbf{b}_{i,m}^*)^{-1} \qquad (3.3.23)$$

Start
Initialize $\mathbf{W}$ with random values
1 $E = 0$
$i = 0$(i: index of output unit)
2 $i \leftarrow i + 1$
$\bar{\mathbf{w}} = \mathbf{w}_i$
$\mathbf{P} = (1/\eta)\,\mathbf{I}$
$k = 0$(k: number of association)
3 $k \leftarrow k + 1$
$\mathbf{x} = \mathbf{x}_k$
$y = y_{i,k}$
$\hat{y} = \sigma(\sum_{j=1}^{n_i} \bar{w}_j\, x_j)$
$c = (1 - \hat{y}^2)^2$
$\delta = c\,(1 + \mathbf{x}^*\,\mathbf{P}\,\mathbf{x})^{-1}$
$\mathbf{P} \leftarrow \mathbf{P} - \delta\,\mathbf{P}\,\mathbf{x}\,\mathbf{x}^*\,\mathbf{P}$
$\varepsilon^o(\lambda) = (1 - \hat{y}^2)(1 - \hat{y})$
$\bar{\mathbf{w}} \leftarrow \bar{\mathbf{w}} + \alpha\,\varepsilon^o\,\mathbf{P}\,\mathbf{x}$
$\hat{y} = \rho(\sum_{j=1} \bar{w}_j\, x_j)$
$E \leftarrow E + \frac{1}{2}\,(y - \hat{y})^2$
if $k < m$; then go to 3
$\mathbf{w}_i = \bar{\mathbf{w}}$
if: $i < n_o$; then: go to 2
if: $E > E_o$; then: go to 1
Stop

Table 3.3: Flowchart of ELEANNE 3.

Assuming that $\mathbf{P}_{i,m-1} = \mathbf{H}_{i,m-1}^{-1}$ is available, $\mathbf{P}_{i,m} = \mathbf{H}_{i,m}^{-1}$ can be evaluated by using the matrix inversion lemma presented in section 3.2. Clearly, from (3.3.22),

$$\mathbf{P}_{i,m} = \mathbf{P}_{i,m-1} - (1 + \mathbf{b}_{i,m}^* \mathbf{P}_{i,m-1} \mathbf{b}_{i,m})^{-1} \mathbf{P}_{i,m-1} \mathbf{b}_{i,m} \mathbf{b}_{i,m}^* \mathbf{P}_{i,m-1} \tag{3.3.24}$$

The equation for the recursive evaluation of $\mathbf{P}_{i,m}$ can be expressed in terms of the input patterns $\mathbf{x}_k$ by substituting $\mathbf{b}_{i,k}$ from (3.3.21) as follows

$$\mathbf{P}_{i,m} = \mathbf{P}_{i,m-1} - \delta_{i,m} \mathbf{P}_{i,m-1} \mathbf{x}_m \mathbf{x}_m^* \mathbf{P}_{i,m-1} \tag{3.3.25}$$

where $\mathbf{P}_{i,0} = (1/\eta)\mathbf{I}$, $\delta_{i,m} = c_{i,m}(1 + c_{i,m} \mathbf{x}_m^* \mathbf{P}_{i,m-1} \mathbf{x}_m)^{-1}$ and $c_{i,m} = (1 - \hat{y}_{i,m}^2)^2$. The update equation for each row of the matrix of synaptic weights can be obtained from (3.3.12) by substituting $\mathbf{H}_{i,m}^{-1} = \mathbf{P}_{i,m}$ and $\mathbf{b}_{i,m}$ from (3.3.21) as

$$\mathbf{w}_{i,m} = \mathbf{w}_{i,m-1} + \alpha\, \varepsilon_{i,m}^o \mathbf{P}_{i,m} \mathbf{x}_m \tag{3.3.26}$$

where

$$\varepsilon_{i,m}^o = (1 - \hat{y}_{i,m}^2)(y_{i,m} - \hat{y}_{i,m}) \tag{3.3.27}$$

This algorithm, called ELEANNE 3, is summarized in the flowchart presented in Table 3.3.

3.3.2 ELEANNE 4

A computationally less demanding algorithm is derived here on the basis of some simplifying assumptions (Karayiannis, 1991a; Karayiannis and Venetsanopoulos, 1990c). According to the above analysis, the Hessian matrix which corresponds to the objective function $E_{i,m}$ can be approximated by

$$\mathbf{H}_{i,m} = \sum_{k=1}^{m} \mathbf{b}_{i,k} \mathbf{b}_{i,k}^* = \sum_{k=1}^{m} c_{i,k} \mathbf{x}_k \mathbf{x}_k^* \;\forall i = 1, 2, \ldots, n_o \tag{3.3.28}$$

where $c_{i,k} = (1 - \hat{y}_{i,k}^2)^2$. The requirement for the recursive evaluation of the n_o matrices $\mathbf{P}_{i,m} = \mathbf{H}_{i,m}^{-1}, i = 1, 2, \ldots, n_o$ is imposed by the existence of n_o different coefficients $c_{i,k}, i = 1, 2, \ldots, n_o$. The simplification of the ELEANNE 3 cannot be attempted by using the crude assumption that $c_{i,k}$ is constant. Such an assumption implies that the Hessian matrix (3.3.28) can be approximated by (3.3.14), that is, the Hessian matrix resulting when the network is trained with respect to its analog output. An alternative approach to simplifying the ELEANNE 3 algorithm can be obtained by considering the assumption that

$$c_{i,k} = c_{i_0,k} = (1 - \hat{y}_{i_0,k}^2)^2 = c_k \, \forall \, i = 1, 2, \ldots, n_o \tag{3.3.29}$$

According to this assumption,

$$\mathbf{H}_{i,m} = \sum_{k=1}^{m} c_k \, \mathbf{x}_k \, \mathbf{x}_k^* = \mathbf{H}_m \, \forall \, i = 1, 2, \ldots, n_o \tag{3.3.30}$$

The replacement of each of the n_o matrices given by (3.3.28) by the "average" matrix (3.3.30), implies that all the rows of the matrix of synaptic weights can be updated with respect to each of the m associations simultaneously. This is the major difference between the algorithm developed here and ELEANNE 3. Each adaptation cycle of the simplified algorithm involves m sequential adaptations of all the synaptic weights of the network, each one corresponding to one of the associations $(\mathbf{y}_k, \mathbf{x}_k), k = 1, 2, \ldots, m$. The reliable estimation of c_k is the crucial problem associated with the above assumption.

Consider the case where $y_{i_0,k} = +1$ and $\hat{y}_{i_0,k} \in [0, 1)$. It can reasonably be assumed that during the initial adaptation cycles the output of the network is far away from the optimum and, therefore, $\hat{y}_{i_0,k} \approx 0$. On the other hand, after a certain number of adaptation cycles, the output of the network approaches the optimum and, therefore, $\hat{y}_{i_0,k} \approx 1$. For the sake of analytical simplicity, assume that $\hat{y}_{i_0,k} \in [0, 1)$ is a linear function of the number ν of adaptation cycles and also $\hat{y}_{i_0,k} = 0$ when $\nu = 1$. In mathematical terms, $\hat{y}_{i_0,k} = \beta_k(\nu - 1)$, $\beta_k > 0$. In fact, the parameter β_k determines the speed of convergence of $\hat{y}_{i_0,k}$ to unity. Clearly, the larger the parameter β_k becomes, the faster $\hat{y}_{i_0,k}$ converges to unity. Following a similar reasoning, it can also be assumed that if $y_{i_0,k} = -1$, $\hat{y}_{i_0,k} = -\beta_k(\nu - 1)$, $\beta_k > 0$. According to these assumptions,

$$\hat{y}_{i_0,k}^2 = \beta_k^2(\nu - 1)^2 \, \forall \, k = 1, 2, \ldots, m \tag{3.3.31}$$

Start
Initialize $\mathbf{W}$ with random values
$\nu = 0$ (ν: number of adaptation cycle)
1 $\nu \leftarrow \nu + 1$
$E = 0$
$\mathbf{P} = (1/\eta)\,\mathbf{I}$
$k = 0$ (k: number of association)
2 $k \leftarrow k + 1$
$\mathbf{x} = \mathbf{x}_k$
$c = \exp(-2\,\beta_k^2\,(\nu - 1)^2)$
$\delta = c\,(1 + c\,\mathbf{x}^*\,\mathbf{P}\,\mathbf{x})^{-1}$
$\mathbf{P} \leftarrow \mathbf{P} - \delta\,\mathbf{P}\,\mathbf{x}\,\mathbf{x}^*\,\mathbf{P}$
$i = 0$ (i: index of output unit)
3 $i \leftarrow i + 1$
$\bar{\mathbf{w}} = \mathbf{w}_i$
$y = y_{i,k}$
$\hat{y} = \sigma(\sum_{j=1}^{n_i} \bar{w}_j\, x_j)$
$\varepsilon^o = (1 - \hat{y}^2)(y - \hat{y})$
$\bar{\mathbf{w}} \leftarrow \bar{\mathbf{w}} + \alpha\,\varepsilon^o\,\mathbf{P}\,\mathbf{x}$
$\hat{y} = \sigma(\sum_{j=1}^{n_i} \bar{w}_j\, x_j)$
$E \leftarrow E + \frac{1}{2}(y - \hat{y})^2$
$\mathbf{w}_i = \bar{\mathbf{w}}$
if $i < n_o$; then go to 3
if $k < m$; then go to 2
if: $E > E_o$; then go to 1
Stop

Table 3.4: Flowchart of ELEANNE 4.

Since $|\hat{y}_{i_o,k}| < 1$, the derivation of the estimate of c_k can be further simplified by using the approximation $\exp(-\hat{y}^2_{i_o,k}) \approx 1 - \hat{y}^2_{i_o,k}$. Then from the definition of c_k in (3.3.29),

$$c_k \approx \exp(-2\hat{y}^2_{i_o,k}) \tag{3.3.32}$$

The combination of (3.3.31) and (3.3.32) indicates that c_k can be estimated during each adaptation cycle by

$$c_k = c_k^\nu = \exp(-2\beta_k^2(\nu - 1)^2) \tag{3.3.33}$$

where ν is the number of the current adaptation cycle and β_k is a positive constant. According to the matrix inversion lemma, $\mathbf{P}^\nu_m = (\mathbf{H}^\nu_m)^{-1}$ can be evaluated recursively through the equation

$$\mathbf{P}^\nu_m = \mathbf{P}^\nu_{m-1} - \delta^\nu_m \, \mathbf{P}^\nu_{m-1} \, \mathbf{x}_m \, \mathbf{x}^*_m \, \mathbf{P}^\nu_{m-1} \tag{3.3.34}$$

where $\mathbf{P}^\nu_0 = (1/\eta)\,\mathbf{I}$ and $\delta^\nu_m = c^\nu_m \, (1 + c^\nu_m \, \mathbf{x}^*_m \, \mathbf{P}^\nu_{m-1} \, \mathbf{x}_m)^{-1}$. During the νth adaptation cycle, the rows of the matrix of synaptic weights are updated using the following set of equations

$$\mathbf{w}_{i,m} = \mathbf{w}_{i,m-1} + \alpha \, \varepsilon^o_{i,m} \, \mathbf{P}^\nu_m \, \mathbf{x}_m \; \forall i = 1, 2, \ldots, n_o \tag{3.3.35}$$

where $\varepsilon^o_{i,m}$ is defined by (3.3.27) and $\mathbf{P}^\nu_m$ is recursively evaluated by (3.3.34). The resulting algorithm, called ELEANNE 4, is summarized in the flowchart presented in Table 3.4.

3.4 Efficient Learning Algorithms for Multi-layered Neural Networks

Consider the neural network with one layer of hidden units, shown in Figure 2.9. Assuming that the input of the network is formed by $x_{j,k}, j = 1, 2, \ldots, n_i$, the corresponding outputs of the hidden units are

$$\hat{h}_{j,k} = \rho(\bar{h}_{j,k}) = \rho(\mathbf{x}^*_k \, \mathbf{v}_j) = \rho(\sum_{\ell=0}^{n_i} v_{j\ell} x_{\ell,k}) \tag{3.4.1}$$

where $\mathbf{x}^*_k = [x_{0,k} \, x_{1,k} \, x_{2,k} \, \ldots \, x_{n_i,k}]$, $x_{0,k} = 1 \, \forall \, k = 1, 2, \ldots, m$, and $\rho(x) = \tanh(x)$. Let $\hat{\mathbf{h}}_k$ be the $(n_h + 1) \times 1$ vector formed by the outputs of the hidden units (3.4.1) as follows

$$\hat{\mathbf{h}}^*_k = [\hat{h}_{0,k} \, \hat{h}_{1,k} \, \hat{h}_{2,k} \, \ldots \, \hat{h}_{n_h,k}] \tag{3.4.2}$$

where $\hat{h}_{0,k} = 1 \, \forall \, k = 1, 2, \ldots, m$. Clearly, $\hat{\mathbf{h}}_k$ is a function of the corresponding input pattern $\mathbf{x}_k$ as well as the synaptic weights of the lower network v_{pq}. The output of the network is formed by the elements

$$\hat{y}_{i,k} = \sigma(\bar{y}_{i,k}) = \sigma(\hat{\mathbf{h}}_k^* \mathbf{w}_i) = \sigma(\sum_{j=0}^{n_h} w_{ij} \hat{h}_{j,k}) \tag{3.4.3}$$

where $\sigma(x) = x$ if the output of the network is analog and $\sigma(x) = \tanh(x)$ if the output of the network consists of binary elements.

Consider also the network with L layers of hidden units, shown in Figure 2.10. In this case, the estimate $\hat{y}_{i,k}$ of the ith element of the expected output $\mathbf{y}_k$ is given by

$$\hat{y}_{i,k} = \sigma(\sum_{j=0}^{n_1} w_{ij} \hat{h}_{j,k}^{(1)}) \; \forall \, i = 1, 2, \ldots, n_o \tag{3.4.4}$$

where $\sigma(x) = x$ if the output of the network is analog and $\sigma(x) = \tanh(x)$ if the output of the network consists of binary elements. The outputs of the units which belong to the hidden layers are given by

$$\hat{h}_{j,k}^{(r)} = \rho(\sum_{\ell=0}^{n_{r+1}} v_{j\ell}^{(r)} \hat{h}_{\ell,k}^{(r+1)}) \; \forall \, r = 1, 2, \ldots, L \tag{3.4.5}$$

where $\hat{h}_{\ell,k}^{(L+1)} = x_{\ell,k} \, \forall \, \ell = 0, 1, \ldots, n_i$.

A multi-layered neural network with one or more layers of hidden units can be trained using the Error Back Propagation algorithm presented in chapter 2 (Rumelhart et al., 1986). According to this algorithm, the network is trained with respect to each association $(\mathbf{y}_k, \mathbf{x}_k)$ by minimizing the objective function E_k, defined by (3.3.2). This minimization is achieved by updating the sets of synaptic weights which connect the units of different layers using the Delta Rule. Such an update scheme propagates back the error between the the expected output and the output provided by the network. This section presents the derivation of alternative learning algorithms for the training of multi-layered neural networks, which can achieve better convergence than the Error Back Propagation algorithm (Karayiannis, 1991a). This derivation is strongly influenced by the analysis presented in section 3.3, which provided second-order learning algorithms for single-layered neural networks.

3.4.1 ELEANNE 5

In order to simplify the analysis which follows, assume that the multi-layered neural network shown in Figure 2.9 consists of two single-layered neural networks; the upper network with synaptic weights w_{pq}, and the lower network with synaptic weights v_{pq}. Clearly, the upper network can be regarded as a single-layered network whose inputs are formed by the outputs of the hidden units. This network can be trained by a learning algorithm based on ELEANNE 3, proposed in section 3.3 for training single-layered neural networks. According to that algorithm,

- each adaptation cycle consists of the adaptation of all the rows of the matrix of synaptic weights $\mathbf{W}$,
- the rows of $\mathbf{W}$ are updated sequentially, and
- the adaptation of each row $\mathbf{w}_i$ consists of m steps, each one corresponding to the adaptation of $\mathbf{w}_i$ with respect to one of the m associations.

These properties of ELEANNE 3 suggest the following strategy for training the neural network shown in Figure 2.9. An adaptation cycle consists of n_o stages, each one corresponding to the rows $\mathbf{w}_i, i = 1, 2, \ldots, n_o$ or, equivalently, to the outputs of the upper network. Each stage consists of the adaptation of the corresponding row of the matrix of synaptic weights $\mathbf{W}$ and all the synaptic weights v_{pq} of the lower network by sequentially considering the m associations. Consider a certain adaptation stage and assume that the corresponding row of $\mathbf{W}$, as well as the synaptic weights v_{pq}, have been updated with respect to the associations $(\mathbf{y}_1, \mathbf{x}_1), (\mathbf{y}_2, \mathbf{x}_2), \ldots, (\mathbf{y}_{k-1}, \mathbf{x}_{k-1})$. The adaptation of the same synaptic weights with respect to the association $(\mathbf{y}_k, \mathbf{x}_k)$ can be achieved by:

- updating the same row of the matrix of synaptic weights $\mathbf{W}$ of the upper network by using ELEANNE 3, and
- appropriately updating all the synaptic weights v_{pq} of the lower network.

The synaptic weights v_{pq} of the lower network could also be updated using ELEANNE 3. According to this algorithm, each row of the matrix of synaptic weights $\mathbf{V}$ should be updated separately. However, the use of ELEANNE 3 for the adaptation of the synaptic weights v_{pq} requires the storage of the n_h matrices which correspond to the n_h rows of $\mathbf{V}$. Also, taking into account the Hessian matrix which corresponds to the particular row of $\mathbf{W}$ considered, this scheme requires the simultaneous storage of $n_h + 1$ matrices. Such an approach considerably increases the arithmetic operations required for each adaptation cycle. The enormous storage and computational requirements imposed by such an algorithm suggest the use of an alternative method such as the Delta Rule for the adaptation of the synaptic weights v_{pq} of the lower network. This hybrid scheme combines the convergence properties of ELEANNE 3 and the computational simplicity of the Delta Rule.

Consider the ith stage of an adaptation cycle and assume that the synaptic weights of the network have been updated with respect to the associations $(\mathbf{y}_1, \mathbf{x}_1), (\mathbf{y}_2, \mathbf{x}_2), \ldots, (\mathbf{y}_{k-1}, \mathbf{x}_{k-1})$. In addition, assume that training with respect to these associations resulted in the estimates $\mathbf{w}_i = \mathbf{w}_{i,k-1}$ and $\mathbf{V} = \mathbf{V}_{k-1}$ of the ith row of the matrix of synaptic weights $\mathbf{W}$ of the upper network and the matrix of synaptic weights of the lower network, respectively. Assuming that $\mathbf{x}_k$ is the input of the lower network, the input of the upper network is the *temporary input pattern* $\hat{\mathbf{h}}_k$. Therefore, $\mathbf{w}_i$ can be updated with respect to the *temporary association* $(\mathbf{y}_k, \hat{\mathbf{h}}_k)$ using ELEANNE 3. It should be emphasized here that the temporary input pattern $\hat{\mathbf{h}}_k$ depends on the constant input pattern $\mathbf{x}_k$ but varies with the synaptic weights v_{pq} of the lower network. According to the analysis presented in section 3.3, $\mathbf{w}_i$ can be updated through the equation

$$\mathbf{w}_{i,k} = \mathbf{w}_{i,k-1} + \alpha\, \varepsilon^{o}_{i,k}\, \mathbf{P}_{i,k}\, \hat{\mathbf{h}}_k \tag{3.4.6}$$

where $\hat{\mathbf{h}}_k$ is defined by (3.4.1) and (3.4.2) and $\varepsilon^{o}_{i,k}$ is defined by (3.3.27). The $(n_h+1) \times (n_h+1)$ matrix $\mathbf{P}_{i,k}$ is evaluated recursively by

$$\mathbf{P}_{i,k} = \mathbf{P}_{i,k-1} - \delta_{i,k}\, \mathbf{P}_{i,k-1}\, \hat{\mathbf{h}}_k\, \hat{\mathbf{h}}_k^*\, \mathbf{P}_{i,k-1} \tag{3.4.7}$$

where $\mathbf{P}_{i,0} = (1/\eta)\,\mathbf{I}$, $\delta_{i,k} = c_{i,k}\,(1 + c_{i,k}\,\hat{\mathbf{h}}_k^*\,\mathbf{P}_{i,k-1}\,\hat{\mathbf{h}}_k)^{-1}$ and $c_{i,k} = (1 - \hat{y}_{i,k}^2)^2$.

To complete the derivation of this algorithm, consider the adaptation of the synaptic weights v_{pq} of the lower network with respect to the association $(\mathbf{y}_k, \mathbf{x}_k)$. The synaptic weights v_{pq} can be updated by minimizing the objective function E_k using the gradient descent method. According to (3.3.2), $E_k = \bar{E}_k^i + E_k^i$, where $\bar{E}_k^i = \frac{1}{2}\sum_{j \neq i}^{n_o}(y_{j,k} - \hat{y}_{j,k})^2$, and

$$E_k^i = \frac{1}{2}\,e_{i,k}^2 = \frac{1}{2}\,(y_{i,k} - \hat{y}_{i,k})^2 \tag{3.4.8}$$

Obviously, $\bar{E}_k^i = \bar{E}_k^i(\mathbf{w}_j\,\forall j \neq i)$ and $E_k^i = E_k^i(\mathbf{w}_i)$. Since only $\mathbf{w}_i$ has been updated, $\bar{E}_k^i$ is unaffected and the adaptation of the synaptic weights v_{pq} can be achieved by simply minimizing E_k^i. It is shown in Appendix A3.4 that the synaptic weights v_{pq} of the lower network can be updated by

$$\mathbf{v}_{p,k} = \mathbf{v}_{p,k-1} + \alpha\,\varepsilon_{pi,k}^h\,\mathbf{x}_k \tag{3.4.9}$$

where

$$\varepsilon_{pi,k}^h = (1 - \hat{h}_{p,k}^2)\,\varepsilon_{i,k}^o\,w_{ip} \tag{3.4.10}$$

where $\varepsilon_{i,k}^o$ is defined by (3.3.27). The resulting algorithm, called ELEANNE 5, is summarized in the flowchart presented in Table 3.5.

A multi-layered neural network with more than one layer of hidden units can be trained by generalizing the previous algorithm. Assuming that $\mathbf{x}_k$ is the input of the network, the input of the upper network is the temporary input pattern $\hat{\mathbf{h}}_k^{(1)}$, whose elements can be obtained from (3.4.5) for $r = 1$. According to the analysis presented in section 3.3, the ith row of the matrix of synaptic weights $\mathbf{V}^{(0)}$ can be updated through the equation

$$\mathbf{v}_{i,k}^{(0)} = \mathbf{v}_{i,k-1}^{(0)} + \alpha\,\varepsilon_{i,k}^{(0)}\,\mathbf{P}_{i,k}\,\hat{\mathbf{h}}_k^{(1)} \tag{3.4.11}$$

where $\varepsilon_{i,k}^{(0)} = \varepsilon_{i,k}^o$ is defined by (3.3.27) and $\hat{y}_{i,k}$ by (3.4.4). The $(n_1 + 1) \times (n_1 + 1)$ matrix $\mathbf{P}_{i,k}$ is evaluated recursively by

$$\mathbf{P}_{i,k} = \mathbf{P}_{i,k-1} - \delta_{i,k}\,\mathbf{P}_{i,k-1}\,\mathbf{h}_k^{(1)}\,\hat{\mathbf{h}}_k^{(1)*}\,\mathbf{P}_{i,k-1} \tag{3.4.12}$$

Start

Initialize $\mathbf{W}$ and $\mathbf{V}$ with random values

1 $E = 0$

$i = 0$ (i: index of output unit)

2 $i \leftarrow i + 1$

$\bar{\mathbf{w}} = \mathbf{w}_i$

$\mathbf{P} = (1/\eta)\,\mathbf{I}$

$k = 0$ (k: number of association)

3 $k \leftarrow k + 1$

$\mathbf{x} = \mathbf{x}_k$

$y_i = y_{i,k}$

$\hat{h}_j = \rho(\sum_{\ell=1}^{n_i} v_{j\ell}\, x_\ell)$

$\hat{y} = \sigma(\sum_{j=1}^{n_h} \bar{w}_j\, \hat{h}_j)$

$c = (1 - \hat{y}^2)^2$

$\varepsilon^o = (1 - \hat{y}^2)(y - \hat{y})$

$\delta = c\,(1 + c\,\hat{\mathbf{h}}^*\,\mathbf{P}\,\hat{\mathbf{h}})^{-1}$

$\mathbf{P} \leftarrow \mathbf{P} - \delta\,\mathbf{P}\,\hat{\mathbf{h}}\,\hat{\mathbf{h}}^*\,\mathbf{P}$

$\bar{\mathbf{w}} \leftarrow \bar{\mathbf{w}} + \alpha\,\varepsilon^o\,\mathbf{P}\,\hat{\mathbf{h}}$

$\varepsilon_j^h = (1 - \hat{h}_j^2)\varepsilon^o\,\bar{w}_j$

$\mathbf{v}_j \leftarrow \mathbf{v}_j + \alpha\,\varepsilon^h\,\mathbf{x}$

$\hat{h}_j = \rho(\sum_{\ell=1}^{n_i} v_{j\ell}\, x_\ell)$

$\hat{y} = \sigma(\sum_{j=1}^{n_h} \bar{w}_j\, \hat{h}_j)$

$E \leftarrow E + \frac{1}{2}\,(y - \hat{y})^2$

if: $k < m$; then: go to 2

$\mathbf{w}_i = \bar{\mathbf{w}}$

if: $k < m$; then: go to 2

if: $E > E_o$; then: go to 1

Stop

Table 3.5: Flowchart of ELEANNE 5.

where $\mathbf{P}_{i,0} = (1/\eta)\,\mathbf{I}$, $\delta_{i,k} = c_{i,k}\,(1 + c_{i,k}\,\hat{\mathbf{h}}_k^{(1)*}\,\mathbf{P}_{i,k-1}\,\hat{\mathbf{h}}_k^{(1)})^{-1}$ and $c_{i,k} = (1 - \hat{y}_{i,k}^2)^2$. The synaptic weights $v_{pq}^{(r)}, r = 1, 2, \ldots, L$ can be updated by minimizing the objective function E_k^i, defined by (3.4.8), using the Delta Rule. It is shown in Appendix A3.4 that each set of synaptic weights $v_{pq}^{(r)}, r = 1, 2, \ldots, L$ can be updated by

$$\mathbf{v}_{p,k}^{(r)} = \mathbf{v}_{p,k-1}^{(r)} + \alpha\,\varepsilon_{pi,k}^{(r)}\,\hat{\mathbf{h}}_k^{(r+1)} \tag{3.4.13}$$

where $\hat{\mathbf{h}}_k^{(L+1)} = \mathbf{x}_k$,

$$\varepsilon_{pi,k}^{(1)} = (1 - \hat{h}_{p,k}^{(1)\,2})\,\varepsilon_{i,k}^{(0)}\,v_{ip}^{(0)} \tag{3.4.14}$$

and

$$\varepsilon_{pi,k}^{(r)} = (1 - \hat{h}_{p,k}^{(r)\,2}) \sum_{\ell=1}^{n_{r-1}} \varepsilon_{\ell i,k}^{(r-1)} v_{\ell p}^{(r-1)} \;\forall\, r = 2, 3, \ldots, L \tag{3.4.15}$$

3.4.2 ELEANNE 6

The ELEANNE 4 algorithm, derived in section 3.3 as a simplified version of the ELEANNE 3, can also be the basis for the development of a learning algorithm for multi-layered neural networks. The structure of this algorithm is closer to that of the Error Back Propagation algorithm. This is a consequence of the simplifying assumptions made in section 3.3. The ELEANNE 3 algorithm allows the adaptation of a single row of the matrix of synaptic weights with respect to each association. Its simplified version, the ELEANNE 4, allows the simultaneous adaptation of all the rows of the matrix of synaptic weights with respect to each association. According to ELEANNE 4,

- each adaptation cycle consists of m steps, each corresponding to one of the m associations, and
- during each step, the rows $\mathbf{w}_i, i = 1, 2, \ldots, n_o$ of the matrix of synaptic weights $\mathbf{W}$ are updated simultaneously.

Consider the neural network with one layer of hidden units, shown in Figure 2.9. Assume that during the νth adaptation cycle the synaptic weights w_{pq} of the upper network and v_{pq} of the lower network have been updated with respect to the associations $(\mathbf{y}_1, \mathbf{x}_1), (\mathbf{y}_2, \mathbf{x}_2), \ldots, (\mathbf{y}_{k-1}, \mathbf{x}_{k-1})$. The adaptation of the same synaptic weights with respect to the association $(\mathbf{y}_k, \mathbf{x}_k)$ can be achieved by:

- updating all the rows of the matrix of synaptic weights $\mathbf{W}$ by using ELEANNE 4, and
- updating the synaptic weights v_{pq} of the lower network on the basis of the Delta Rule.

Consider here the adaptation of the matrix of synaptic weights $\mathbf{W}$ of the upper network with respect to the association $(\mathbf{y}_k, \mathbf{x}_k)$, during the νth adaptation cycle. Assuming that $\mathbf{x}_k$ is the input of the lower network, the input of the upper network is the temporary input pattern $\hat{\mathbf{h}}_k$, defined by the combination of (3.4.1) and (3.4.2). The matrix of synaptic weights $\mathbf{W}$ of the upper network can be updated with respect to the temporary association $(\mathbf{y}_k, \hat{\mathbf{h}}_k)$ using ELEANNE 4. According to the analysis presented in section 3.3, the n_o rows of the matrix of synaptic weights $\mathbf{W}$ of the upper network are updated with respect to the temporary association $(\mathbf{y}_k, \hat{\mathbf{h}}_k)$ through the equations

$$\mathbf{w}_{i,k} = \mathbf{w}_{i,k-1} + \alpha\, \varepsilon_{i,k}^{o}\, \mathbf{P}_k^{\nu}\, \hat{\mathbf{h}}_k \;\forall\, i = 1, 2, \ldots, n_o \tag{3.4.16}$$

where $\hat{\mathbf{h}}_k$ is the temporary input pattern (3.4.2) whose elements are given by (3.4.1) and $\varepsilon_{i,k}^{o}$ is given by (3.3.27). $\mathbf{P}_k = \mathbf{P}_k^{\nu}$ is evaluated in terms of $\mathbf{P}_{k-1} = \mathbf{P}_{k-1}^{\nu}$ by the following recursive equation

$$\mathbf{P}_k^{\nu} = \mathbf{P}_{k-1}^{\nu} - \delta_k^{\nu}\, \mathbf{P}_{k-1}^{\nu}\, \hat{\mathbf{h}}_k\, \hat{\mathbf{h}}_k^{*}\, \mathbf{P}_{k-1}^{\nu} \tag{3.4.17}$$

where $\mathbf{P}_0^{\nu} = (1/\eta)\,\mathbf{I}$, $\delta_k^{\nu} = c_k^{\nu}\,(1 + c_k^{\nu}\, \hat{\mathbf{h}}_k^{*}\, \mathbf{P}_{k-1}^{\nu}\, \hat{\mathbf{h}}_k)^{-1}$ and c_k^{ν} is defined in terms of the adaptation cycle ν by (3.3.33).

The synaptic weights v_{pq} can be updated by minimizing the objective function E_k, defined by (3.3.2), using the gradient descent method. It was shown in chapter 2 that the synaptic weights of the lower network can be updated by

$$\mathbf{v}_{p,k} = \mathbf{v}_{p,k-1} + \alpha\, \varepsilon_{p,k}^{h}\, \mathbf{x}_k \tag{3.4.18}$$

Start

Initialize **W** and **V** with random values

$\nu = 0$ (ν: number of adaptation cycle)

1 $\nu \leftarrow \nu + 1$

$E = 0$

$\mathbf{P} = (1/\eta)\,\mathbf{I}$

$k = 0$ (k: number of association)

2 $k \leftarrow k + 1$

$\mathbf{x} = \mathbf{x}_k$

$\mathbf{y} = \mathbf{y}_k$

$\hat{h}_j = \rho(\sum_{\ell=1}^{n_i} v_{j\ell}\, x_\ell)$

$c = \exp(-2\,\beta_k^2\,(\nu - 1)^2)$

$\delta = c\,(1 + c\,\hat{\mathbf{h}}^*\,\mathbf{P}\,\hat{\mathbf{h}})^{-1}$

$\mathbf{P} \leftarrow \mathbf{P} - \delta\,\mathbf{P}\,\hat{\mathbf{h}}\,\hat{\mathbf{h}}^*\,\mathbf{P}$

$\hat{y}_i = \sigma(\sum_{j=1}^{n_h} w_{ij}\,\hat{h}_j)$

$\varepsilon_i^o = (1 - \hat{y}_i^2)(y_i - \hat{y}_i)$

$\mathbf{w}_i \leftarrow \mathbf{w}_i + \alpha\,\varepsilon_i^o\,\mathbf{P}\,\hat{\mathbf{h}}$

$\varepsilon_j^h = (1 - \hat{h}_j^2)\sum_{i=1}^{n_o} \varepsilon_i^o\, w_{ij}$

$\mathbf{v}_j \leftarrow \mathbf{v}_j + \alpha\,\varepsilon_j^h\,\mathbf{x}$

$\hat{h}_j = \rho(\sum_{\ell=1}^{n_i} v_{j\ell}\, x_\ell)$

$\hat{y}_i = \sigma(\sum_{j=1}^{n_h} w_{ij}\,\hat{h}_j)$

$E \leftarrow E + \frac{1}{2}\sum_{i=1}^{n_o}(y_i - \hat{y}_i)^2$

if: $k < m$; then: go to 2

if: $E > E_o$; then: go to 1

Stop

Table 3.6: Flowchart of ELEANNE 6.

where

$$\varepsilon^h_{p,k} = (1 - \hat{h}^2_{p,k}) \sum_{i=1}^{n_o} \varepsilon^o_{i,k}\, w_{ip} \tag{3.4.19}$$

and $\varepsilon^o_{i,k}$ is defined by (3.3.27). The resulting algorithm, called ELEANNE 6, is summarized in the flowchart presented in Table 3.6.

The above analysis indicates that the update equations for the synaptic weights v_{pq} of the lower network are identical to those of the Error Back Propagation algorithm (Rumelhart et al., 1986). This is due to the fact that the simplified algorithm used for the adaptation of the synaptic weights w_{pq} of the upper network allows for the simultaneous evaluation of all the rows of the matrix of synaptic weights $\mathbf{W}$. The generalization of this algorithm to networks with more than one layer of hidden units is straightforward and can be performed as in chapter 2.

3.4.3 ELEANNE 7

The ELEANNE 5 and ELEANNE 6 algorithms were devised for training multi-layered neural networks with binary outputs. However, feed-forward neural networks with analog output have recently attracted the attention of many researchers (Hertz et al., 1991). For example, such a neural network with analog output has been used in signal and image compression (Cottrell et al., 1987; Boulard and Kamp, 1988). The increasing popularity of this type of feed-forward neural networks is the motivation for the development of ELEANNE 7, an algorithm which can be used for training neural networks with either analog or binary output. This algorithm can be derived by simply combining the results of the analysis presented in section 3.3 and the strategy which provided the basis for the development of the ELEANNE 6.

Consider the feed-forward neural network with one layer of hidden units shown in Figure 2.9. The matrices $\mathbf{H}_{i,m}, i = 1, 2, \ldots, n_o$ can be evaluated in terms of the temporary input patterns $\hat{\mathbf{h}}_k$ on the basis of the analysis presented in section 3.3. If the output of this network is analog, i.e., $\sigma(x) = x$, these matrices are given by

$$\mathbf{H}_{i,m} = \mathbf{H}_m = \sum_{k=1}^{m} \hat{\mathbf{h}}_k \, \hat{\mathbf{h}}_k^* \;\forall\, i = 1, 2, \ldots, n_o \tag{3.4.20}$$

It was also shown in section 3.3 that if the output of the network is binary, i.e., $\sigma(x) = \tanh(x)$, these matrices can be approximated by

$$\mathbf{H}_{i,m} = \sum_{k=1}^{m} c_{i,k} \, \hat{\mathbf{h}}_k \, \hat{\mathbf{h}}_k^* \;\forall\, i = 1, 2, \ldots, n_o \tag{3.4.21}$$

where $c_{i,k} = (1 - \hat{y}_{i,k}^2)^2$. The computational burden imposed by the existence of n_o distinct matrices can be significantly reduced if each of these matrices is replaced by their average, that is,

$$\bar{\mathbf{H}}_m = \frac{1}{n_o} \sum_{i=1}^{n_o} \mathbf{H}_{i,m} = \sum_{k=1}^{m} \bar{c}_k \, \hat{\mathbf{h}}_k \, \hat{\mathbf{h}}_k^* \tag{3.4.22}$$

where

$$\bar{c}_k = \frac{1}{n_o} \sum_{i=1}^{n_o} c_{i,k} = \frac{1}{n_o} \sum_{i=1}^{n_o} (1 - \hat{y}_{i,k}^2)^2 \tag{3.4.23}$$

Finally, the efficient adaptation of the synaptic weights requires the recursive evaluation of $\bar{\mathbf{P}}_k = \bar{\mathbf{H}}_k^{-1}$, where

$$\bar{\mathbf{H}}_m = \sum_{k=1}^{m} \bar{c}_k \, \hat{\mathbf{h}}_k \, \hat{\mathbf{h}}_k^* \tag{3.4.24}$$

where

$$\bar{c}_k = \begin{cases} 1 & \text{if the output is analog} \\ \frac{1}{n_o} \sum_{i=1}^{n_o} (1 - \hat{y}_{i,k}^2)^2 & \text{if the output is binary} \end{cases} \tag{3.4.25}$$

The training of the multi-layered neural network considered here involves the adaptation of two sets of synaptic weights, that is, the synaptic weights which connect the hidden units and the output units and the synaptic weights which connect the input of the network and the hidden units. The elements of the matrix $\bar{\mathbf{H}}_m$ are obtained from the second-order partial derivatives of the objective function with respect to the synaptic weights w_{pq} which connect the hidden and the output units. It should be emphasized here that if

Start

Initialize $\mathbf{W}$ and $\mathbf{V}$ with random values

1 $\quad E = 0$

$\quad \mathbf{P} = (1/\eta)\,\mathbf{I}$

$\quad\quad k = 0$ (k: number of association)

$\quad$ 2 $\quad k \leftarrow k + 1$

$\quad\quad \mathbf{x} = \mathbf{x}_k$

$\quad\quad \mathbf{y} = \mathbf{y}_k$

$\quad\quad \hat{h}_j = \rho(\sum_{\ell=1}^{n_i} v_{j\ell}\, x_\ell)$

$\quad\quad \bar{c} = 1$ (analog output)

$\quad\quad c_i = (1 - \hat{y}_i^2)^2$ (binary output)

$\quad\quad \bar{c} = (1/n_o) \sum_{i=1}^{n_o} c_i$ (binary output)

$\quad\quad \bar{\delta} = \bar{c}\,(1 + \bar{c}\,\hat{\mathbf{h}}^*\, \bar{\mathbf{P}}\, \hat{\mathbf{h}})^{-1}$

$\quad\quad \bar{\mathbf{P}} \leftarrow \bar{\mathbf{P}} - \bar{\delta}\, \bar{\mathbf{P}}\, \hat{\mathbf{h}}\, \hat{\mathbf{h}}^*\, \bar{\mathbf{P}}$

$\quad\quad \hat{y}_i = \sigma(\sum_{j=1}^{n_h} w_{ij}\, \hat{h}_j)$

$\quad\quad \varepsilon_i^o = (1 - \hat{y}_i^2)(y_i - \hat{y}_i)$

$\quad\quad \mathbf{w}_i \leftarrow \mathbf{w}_i + \alpha\, \varepsilon_i^o\, \bar{\mathbf{P}}\, \hat{\mathbf{h}}$

$\quad\quad \varepsilon_j^h = (1 - \hat{h}_j^2) \sum_{i=1}^{n_o} \varepsilon_i^o\, w_{ij}$

$\quad\quad \mathbf{v}_j \leftarrow \mathbf{v}_j + \alpha\, \varepsilon_j^h\, \mathbf{x}$

$\quad\quad \hat{h}_j = \rho(\sum_{\ell=1}^{n_i} v_{j\ell}\, x_\ell)$

$\quad\quad \hat{y}_i = \sigma(\sum_{j=1}^{n_h} w_{ij}\, \hat{h}_j)$

$\quad\quad E \leftarrow E + \frac{1}{2} \sum_{i=1}^{n_o} (y_i - \hat{y}_i)^2$

$\quad\quad$ if: $k < m$; then: go to 2

$\quad$ if: $E > E_o$; then: go to 1

Stop

Table 3.7: Flowchart of ELEANNE 7.

a multi-layered neural network is considered, $\bar{\mathbf{H}}_m$ is a submatrix of the true Hessian matrix, which is formed by the second-order partial derivatives of the objective function with respect to all the synaptic weights of the network.

The previous discussion suggests that this algorithm can be devised on the basis of the same strategy that provided the ELEANNE 6. Suppose that during a certain adaptation cycle the synaptic weights of the network have been updated with respect to the associations $(\mathbf{y}_1, \mathbf{x}_1)$, $(\mathbf{y}_2, \mathbf{x}_2)$, $\ldots$, $(\mathbf{y}_{k-1}, \mathbf{x}_{k-1})$, resulting in the estimates $\mathbf{W} = \mathbf{W}_{k-1}$ and $\mathbf{V} = \mathbf{V}_{k-1}$. The rows $\mathbf{w}_i, i = 1, 2, \ldots, n_o$ of $\mathbf{W}$ can be updated with respect to the association $(\mathbf{y}_k, \mathbf{x}_k)$ by

$$\mathbf{w}_{i,k} = \mathbf{w}_{i,k-1} + \alpha\, \varepsilon^o_{i,k}\, \bar{\mathbf{P}}_k\, \hat{\mathbf{h}}_k \;\forall\, i = 1, 2, \ldots, n_o \tag{3.4.26}$$

where $\hat{\mathbf{h}}_k$ is the temporary input pattern and

$$\varepsilon^o_{i,k} = \begin{cases} y_{i,k} - \hat{y}_{i,k} & \text{if the output is analog} \\ (1 - \hat{y}^2_{i,k})(y_{i,k} - \hat{y}_{i,k}) & \text{if the output is binary} \end{cases} \tag{3.4.27}$$

$\bar{\mathbf{P}}_k$ can be evaluated recursively in terms of $\bar{\mathbf{P}}_{k-1}$ by

$$\bar{\mathbf{P}}_k = \bar{\mathbf{P}}_{k-1} - \bar{\delta}_k\, \bar{\mathbf{P}}_{k-1}\, \hat{\mathbf{h}}_k\, \hat{\mathbf{h}}^*_k\, \bar{\mathbf{P}}_{k-1} \tag{3.4.28}$$

where $\bar{\mathbf{P}}_0 = (1/\eta)\,\mathbf{I}$, $\bar{\delta}_k = \bar{c}_k\,(1 + \bar{c}_k\, \hat{\mathbf{h}}^*_k\, \bar{\mathbf{P}}_{k-1}\, \hat{\mathbf{h}}_k)^{-1}$ and $\bar{c}_k$ is given by (3.4.25).

After the adaptation of all the synaptic weights by (3.4.26), the synaptic weights v_{pq} of the lower network can be updated with respect to the same association $(\mathbf{y}_k, \mathbf{x}_k)$ by the combination of (3.4.18) and (3.4.19). The resulting algorithm, called ELEANNE 7, is summarized in the flowchart presented in Table 3.7. This algorithm can easily be generalized to feed-forward neural networks with more than one layer of hidden units.

3.5 Computational Considerations

The convergence of a learning algorithm is judged on the basis of the number of adaptation cycles required by the algorithm to reach a predetermined small value of the total error E. The efficiency of

Algorithm	Multiplications per adaptation cycle
Back Propagation	$m(4n_hn_i + 4n_hn_o + 4n_o + 2n_h + 1)$
ELEANNE 5	$mn_o(3n_h^2 + 4n_hn_i + 7n_h + 8)$
ELEANNE 6 & 7	$m(3n_h^2 + 4n_hn_i + 4n_hn_o + 4n_o + 3n_h + 7)$

Table 3.8: Number of multiplications per adaptation cycle required by the Error Back Propagation, ELEANNE 5, ELEANNE 6, and ELEANNE 7 algorithms.

Algorithms	Additions per adaptation cycle
Back Propagation	$m(3n_hn_i + 4n_on_h + 2n_o - n_h + 1)$
ELEANNE 5	$mn_o(2n_h^2 + 3n_hn_i + 3n_h + 4)$
ELEANNE 6 & 7	$m(2n_h^2 + 3n_hn_i + 4n_on_h + 2n_o - n_h + 2)$

Table 3.9: Number of additions per adaptation cycle required by the Error Back Propagation, ELEANNE 5, ELEANNE 6, and ELEANNE 7 algorithms.

the algorithm depends on the number of arithmetic operations required for each adaptation cycle. The comparison of the algorithms for single-layered neural networks in terms of their computational requirements can be found in (Karayiannis, 1991a).

This section presents an evaluation of the efficient learning algorithms for multi-layered neural networks in terms of the number of arithmetic operations required for each adaptation cycle. Table 3.8 shows the number of multiplications per adaptation cycle required by the Error Back Propagation (M_{EBP}), the ELEANNE 5 (M_{E5}), the ELEANNE 6 (M_{E6}), and the ELEANNE 7 (M_{E7}) algorithms. Table 3.9 shows the number of additions per adaptation cycle required by Error Back Propagation (A_{EBP}), ELEANNE 5 (A_{E5}), ELEANNE 6 (A_{E6}), and ELEANNE 7 (A_{E7}).

Under the assumption that $n_h\, n_i \gg n_o$ and/or $n_h\, n_o \gg n_o (n_h \gg 1)$, Table 3.8 shows that

$$\frac{M_{E5}}{M_{E6}} \approx \frac{M_{E5}}{M_{E7}} \approx n_o \frac{3r + 4 + 7/n_i}{3r + 4 + (4n_o + 3)/n_i} \tag{3.6.1}$$

where $r = n_h/n_i$. If $n_o = 1$, (3.6.1) implies that $M_{E5} \approx M_{E6} \approx M_{E7}$.

Under the assumption that $n_i \gg n_o > 1$, (3.6.1) can be simplified to $M_{E5} \approx n_o M_{E6} \approx n_o M_{E7}$.

Under the same assumptions, Table 3.9 shows that

$$\frac{A_{E5}}{A_{E6}} \approx \frac{A_{E5}}{A_{E7}} \approx n_o \frac{2r + 3 + 3/n_i}{2r + 3 + (4n_o - 1)/n_i} \tag{3.6.2}$$

If $n_o = 1$, (3.6.2) implies that $A_{E5} \approx A_{E6} \approx A_{E7}$. Under the assumption that $n_i \gg n_o > 1$, (3.6.2) can be simplified to $A_{E5} \approx n_o A_{E6} \approx n_o A_{E7}$.

Among the algorithms proposed for training multi-layered neural networks, ELEANNE 6 and ELEANNE 7 are computationally less demanding. The analysis which follows compares the number of multiplications per adaptation cycle required by the Error Back Propagation, the ELEANNE 6, and the ELEANNE 7 algorithms. In practical situations, $n_h n_o \gg n_o (n_h \gg 1)$ and/or $n_h n_i \gg n_o$. Therefore, M_{EBP} can be approximated by $M_{EBP} \approx 2 m n_h (2n_i + 2n_o + 1)$. If, in addition, $n_h^2 \gg 1$, Table 3.8 shows that

$$\frac{M_{E6}}{M_{EBP}} \approx \frac{M_{E7}}{M_{EBP}} \approx 1 + \frac{3n_h + 1}{2(2n_i + 2n_o + 1)} \tag{3.6.3}$$

Let the number n_h of hidden units be related to the number n_i of inputs by $n_h = r\, n_i$. Assuming that $n_i \gg 1$, (3.6.3) simplifies to

$$\frac{M_{E6}}{M_{EBP}} \approx \frac{M_{E7}}{M_{EBP}} \approx 1 + \frac{3r}{4(1 + (n_o/n_i))} \tag{3.6.4}$$

This equation indicates that the ratio $M_{E6}/M_{EBP}(M_{E7}/M_{EBP})$ depends on the number of input, hidden, and output units. When the ratio n_o/n_i is constant, $M_{E6}/M_{EBP}(M_{E7}/M_{EBP})$ is an increasing function of $r = n_h/n_i$. Assuming that r is constant and $n_i \gg 1$, the maximum value of $M_{E6}/M_{EBP}(M_{E7}/M_{EBP})$ corresponds to $n_o = 1$. In this case, $M_{E6} \approx M_{E7} \approx [1 + \frac{3}{4} r] M_{EBP}$. When r is constant, the ratio $M_{E6}/M_{EBP}(M_{E7}/M_{EBP})$ decreases when n_o/n_i increases. For example, when $n_o = n_i$, $M_{E6} \approx M_{E7} \approx [1 + \frac{3}{8} r] M_{EBP}$. The ratio $M_{E6}/M_{EBP}(M_{E7}/M_{EBP})$ is plotted as a function of the ratio n_o/n_i for various values of $r = n_h/n_i$ in Figure 3.1. The additional computational burden due to the use of ELEANNE 6 or

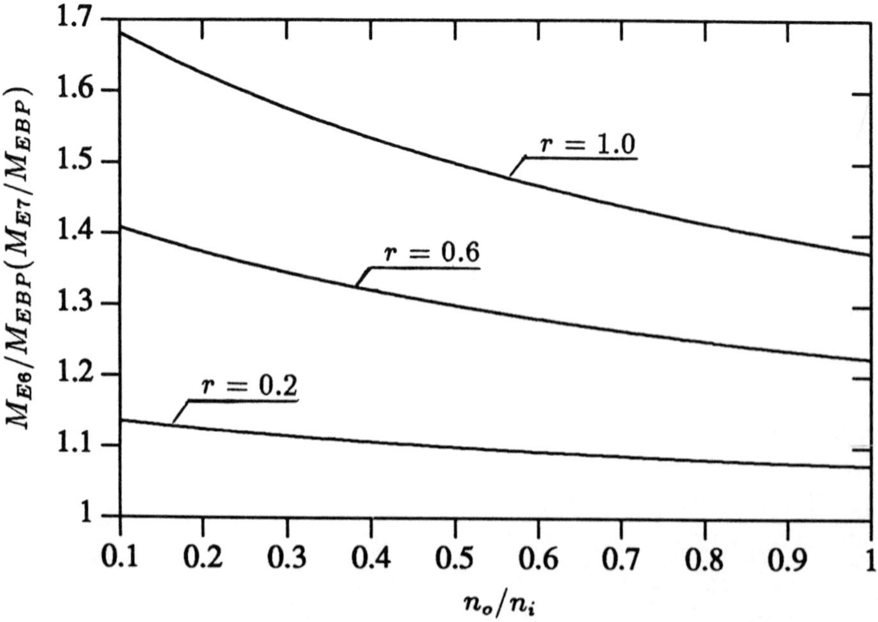

Figure 3.1: The ratio M_{E6}/M_{EBP} of the number of multiplications per adaptation cycle required by the ELEANNE 6 and Error Back Propagation (EBP) algorithms as a function of the ratio n_o/n_i for various $r = n_h/n_i$.

ELEANNE 7 is illustrated by the following example. If $n_o = n_i/2$ and $r = n_h/n_i = 2/3$, $M_{E6} \approx M_{E7} \approx M_{EBP} + \frac{1}{3} M_{EBP}$. The additional number of multiplications per adaptation cycle required by ELEANNE 6 or ELEANNE 7 is a relatively small portion of the number of multiplications per adaptation cycle required by the Error Back Propagation algorithm.

A similar analysis is now carried out for the number of additions per adaptation cycle required by the Error Back Propagation algorithm, ELEANNE 6, and ELEANNE 7. Once again, provided that either one of the conditions $n_h\, n_i \gg n_o$ and/or $n_o\, n_h \gg n_o (n_h \gg 1)$, A_{EBP} can be approximated by $A_{EBP} \approx m\, n_h\, (3n_i + 4n_o - 1)$. If, in addition, $n_h^2 \gg 1$, Table 3.9 shows that

$$\frac{A_{E6}}{A_{EBP}} \approx \frac{A_{E7}}{A_{EBP}} \approx 1 + \frac{2n_h}{3n_i + 4n_o - 1} \tag{3.6.5}$$

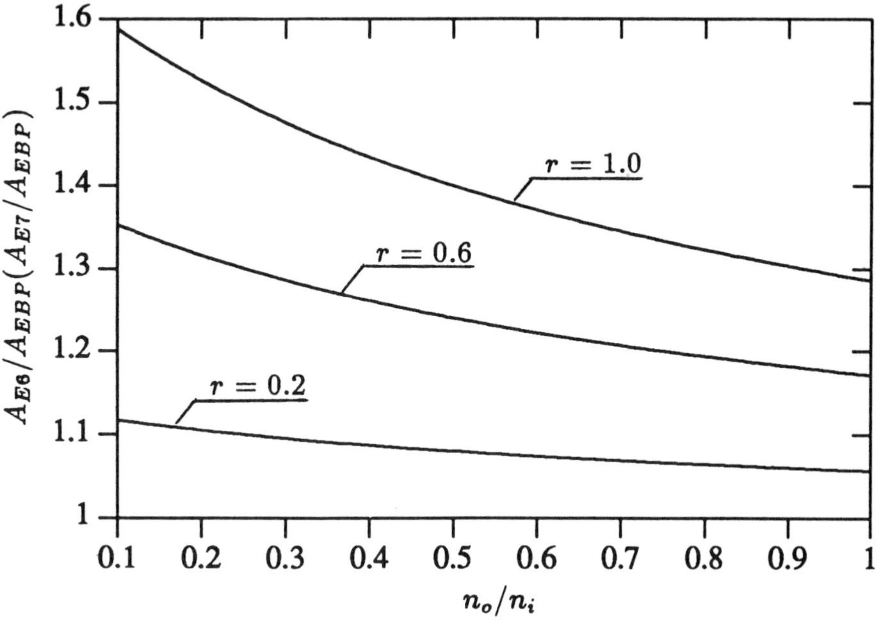

Figure 3.2: The ratio A_{E6}/A_{EBP} of the number of additions per adaptation cycle required by the ELEANNE 6 and Error Back Propagation (EBP) algorithms as a function of n_o/n_i for various $r = n_h/n_i$.

Assuming once again that $n_h = r\, n_i$ and $n_i \gg 1$, (3.6.5) simplifies to

$$\frac{A_{E6}}{A_{EBP}} \approx \frac{A_{E7}}{A_{EBP}} \approx 1 + \frac{2r}{3 + 4(n_o/n_i)} \tag{3.6.6}$$

When the ratio n_o/n_i is constant, $A_{E6}/A_{EBP}(A_{E7}/A_{EBP})$ is also an increasing function of the ratio $r = n_h/n_i$. Assuming that $r = n_h/n_i$ is constant, the ratio $A_{E6}/A_{EBP}(A_{E7}/A_{EBP})$ attains its maximum value when $n_o = 1$. In this case, $A_{E6} \approx A_{E7} \approx [1+\frac{2}{3}\,r]A_{EBP}$. Figure 3.2 shows the ratio A_{E6}/A_{EBP} as a function of the ratio n_o/n_i for various values of $r = n_h/n_i$. When $n_o = n_i$, $A_{E6} \approx A_{E7} \approx [1 + \frac{2}{7}\,r]A_{EBP}$. Considering again the typical situation that $n_o = n_i/2$ and $r = n_h/n_i = 2/3$, $A_{E6} \approx A_{E7} \approx A_{EBP} + \frac{4}{15}\,A_{EBP}$. Once again, the number of additional additions per adaptation cycle required by ELEANNE 6 or ELEANNE 7 is a relatively small portion of the total number of additions per adaptation cycle required by the

Error Back Propagation algorithm.

3.6 Experimental Results

The overall efficiency of a learning algorithm is evaluated on the basis of a criterion comprising the convergence rate achieved by the algorithm during the training and the computational requirements imposed by the algorithm. The evaluation of the proposed learning algorithms in terms of the arithmetic operations involved in each adaptation cycle was presented in section 3.5. This section presents some experimental results using the learning algorithms presented in this chapter to train multi-layered neural networks. An experimental evaluation of the learning algorithms proposed for single-layered neural networks can be found in (Karayiannis, 1991a; Karayiannis and Venetsanopoulos, 1990b).

The first set of experiments compares the ELEANNE 7 and Error Back Propagation algorithms used to train a multi-layered neural network with one hidden layer to map its analog input to its output. Such a network performs auto-association, since its input and output coincide. It was suggested that when $n_h < n_o = n_i$ such a network can be used for signal and image compression, where the output of the hidden layer provides the compressed signal (Cottrell et al., 1987). A remarkable observation is that in this application the nonlinearity of the hidden units does not considerably affect the training task (Cottrell et al., 1987). In fact, the outputs of the hidden units are related to the principal components of the input signal (Boulard and Kamp, 1988). In this set of experiments, a discrete signal of length 800 was normalized between -1 and $+1$ and then divided into 25 blocks of length 32. The training set formed by these blocks was used to train a neural network with 32 inputs, 24 hidden units, and 32 output units. Figure 3.3 shows the error in dB when this network was trained by ELEANNE 7 and Error Back Propagation. The error curves were obtained by averaging error curves resulting from five trials, each starting with a different set of synaptic weights generated by a random number generator producing values between -0.5 and $+0.5$. The learning rate in these experiments was the highest value which still guaranteed that the

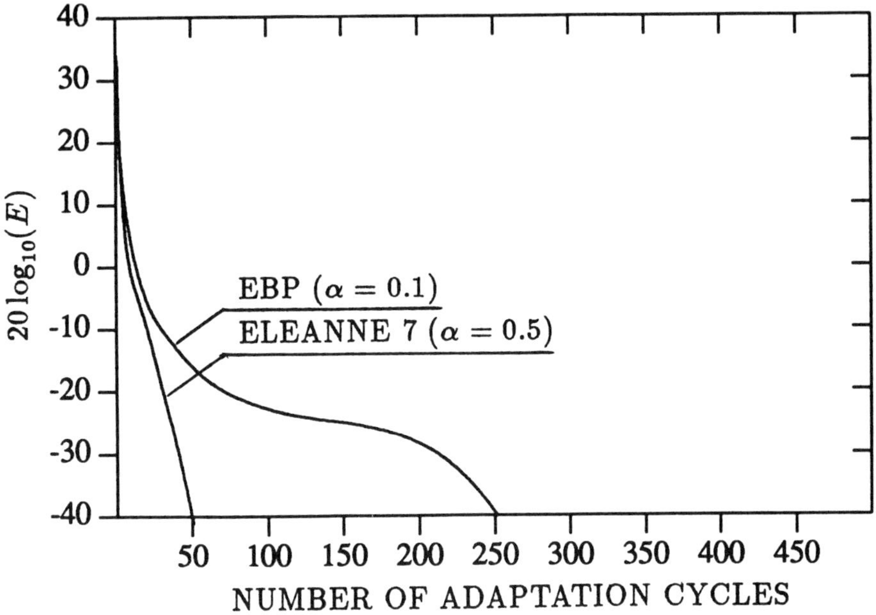

Figure 3.3: Training a multi-layered neural network with analog output using the Error Back Propagation (EBP) and the ELEANNE 7 algorithms. $20\log_{10}(E)$ as a function of the number of adaptation cycles.

error decreased monotonically during the training. The parameter η used for initializing the recursive inversion of the Hessian matrix required by ELEANNE 7 was $\eta = 1$. A comparison of the curves shown in Figure 3.3 indicates that the ELEANNE 7 algorithm converges to the target error value $E = 0.01$ much faster than Error Back Propagation. In fact, the convergence of the Error Back Propagation algorithm slows significantly after the initial adaptation cycles. In contrast, the convergence rate of the ELEANNE 7 remains almost unchanged throughout the training. This observation reveals the efficiency of ELEANNE 7 in training tasks where the learning algorithm must reach a very small target value of the error. It should be emphasized that the selection of a very small value of the total error is particularly important if the output of the trained neural network is analog. In this experiment, $n_i = n_o = 32$ and $r = n_h/n_i = 3/4$. According to section 3.5, $M_{E7} \approx \frac{4}{3} M_{EBP}$ and

$A_{E7} \approx \frac{17}{14} A_{EBP}$. The performance of the algorithms, in conjunction with their computational requirements, indicates that ELEANNE 7 performs this training task faster than the Error Back Propagation algorithm.

The remainder of this section presents a comparison of the algorithms presented in this chapter for feed-forward neural networks with binary output and the Error Back Propagation algorithm used to train a neural network to perform the Exclusive OR operation (Rumelhart et al., 1986). Such a network consists of two inputs, two hidden units and one output unit (Rumelhart et al., 1986; Hertz et al., 1991). The network was trained using the four associations $(\mathbf{y}_k, \mathbf{x}_k), k = 1,2,3,4$, where $\mathbf{x}_1^* = (-1,-1), y_1 = -1$, $\mathbf{x}_2^* = (-1,+1), y_2 = +1$, $\mathbf{x}_3^* = (+1,-1), y_3 = +1$ and $\mathbf{x}_4^* = (+1,+1), y_4 = -1$. Each adaptation cycle involved the presentation of 20 combinations of the associations $(y_k, \mathbf{x}_k), k = 1,2,3,4$ (Kollias and Anastassiou, 1989). Figure 3.4 shows the error in dB as a function of the number of adaptation cycles when the network was trained by the Error Back Propagation algorithm, ELEANNE 5, ELEANNE 6, and ELEANNE 7. The curves shown were obtained by averaging error curves resulting from five trials, each starting with a different initial set of synaptic weights generated by a random number generator producing values between -0.9 and $+0.9$. The learning rate used in each experiment was chosen in such a way that there were no temporary oscillations of the error during the training. This experiment emphasizes the importance of the selection of the parameter η used in the recursive inversion of the Hessian matrix. It was experimentally verified that the parameter η should be relatively large during the initial adaptation cycles in order to guarantee the stability of the recursive inversion of the Hessian matrix. According to the analysis presented in section 3.2, this parameter must be reduced during the training. In this experiment, η was determined during each adaptation cycle by $1/\eta = 1/\alpha + \lambda(1/\eta_0 - 1/\alpha)$, where λ was determined in terms of the current value of the total error E as $\lambda = \exp(-1/E^2)$ and $\eta_0 = 1$. This rule guarantees that during the initial adaptation cycles, where the error E is relatively large, $\lambda \approx 1$ and hence $\eta \approx \eta_0$. After the initial adaptation cycles, the error reduces and $\lambda \approx 0$. In this case, the parameter η approaches the learning rate, i.e., $\eta \approx \alpha$.

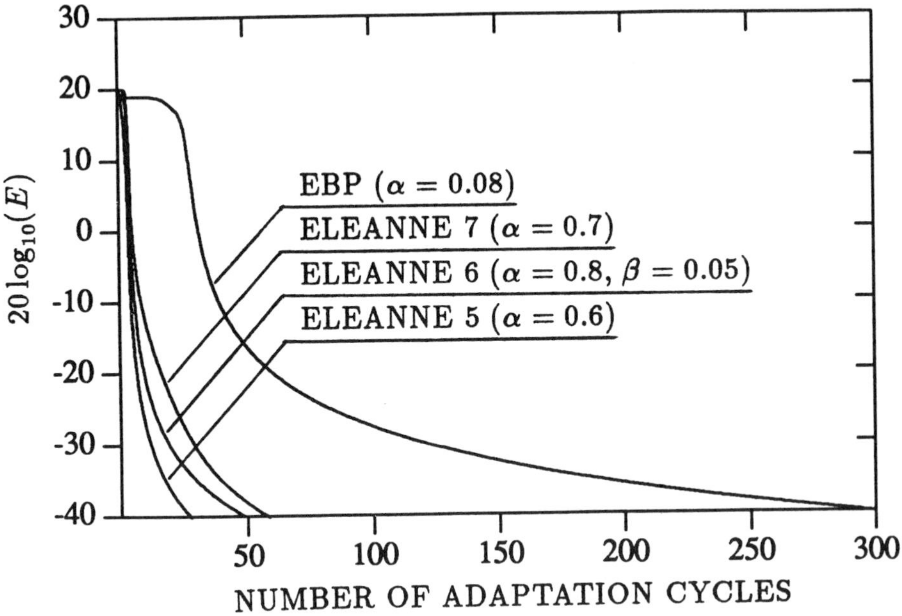

Figure 3.4: Training a neural network with one layer of hidden units to perform the Exclusive OR function by the Error Back Propagation (EBP), the ELEANNE 5, the ELEANNE 6, and the ELEANNE 7 algorithms. $20\log_{10}(E)$ as a function of the number of adaptation cycles.

According to Figure 3.4, the algorithms proposed in this chapter converge considerably faster than the Error Back Propagation algorithm, with ELEANNE 5 converging to a small target value of the error the fastest. Since the network trained in this experiment contains only one output unit, ELEANNE 5 performs approximately the same number of multiplications and additions per adaptation cycle as ELEANNE 6 or ELEANNE 7. The time required by the algorithms to perform this training task depends on the number of arithmetic operations performed by the algorithms in each adaptation cycle. In this experiment, $n_i = 2$, $n_o = 1$, and $r = n_h/n_i = 1$. According to section 3.5, $M_{E5} \approx M_{E6} \approx M_{E7} \approx \frac{3}{2} M_{EBP}$ and also $A_{E5} \approx A_{E6} \approx A_{E7} \approx \frac{7}{5} A_{EBP}$. This is an indication that all the algorithms in this chapter perform this training task faster than the Error Back Propagation algorithm.

In the next experiment, a multi-layered neural network was trained to function as an n to n encoder. This problem was considered in the first experiments involving the Error Back Propagation algorithm (Rumelhart et al., 1986) and it was subsequently used as a benchmark (Hertz et al., 1991). The n to n encoder is implemented by a neural network with n inputs and n output units trained to map n input patterns into the output. Each of the n patterns contains only one active element, represented in this experiment by $+1$, while all the other elements are inactive, represented by -1. The number of hidden units of an n to n encoder is typically smaller than n and at least equal to $\log_2(n)$. In this experiment the network consisted of 10 inputs, 5 hidden units and 10 output units. Figure 3.5 shows the error in dB as a function of the number of adaptation cycles when the network was trained by the Error Back Propagation, the ELEANNE 5, the ELEANNE 6, and the ELEANNE 7 algorithms. The curves shown were obtained by averaging five error curves as indicated earlier. Once again, the initial sets of synaptic weights were provided by a random number generator producing numbers between -0.9 and $+0.9$. The learning rate for each algorithm was also chosen sufficiently small to guarantee a smooth reduction of the error during the training. The parameter η used in the initialization of the recursive inversion of the Hessian matrix was evaluated during the training in terms of the learning rate α by $1/\eta = 1/\alpha + \lambda(1/\eta_0 - 1/\alpha)$, where $\lambda = \exp(-1/E^2)$ and $\eta_0 = 1/3$. Figure 3.5 shows that all the learning algorithms presented in this chapter achieve faster convergence than the Error Back Propagation algorithm. However, these algorithms are more demanding than the Error Back Propagation in terms of the number of arithmetic operations required per adaptation cycle. Figure 3.5 indicates that ELEANNE 5 converges faster than both ELEANNE 6 and ELEANNE 7. However, there is a tradeoff between the number of arithmetic operations per adaptation cycle required by each algorithm and the corresponding rate of convergence. The comparison of ELEANNE 5 and ELEANNE 6 in section 3.5 indicated that ELEANNE 5 performs almost n_o times as many arithmetic operations per adaptation cycle as ELEANNE 6, and ELEANNE 7 requires the same number of arithmetic oper-

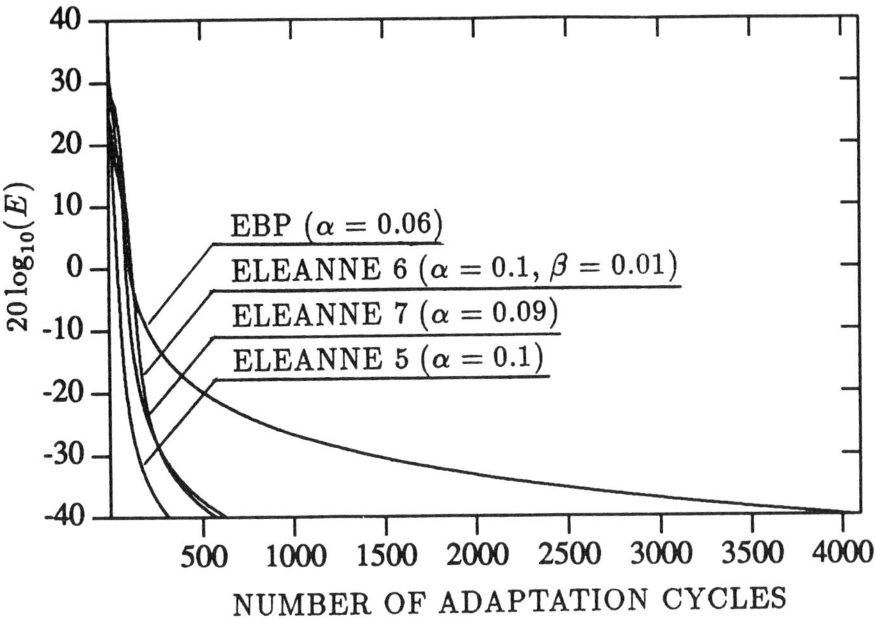

Figure 3.5: Training a neural network with one layer of hidden units to function as an n to n encoder by the Error Back Propagation (EBP), the ELEANNE 5, the ELEANNE 6, and the ELEANNE 7 algorithms. $20\log_{10}(E)$ as a function of the number of adaptation cycles.

ations per adaptation cycle as ELEANNE 6. In this experiment, $n_i = n_o = 10$ and $r = n_h/n_i = 1/2$. According to section 3.5, $M_{E6} \approx M_{E7} \approx \frac{5}{4} M_{EBP}$ and also $A_{E6} \approx A_{E7} \approx \frac{8}{7} A_{EBP}$. This is an indication that the ELEANNE 6 and the ELEANNE 7 converge to the target error value faster than the Error Back Propagation algorithm.

3.7 Discussion

This chapter presented the development and evaluation of efficient learning algorithms for neural networks. The use of optimal least-squares techniques for training a single-layered neural network with respect to its analog output provided two recursive least-squares learning algorithms, namely the ELEANNE 1 and the ELEANNE 2 algorithms. A new optimization strategy for training single-layered neural networks was also proposed. The application of this strategy to the training of a single-layered neural network with respect to its analog output also led to ELEANNE 2. The application of the same strategy to the training of a single-layered neural network with respect to its binary output led to the ELEANNE 3 algorithm. According to this algorithm, the rows of the matrix of synaptic weights are updated separately. ELEANNE 4 resulted from a simplification of ELEANNE 3 which permitted the entire matrix of synaptic weights to be updated on each step. The learning algorithms developed for single-layered neural networks were the basis for the derivation of efficient learning algorithms for training multi-layered neural networks. The ELEANNE 5, ELEANNE 6, and ELEANNE 7 algorithms were derived from ELEANNE 3 and ELEANNE 4, respectively. The last part of this chapter presented an experimental comparison of the proposed algorithms and existing ones, evaluated the effect of the free parameters of the algorithms on their performance, and studied the effect of the architecture of multi-layered neural networks on the convergence of the learning algorithms used for their training. It was experimentally verified that the algorithms proposed in this chapter converged faster than the ones already existing in the literature. The simplified algorithms, ELEANNE 4, ELEANNE 6, and ELEANNE 7, provide an attractive alternative for training neural networks in practical applications.

Appendix A3.1

The derivation of the ELEANNE 1 is completed here by determining the update equation for the matrix of synaptic weights $\mathbf{W}_m$ in the case where $\mathbf{x}_m \in \mathcal{C}(\mathbf{X}_{m-1})$, that is, $\mathbf{x}_m$ is a linear combination of the column vectors $\mathbf{x}_k, k = 1, 2, \ldots, m-1$ of $\mathbf{X}_{m-1}$. The update equation for the matrix of synaptic weights $\mathbf{U}_m$ can be provided by a closer consideration of the meaning of $\mathbf{x}_m \in \mathcal{C}(\mathbf{X}_{m-1})$. The column space of $\mathbf{X}_{m-1}$ consists of all vectors in $\mathcal{R}^{n_i}$ which can be written as a linear combination of its columns $\mathbf{x}_k, k = 1, 2, \ldots, m-1$. In other words, the column space $\mathcal{C}(\mathbf{X}_{m-1})$ of $\mathbf{X}_{m-1}$ consists of all vectors

$\mathbf{x} \in \mathcal{R}^{n_i}$, which can be written as $\mathbf{x} = \mathbf{X}_{m-1}\mathbf{w}$, for some $\mathbf{w} \in \mathcal{R}^{m-1}$
If $\mathbf{x}_m \in \mathcal{C}(\mathbf{X}_{m-1})$, there exists a $\mathbf{w}_m \in \mathcal{R}^{m-1}$, such that

$$\mathbf{x}_m = \mathbf{X}_{m-1}\,\mathbf{w}_m \tag{A3.1.1}$$

Taking into account the definition of $\mathbf{U}_{m-1}$, (A3.1.1) provides

$$\mathbf{U}_{m-1}\,\mathbf{x}_m = \mathbf{X}_{m-1}\,\mathbf{X}_{m-1}^{+}\,\mathbf{X}_{m-1}\,\mathbf{w}_m = \mathbf{X}_{m-1}\,\mathbf{w}_m = \mathbf{x}_m \tag{A3.1.2}$$

According to (A3.1.2), the matrix of synaptic weights $\mathbf{U}_{m-1}$ which corresponds to the $m-1$ associations $(\mathbf{x}_k, \mathbf{x}_k), k = 1, 2, \ldots, m-1$ achieves perfect association of the new input pattern $\mathbf{x}_m$ with itself. In other words, the association $(\mathbf{x}_m, \mathbf{x}_m)$ is already stored in the matrix of synaptic weights $\mathbf{U}_{m-1}$. This is a verification that if $\mathbf{x}_m \in \mathcal{C}(\mathbf{X}_{m-1})$, $\mathbf{U}_m = \mathbf{U}_{m-1}$. Consider now the update equation for the matrix of synaptic weights $\mathbf{W}_m^h$ under the assumption that $\mathbf{x}_m \in \mathcal{C}(\mathbf{X}_{m-1})$. In this case, $\mathbf{z}_m$ depends on the particular definition of the generalized inverse (Rao and Mitra, 1971). For example, if the minimum norm generalized inverse is considered,

$$\mathbf{z}_m = \frac{(\mathbf{X}_{m-1}^{+})^{*}\,\mathbf{X}_{m-1}^{+}\,\mathbf{x}_m}{1 + ||\mathbf{X}_{m-1}^{+}\,\mathbf{x}_m||^2} \tag{A3.1.3}$$

It can be verified that (3.2.7) amounts to

$$\mathbf{y}_m - \mathbf{W}_m\,\mathbf{x}_m = (1 - \mathbf{z}_m^{*}\,\mathbf{x}_m)(\mathbf{y}_m - \mathbf{W}_{m-1}\,\mathbf{x}_m) \tag{A3.1.4}$$

If $\mathbf{x}_m \notin \mathcal{C}(\mathbf{X}_{m-1})$, $\mathbf{z}_m$ is given by (3.2.6). It can be shown that in this case

$$1 - \mathbf{z}_m^{*}\,\mathbf{x}_m = 0 \tag{A3.1.5}$$

The combination of (A3.1.4) and (A3.1.5) indicates that if $\mathbf{x}_m \in \mathcal{C}(\mathbf{X}_{m-1})$

$$\mathbf{y}_m = \mathbf{W}_m\,\mathbf{x}_m \tag{A3.1.6}$$

This is exactly the condition proposed for the evaluation of $\mathbf{W}_m$ by (3.2.1), which corresponds to the mth association $(\mathbf{y}_m, \mathbf{x}_m)$. Regardless of the definition of the generalized inverse, it can be shown that if $\mathbf{x}_m \in \mathcal{C}(\mathbf{X}_{m-1})$

$$1 - \mathbf{z}_m^{*}\,\mathbf{x}_m \neq 0 \tag{A3.1.7}$$

For example, in the case where $\mathbf{z}_m$ is given by (A3.1.3), $1 - \mathbf{z}_m^* \mathbf{x}_m = (1 + ||\mathbf{X}_{m-1} \mathbf{x}_m||^2)^{-1} \neq 0$. The combination of (A3.1.4) and (A3.1.7) indicates that the condition $\mathbf{y}_m = \mathbf{W}_m \mathbf{x}_m$ holds if and only if

$$\mathbf{y}_m = \mathbf{W}_{m-1} \mathbf{x}_m \tag{A3.1.8}$$

The combination of (A3.1.6) and (A3.1.8) indicates that if $\mathbf{x}_m \in \mathcal{C}(\mathbf{X}_{m-1})$,

$$\mathbf{W}_m = \mathbf{W}_{m-1} \tag{A3.1.9}$$

In conclusion, if the new key pattern $\mathbf{x}_m$ is a linear combination of the input patterns belonging to the already stored associations, the new association $(\mathbf{y}_m, \mathbf{x}_m)$ has no effect on the matrix of synaptic weights.

Appendix A3.2

The derivation of the ELEANNE 2 algorithm is completed here by determining the update equation for the matrix of synaptic weights $\mathbf{W}$ with respect to the association $(\mathbf{y}_m, \mathbf{x}_m)$. It was shown in section 3.2 that $\mathbf{P}_m = (\mathbf{X}_m \mathbf{X}_m^*)^{-1}$ can be evaluated in terms of $\mathbf{P}_{m-1} = (\mathbf{X}_{m-1} \mathbf{X}_{m-1}^*)^{-1}$ by

$$\mathbf{P}_m = \mathbf{P}_{m-1} - (1 + \mathbf{x}_m^* \mathbf{P}_{m-1} \mathbf{x}_m)^{-1} \mathbf{P}_{m-1} \mathbf{x}_m \mathbf{x}_m^* \mathbf{P}_{m-1} \tag{A3.2.1}$$

In addition, since $\mathbf{y}_{i,m}^* = [\mathbf{y}_{i,m-1}^* \, y_{i,m}]$,

$$\mathbf{X}_m \mathbf{y}_{i,m} = \mathbf{X}_{m-1} \mathbf{y}_{i,m-1} + \mathbf{x}_m y_{i,m} \tag{A3.2.2}$$

Substituting (A3.2.1) and (A3.2.2) into (3.2.17) gives

$$\begin{aligned}
\mathbf{w}_{i,m} &= \mathbf{P}_m \mathbf{X}_m \mathbf{y}_{i,m} \\
&= [\mathbf{P}_{m-1} - (1 + \mathbf{x}_m^* \mathbf{P}_{m-1} \mathbf{x}_m)^{-1} \mathbf{P}_{m-1} \mathbf{x}_m \mathbf{x}_m^* \mathbf{P}_{m-1}] \\
&\cdot\ (\mathbf{X}_{m-1} \mathbf{y}_{i,m-1} + \mathbf{x}_m y_{i,m}) \\
&= \mathbf{P}_{m-1} \mathbf{X}_{m-1} \mathbf{y}_{i,m-1} + \mathbf{P}_{m-1} \mathbf{x}_m y_{i,m} - (1 + \mathbf{x}_m^* \mathbf{P}_{m-1} \mathbf{x}_m)^{-1} \\
&\cdot\ [\mathbf{P}_{m-1} \mathbf{x}_m \mathbf{x}_m^* \mathbf{P}_{m-1} \mathbf{X}_{m-1} \mathbf{y}_{i,m-1} \\
&+\ \mathbf{P}_{m-1} \mathbf{x}_m \mathbf{x}_m^* \mathbf{P}_{m-1} \mathbf{x}_m y_{i,m}] \\
&= \mathbf{w}_{i,m-1} + (1 + \mathbf{x}_m^* \mathbf{P}_{m-1} \mathbf{x}_m)^{-1}
\end{aligned}$$

$$
\begin{aligned}
\cdot \quad & [(1 + \mathbf{x}_m^* \mathbf{P}_{m-1} \mathbf{x}_m) \mathbf{P}_{m-1} \mathbf{x}_m y_{i,m} \\
- \quad & \mathbf{P}_{m-1} \mathbf{x}_m \mathbf{x}_m^* \mathbf{P}_{m-1} \mathbf{X}_{m-1} \mathbf{y}_{i,m-1} \\
- \quad & (\mathbf{x}_m^* \mathbf{P}_{m-1} \mathbf{x}_m) \mathbf{P}_{m-1} \mathbf{x}_m y_{i,m}]
\end{aligned}
\tag{A3.2.3}
$$

Finally, the update equation for each row of the matrix of synaptic weights is obtained by simplifying (A3.2.3) as

$$\mathbf{w}_{i,m} = \mathbf{w}_{i,m-1} + (1 + \mathbf{x}_m^* \, \mathbf{P}_{m-1} \, \mathbf{x}_m)^{-1} \, \mathbf{P}_{m-1} \, \mathbf{x}_m \, e_{i,m} \tag{A3.2.4}$$

where $\mathbf{P}_m$ can be evaluated recursively using (A3.2.1).

According to the above analysis, the matrix $\mathbf{P}_k$ is independent of the particular row of the matrix of synaptic weights considered. Therefore, the appropriate combination of the update equations defined by (A3.2.4) can provide an update equation for the entire matrix of synaptic weights $\mathbf{W}$. Since $\mathbf{P}_k$ is a symmetric matrix, (A3.2.4) can equivalently be written as follows

$$\mathbf{w}_{i,m}^* = \mathbf{w}_{i,m-1}^* + (1 + \mathbf{x}_m^* \, \mathbf{P}_{m-1} \, \mathbf{x}_m)^{-1} \, e_{i,m} \, \mathbf{x}_m^* \, \mathbf{P}_{m-1} \tag{A3.2.5}$$

The combination of the n_o equations defined by (A3.2.4) provides the update equation for the matrix of synaptic weights $\mathbf{W}$ as follows

$$\mathbf{W}_m = \mathbf{W}_{m-1} + (1 + \mathbf{x}_m^* \, \mathbf{P}_{m-1} \, \mathbf{x}_m)^{-1} \, (\mathbf{y}_m - \mathbf{W}_{m-1} \, \mathbf{x}_m) \, \mathbf{x}_m^* \, \mathbf{P}_{m-1} \tag{A3.2.6}$$

Appendix A3.3

This Appendix presents the evaluation of the vector $\mathbf{b}_{i,m}$ and the Hessian matrix $\mathbf{H}_{i,m}$ for the derivation of the ELEANNE 3 algorithm, in which $\hat{y}_{i,k} = \sigma(\bar{y}_{i,k})$. Since $e_{i,k} = y_{i,k} - \hat{y}_{i,k}$,

$$\frac{\partial e_{i,k}}{\partial w_{iq}} = -\frac{\partial \hat{y}_{i,k}}{\partial w_{iq}} = -\frac{\partial \sigma(\bar{y}_{i,k})}{\partial \bar{y}_{i,k}} \frac{\partial \bar{y}_{i,k}}{\partial w_{iq}} = -\sigma'(\bar{y}_{i,k}) \, x_{q,k} \tag{A3.3.1}$$

If $\sigma(x) = \tanh(x)$, it can easily be verified that $\sigma'(x) = (1 - \sigma(x)^2)$. Therefore, (A3.3.1) becomes

$$\frac{\partial e_{i,k}}{\partial w_{iq}} = -(1 - \sigma(\bar{y}_{i,k})^2) \, x_{q,k} = -(1 - \hat{y}_{i,k}^2) \, x_{q,k} \tag{A3.3.2}$$

Substituting (A3.3.2) into the definition of $\mathbf{b}_{i,k}$ in (3.3.8),

$$\mathbf{b}_{i,k} = -(1 - \hat{y}_{i,k}^2)\,\mathbf{x}_k \qquad \text{(A3.3.3)}$$

In the case where $\hat{y}_{i,k} = \bar{y}_{i,k}$, the resulting form of the Hessian matrix is such that its inverse can be evaluated recursively and thus a recursive algorithm can be developed. Unfortunately, this highly desirable property is not present when $\hat{y}_{i,k} = \tanh(\bar{y}_{i,k})$, since $\partial^2 e_{i,k}/\partial w_{ip}\partial w_{iq} \neq 0$. In this case,

$$\frac{\partial^2 e_{i,k}}{\partial w_{ip}\partial w_{iq}} = -x_{q,k}\,\frac{\partial}{\partial w_{ip}}(1 - \hat{y}_{i,k}^2) = -2\,x_{q,k}\,\hat{y}_{i,k}\,\frac{\partial e_{i,k}}{\partial w_{ip}} \qquad \text{(A3.3.4)}$$

Substituting (A3.3.2) into (A3.3.4) and multiplying by $e_{i,k}$ gives

$$e_{i,k}\,\frac{\partial^2 e_{i,k}}{\partial w_{ip}\partial w_{iq}} = 2\,x_{p,k}\,x_{q,k}\,\hat{y}_{i,k}\,(1 - \hat{y}_{i,k}^2)\,e_{i,k} \qquad \text{(A3.3.5)}$$

Comparing (A3.3.2) and (A3.3.5) suggests the form

$$e_{i,k}\,\frac{\partial^2 e_{i,k}}{\partial w_{ip}\partial w_{iq}} = \gamma_{i,k}\,\frac{\partial e_{i,k}}{\partial w_{ip}}\frac{\partial e_{i,k}}{\partial w_{iq}} \qquad \text{(A3.3.6)}$$

where

$$\gamma_{i,k} = \frac{2\,\hat{y}_{i,k}\,(y_{i,k} - \hat{y}_{i,k})}{(1 - \hat{y}_{i,k})(1 + \hat{y}_{i,k})} = 2\,\frac{y_{i,k}\,\hat{y}_{i,k} - \hat{y}_{i,k}^2}{1 - \hat{y}_{i,k}^2} \qquad \text{(A3.3.7)}$$

The substitution of (A3.3.6) into (3.3.13) suggests that the Hessian matrix is of the form

$$\mathbf{H}_{i,m} = \sum_{k=1}^{m}(1 + \gamma_{i,k})\,\mathbf{b}_{i,k}\,\mathbf{b}_{i,k}^* \qquad \text{(A3.3.8)}$$

Assuming that $1 + \gamma_{i,k} > 0\,\forall\,k = 1, 2, \ldots, m$, the Hessian matrix defined by (A3.3.8) is positive semi-definite. This can easily be verified by observing that, if $1 + \gamma_{i,k} > 0\,\forall\,k = 1, 2, \ldots, m$, (A3.3.8) can equivalently be written as $\mathbf{H}_{i,m} = \sum_{k=1}^{m} \bar{\mathbf{b}}_{i,k}\,\bar{\mathbf{b}}_{i,k}^*$, where $\bar{\mathbf{b}}_{i,k} = (1 + \gamma_{i,k})^{1/2}\mathbf{b}_{i,k}$. However, there is no guarantee that $1 + \gamma_{i,k} > 0$. The positive semi-definiteness of the matrix defined by (A3.3.8) is now investigated by examining the sign of $1 + \gamma_{i,k} \,\forall\, k = 1, 2, \ldots, m$. Define the two sets $\mathcal{S}_+(i) = \{k : y_{i,k} = +1\}$ and $\mathcal{S}_-(i) = \{k : y_{i,k} = -1\}$. Assuming that $k \in \mathcal{S}_+(i)$, (A3.3.7) gives

$$1 + \gamma_{i,k} = \frac{1 + 3\,\hat{y}_{i,k}}{1 + \hat{y}_{i,k}} = 2 - \frac{1 - \hat{y}_{i,k}}{1 + \hat{y}_{i,k}} \tag{A3.3.9}$$

According to (A3.3.9), $1+\gamma_{i,k} > 0$ in this case only if $-1/3 < \hat{y}_{i,k} < 1$. Since $y_{i,k} = 1$, $1 + \gamma_{i,k} > 0$ only if the estimate $\hat{y}_{i,k}$ provided by the network is not far away from the target $y_{i,k}$. Similarly, when $k \in \mathcal{S}_-(i)$,

$$1 + \gamma_{i,k} = \frac{3\,\hat{y}_{i,k} - 1}{\hat{y}_{i,k} - 1} = 2 + \frac{\hat{y}_{i,k} + 1}{\hat{y}_{i,k} - 1} \tag{A3.3.10}$$

In this case, (A3.3.10) indicates that $1+\gamma_{i,k} > 0$ only if $-1 < \hat{y}_{i,k} < 1/3$. Once again, $1 + \gamma_{i,k} > 0$ only if the estimate $\hat{y}_{i,k}$ provided by the network is close to the target $y_{i,k} = -1$. In conclusion, the matrix defined by (A3.3.8) is positive semi-definite only if it is evaluated near the optimum. Since in many practical situations the algorithm is initialized randomly, there is no guarantee that the initial estimate will be close to the optimum. The previous analysis indicates that the Hessian matrix (A3.3.8) should be approximated by a matrix that is positive semi-definite regardless of the particular value $\hat{y}_{i,k}$, $k = 1, 2, \ldots, m$.

Consider the approximation of $1+\gamma_{i,k}$ under the assumption that the current estimate is far away from the optimum. In this case, it can be assumed that $\hat{y}_{i,k} \approx 0 \,\forall\, k = 1, 2, \ldots, m$ and, therefore, $1 + \gamma_{i,k} \approx 1 \,\forall\, k = 1, 2, \ldots, m$. As a result, the Hessian matrix can be approximated as

$$\mathbf{H}_{i,m} \approx \sum_{k=1}^{m} \mathbf{b}_{i,k}\, \mathbf{b}_{i,k}^{*} \tag{A3.3.11}$$

Consider now the same approximation under the assumption that the current estimate is close to the optimum. In this case, it can be reasonably assumed that $y_{i,k}\,\hat{y}_{i,k} \approx 1$. The combination of this assumption with (A3.3.7) indicates that $1+\gamma_{i,k} \approx 3 \,\forall\, k = 1, 2, \ldots, m$. Therefore, the Hessian matrix can be approximated as

$$\mathbf{H}_{i,m} \approx 3 \sum_{k=1}^{m} \mathbf{b}_{i,k}\, \mathbf{b}_{i,k}^{*} \tag{A3.3.12}$$

The comparison of (A3.3.11) and (A3.3.12) indicates that the two alternative approximations of the Hessian matrix are basically of the same form. Since the constant 3 appearing in (A3.3.12) can be incorporated in the learning rate α involving in the update equation, the Hessian matrix is finally approximated by (A3.3.11).

Appendix A3.4

The derivation of the ELEANNE 5 algorithm is completed here by determining the adaptation of the synaptic weights v_{pq} of the lower network with respect to the association $(\mathbf{y}_k, \mathbf{x}_k)$. The synaptic weights v_{pq} can be updated by minimizing the objective function E_k^i, defined by (3.4.8). According to the gradient descent method, v_{pq} can be updated by the following equation

$$\mathbf{v}_{p,k} - \mathbf{v}_{p,k-1} = -\alpha \frac{\partial E_k^i}{\partial \mathbf{v}_p} = -\alpha\, e_{i,k} \frac{\partial e_{i,k}}{\partial \mathbf{v}_p} \tag{A3.4.1}$$

From the definition of $\hat{y}_{i,k}$ in (3.4.3),

$$\frac{\partial e_{i,k}}{\partial \mathbf{v}_p} = -\frac{\partial \sigma(\bar{y}_{i,k})}{\partial \bar{y}_{i,k}} \frac{\partial \bar{y}_{i,k}}{\partial \mathbf{v}_p} = -\sigma'(\bar{y}_{i,k}) \sum_{j=1}^{n_h} w_{ij} \frac{\partial \hat{h}_{j,k}}{\partial \mathbf{v}_p} \tag{A3.4.2}$$

Then, using the definition of $\hat{h}_{j,k}$ in (3.4.1),

$$\frac{\partial \hat{h}_{j,k}}{\partial \mathbf{v}_p} = \frac{\partial \rho(\bar{h}_{j,k})}{\partial \bar{h}_{j,k}} \frac{\partial \bar{h}_{j,k}}{\partial \mathbf{v}_p} = \rho'(\bar{h}_{j,k})\, \mathbf{x}_k\, \delta_{jp} \tag{A3.4.3}$$

where δ_{jp} is defined as $\delta_{jp} = 1$ if $j = p$ and $\delta_{jp} = 0$ if $j \neq p$. Assuming that $\rho(x) = \tanh(x)$, $\rho'(x) = (1 - \rho(x)^2)$. The combination of (A3.4.2) and (A3.4.3) with (A3.4.1) provides the following update equation for the synaptic weights v_{pq} of the lower network

$$\mathbf{v}_{p,k} = \mathbf{v}_{p,k-1} + \alpha\, \varepsilon_{pi,k}^h\, \mathbf{x}_k \tag{A3.4.4}$$

where

$$\varepsilon_{pi,k}^h = (1 - \hat{h}_{p,k}^2)\, \varepsilon_{i,k}^o\, w_{ip} \tag{A3.4.5}$$

where $\varepsilon_{i,k}^o$ is defined by (3.3.27).

A similar analysis provides the update equation for the synaptic weights of the neural network with L layers of hidden units shown in Figure 2.10. The synaptic weights $v_{pq}^{(r)}, r = 1, 2, \ldots, L$ can be updated by minimizing the objective function E_k^i, defined by (3.4.8). According to the gradient descent method, the synaptic weights $v_{pq}^{(r)}$ can be updated through the equation

$$\mathbf{v}_{p,k}^{(r)} - \mathbf{v}_{p,k-1}^{(r)} = -\alpha \frac{\partial E_k^i}{\partial \mathbf{v}_p^{(r)}} = -\alpha \, e_{i,k} \frac{\partial e_{i,k}}{\partial \mathbf{v}_p^{(r)}} \tag{A3.4.6}$$

In the case where $r = 1$,

$$\frac{\partial e_{i,k}}{\partial \mathbf{v}_p^{(1)}} = -(1 - \hat{y}_{i,k}^2) \sum_{j=1}^{n_1} w_{ij} \frac{\partial \hat{h}_{j,k}^{(1)}}{\partial \mathbf{v}_p^{(1)}} \tag{A3.4.7}$$

where

$$\frac{\partial \hat{h}_{j,k}^{(1)}}{\partial \mathbf{v}_p^{(1)}} = (1 - \hat{h}_{j,k}^{(1)\,2}) \, \hat{\mathbf{h}}_k^{(2)} \, \delta_{jp} \tag{A3.4.8}$$

For $r = 1$, the combination of (A3.4.7) and (A3.4.8) with (A3.4.6) gives

$$\mathbf{v}_{p,k}^{(1)} = \mathbf{v}_{p,k-1}^{(1)} + \alpha \, \varepsilon_{pi,k}^{(1)} \, \hat{\mathbf{h}}_k^{(2)} \tag{A3.4.9}$$

where

$$\varepsilon_{pi,k}^{(1)} = (1 - \hat{h}_{p,k}^{(1)\,2}) \, \varepsilon_{i,k}^{o} \, w_{ip} \tag{A3.4.10}$$

and $\varepsilon_{i,k}^{o}$ is defined by (3.3.27). A similar analysis provides the following the following update equation for the synaptic weights $v_{pq}^{(r)}, r = 2, 3, \ldots, L$

$$\mathbf{v}_{p,k}^{(r)} = \mathbf{v}_{p,k-1}^{(r)} + \alpha \, \varepsilon_{pi,k}^{(r)} \, \hat{\mathbf{h}}_k^{(r+1)} \tag{A3.4.11}$$

where $\hat{\mathbf{h}}_k^{(L+1)} = \mathbf{x}_k$, and

$$\varepsilon_{pi,k}^{(r)} = (1 - \hat{h}_{p,k}^{(r)\,2}) \sum_{\ell=1}^{n_{r-1}} \varepsilon_{\ell i,k}^{(r-1)} \, v_{\ell p}^{(r-1)} \, \forall \, r = 2, 3, \ldots, L \tag{A3.4.12}$$

Chapter 4

Fast Learning Algorithms for Neural Networks

4.1 Introduction

When neural networks regained popularity in the early eighties, the dominant trend among researchers was an eagerness to develop impressive neural network applications and then quickly produce and market neural network systems. This trend was mainly motivated, and also justified, by a desire to establish neural networks as a viable and realistic alternative for developing speech and image recognition systems, and trainable control devices. During the early stages of this revitalized interest in neural network research, the existing learning algorithms were satisfactory. As research has moved from state-of-the-art paradigms to real-world applications, the associated training time and computing requirements have become an increasingly important consideration in the comparison of neural networks with alternative, competing techniques. The availability of fast and efficient learning algorithms is crucial for the future evolution of this research field.

The development of fast and efficient learning algorithms for neural networks has been a subject of considerable interest over the past few years. As a result, several new algorithms have been proposed for training neural networks with various architectures. For instance, the outer-product rule is a simple but suboptimal learning scheme traditionally used for spin glass models of neural networks. An optimal learning algorithm for this type of networks was reported recently in (Abbot and Kepler, 1989). The AdaTron is another algorithm proposed for spin glass models of neural networks in (Anlauf and Biehl, 1989). This algorithm exponentially relaxes to the network of optimal stability using the concept of adaptive learning. The popularity of multi-layered neural networks motivated several researchers to focus on heuristic techniques for accelerating the Error Back Propagation algorithm or for the development of alternative learning algorithms. Accelerated versions of the Error Back Propagation algorithm were attempted by adapting the learning rate during training or by using various other heuristics to improve the convergence of the algorithm. The development of alternative learning algorithms for multi-layered neural networks was attempted by modifying the optimization strategy and/or employing adaptation rules other than the gradient descent, or by considering alternative criteria to the frequently used quadratic error criterion.

Hush and Salas proposed a simple method for improving the convergence rate of the Error Back Propagation algorithm, referred to as the gradient reuse algorithm (Hush and Salas, 1988). The basic idea of this method is that gradients which are computed during training are reused until the resulting weight updates no longer lead to a reduction in the error. Dahl proposed a method for improving the rate of convergence of the Delta Rule (Dahl, 1987). This method computes the gradient of an error function evaluated by a group of associations. Instead of simply incrementing the weights in the network by some fraction of the components of the gradient, the error function is evaluated at various points along the gradient, and the weights are set to those values which minimize the error function along the gradient. Jacobs suggested that the rate of convergence of the Error Back Propagation algorithm can be accelerated by assigning a learning rate to each of the synaptic weights of the network and appropriately adapting these learning rates during the training of the network (Jacobs, 1988). Silva and Almeida proposed a similar technique for the acceleration of the Error Back Propagation algorithm (Silva and Almeida, 1990). This technique is also based on individual adaptation of the learning rate of each synaptic weight of the network. Battiti presented another attempt to speed up the convergence of the Error Back Propagation algorithm through adaptation of the learning rate (Battiti, 1989). In this paper, the learning rate was adapted in order to reduce the energy value of the gradient direction in a close-to-optimal way. Battiti also proposed an alternative acceleration technique, which is derived from the conjugate gradient method with inexact linear searches (Battiti, 1989). Based on heuristics, other researchers modified the Error Back Propagation algorithm in an attempt to accelerate its convergence (Vogl et al., 1988; Fahlman, 1989).

There are two major issues associated with the development of learning algorithms for neural networks: the formulation of the training problem considered, and the optimization strategy used in performing this optimization. Parker suggested the use of second-order methods for training of neural networks (Parker, 1987). Second-order methods were also used by other researchers for the development of learning algorithms for feed-forward neural networks that converge faster than the Error Back Propagation algorithm (Kollias and Anastassiou, 1989; Becker and le Cun, 1989; Singhal and Wu, 1989; Scalero and Tepedelenioglu, 1992). An alternative treatment of the optimization problem resulting from the conventional formulation of the training problem provided the family of Efficient LEarning Algorithms for Neural NEtworks (ELEANNE), presented in chapter 3.

It has been recently suggested that feed-forward neural networks with binary outputs can be trained by maximizing the relative entropy of the expected output with respect to the estimate provided by the network (Solla et al., 1988; Baum and Wilczek, 1988). A similar criterion was also used by Hopfield to compare layered network learning with Boltzmann machine learning (Hopfield, 1987). It was experimentally verified that the learning algorithm resulting from the maximization of the relative entropy converges faster than the conventional Error Back Propagation algorithm (Solla et al., 1988). However, the networks trained on the basis of the relative entropy criterion are inferior to those trained by the Error Back Propagation algorithm in terms of their generalization ability (Solla et al.,

1988).

This chapter presents the development of fast learning algorithms for feed-forward neural networks by focusing on the criterion which is used for their training. The idea underlying the development of all these algorithms is the appropriate relaxation of the quadratic error criterion during training. The resulting generalized training criterion leads to a variety of fast learning algorithms for neural networks. The form and properties of the resulting algorithms depend on the particular way each deals with the optimization problem.

This chapter is organized as follows: Section 4.2 presents a generalized criterion for training feed-forward neural networks. This criterion leads to fast learning algorithms for single-layered neural networks, which are presented in section 4.3. Section 4.4 presents the development of fast learning algorithms for multi-layered neural networks. The proposed algorithms are compared to existing learning algorithms in section 4.5. Finally, section 4.6 contains concluding remarks.

4.2 A Generalized Training Criterion

Consider a feed-forward neural network trained with respect to the associations $(\mathbf{y}_k, \mathbf{x}_k), k = 1, 2, \ldots, m$, where $\mathbf{x}_k^* = [x_{1,k}\, x_{2,k} \ldots x_{n_i,k}]$ are the inputs of the network and $\mathbf{y}_k^* = [y_{1,k}\, y_{2,k} \ldots y_{n_o,k}]$ are the corresponding outputs. The training is frequently based on the minimization of the following objective function

$$E = \frac{1}{2} \sum_{k=1}^{m} \sum_{i=1}^{n_o} e_{i,k}^2 = \frac{1}{2} \sum_{k=1}^{m} \sum_{i=1}^{n_o} (y_{i,k} - \hat{y}_{i,k})^2 \qquad (4.2.1)$$

where $y_{i,k}$ is the expected output and $\hat{y}_{i,k}$ the corresponding estimate provided by the network. If the output of the network is binary, each $y_{i,k}$ can take the values $+1$ and -1. If the output of the network is analog, the elements $y_{i,k}, i = 1, 2, \ldots, n_o$, which form the output patterns $\mathbf{y}_k, k = 1, 2, \ldots, m$, are normalized between -1 and $+1$.

A characteristic common to learning algorithms resulting from the minimization of the quadratic error criterion (4.2.1) is the fact that the adaptation of the synaptic weights is mainly determined by the estimation error, i.e., the error between the expected output and the estimate provided by the network. The reduction of the total error resulting by such an adaptation of the synaptic weights is significant during the initial adaptation cycles, i.e., when the estimation error is relatively large. However, when the estimation error decreases with training, the convergence of the algorithm becomes very slow. In fact, this is a shortcoming of all existing algorithms, regardless of the optimization strategy upon which they are based. Several researchers attempted to avoid this problem by introducing criteria which indicate when training can be stopped without noticeable effect on the performance of the resulting neural network. The basic argument in these approaches is that the network is 'overtrained' if the learning algorithm continues adapting the synaptic weights beyond a certain number of adaptation cycles. Although this trick is often justified for practical reasons, it can hardly be considered a sound solution to the problem of developing fast learning algorithms for neural networks. Moreover, the combination of learning algorithms with such criteria frequently increases their computational complexity and complicates their application.

The development of fast learning algorithms for feed-forward neural networks presented here is based on the minimization of an alternative criterion after the initial adaptation cycles (Karayiannis, 1991a; Karayiannis, 1992a; Karayiannis and Venetsanopoulos, 1991; Karayiannis and Venetsanopoulos, 1992b). Consider that a feed-forward neural network is trained by minimizing the generalized objective function

$$\begin{aligned} G(\lambda) &= \lambda E + (1-\lambda) E' \\ &= \lambda \sum_{k=1}^{m} \sum_{i=1}^{n_o} \phi_2(e_{i,k}) + (1-\lambda) \sum_{k=1}^{m} \sum_{i=1}^{n_o} \phi_1(e_{i,k}) \qquad (4.2.2) \end{aligned}$$

where $\phi_2(x) = \frac{1}{2}\, x^2$, $\phi_1(.)$ is a positive definite, convex, continuous and differentiable everywhere function, and $\lambda \in [0,1]$. If $\lambda = 1$, (4.2.2) coincides with the original objective function (4.2.1). If $\lambda = 0$, the training of the network is based on the minimization of $G(0) = \sum_{k=1}^{m} \sum_{i=1}^{n_o} \phi_1(e_{i,k})$. When $0 < \lambda < 1$, (4.2.2) defines a variety of alternative objective functions between the two extremes corresponding to $\lambda = 0$ and $\lambda = 1$. The convergence rate of the learning algorithms based on the minimization of $G(1) = E$ reduces significantly after the initial adaptation cycles. The form of (4.2.2) indicates that the minimization of an alternative objective function after the initial adaptation cycles can be achieved by reducing λ from unity to zero during the training of the network. Assuming that the training of a network is based on the generalized criterion (4.2.2), the crucial problem is the selection of the objective function $G(0) = E'$. It was recently found that $\phi_1(.)$ should be selected in such a way that $G(0) = E'$ approximates asymptotically the absolute error criterion (Karayiannis, 1991a). This requirement is satisfied by the following function

$$\phi_1(x) = \frac{1}{\beta} \ln[\cosh \beta x] \tag{4.2.3}$$

for large β (Karayiannis and Venetsanopoulos, 1990a). Since the minimization of $E' = G(0)$ becomes dominant after the initial adaptation cycles, the properties of $\phi_1(e_{i,k})$ are studied here for the case where $\hat{y}_{i,k}$ is close to $y_{i,k}$. Under this assumption, the behavior of the nonlinearity $\phi_1(.)$ depends on the value of β. If β is relatively small, $\cosh(\beta x) \approx 1 + \frac{1}{2}(\beta x)^2$ and also $\frac{1}{\beta}\ln[\cosh \beta x] \approx \frac{\beta}{2} x^2$. In this case, $\phi_1(e_{i,k}) \approx \frac{\beta}{2} e_{i,k}^2$ and $G(0)$ is similar to the objective function (4.2.1). As β increases, $\phi_1(e_{i,k})$ asymptotically approaches $|e_{i,k}|$. In this case, $G(0) = E'$ approaches the absolute error criterion.

The criterion proposed here can be the basis for training neural networks whose output is either analog or binary. If the output of the network is binary, a function simpler than (4.2.3) can be used. Consider that the binary output of the network consists of elements taking the values $+1$ and -1. The Hamming distance between the expected output $\mathbf{y}_k^* = [y_{1,k}\, y_{2,k} \ldots y_{n_o,k}]$ and the estimate $\hat{\mathbf{y}}_k^* = [\hat{y}_{1,k}\, \hat{y}_{2,k} \ldots \hat{y}_{n_o,k}]$ provided by the network in this case is simply the number of components of the two vectors which are different, and is given by (Hertz et al., 1991)

$$\begin{aligned} d(\mathbf{y}_k, \hat{\mathbf{y}}_k) &= \frac{1}{2}\sum_{i=1}^{n_o}[(1 - \hat{y}_{i,k})(1 + y_{i,k}) + (1 - y_{i,k})(1 + \hat{y}_{i,k})] \\ &= \sum_{i=1}^{n_o}(1 - y_{i,k}\, \hat{y}_{i,k}) \end{aligned} \tag{4.2.4}$$

Consider that a feed-forward neural network with binary output is trained by minimizing the generalized objective function (4.2.2), where E is the quadratic error criterion defined by (4.2.1) and

$$E' = \sum_{k=1}^{m} d(\mathbf{y}_k, \hat{\mathbf{y}}_k) = \sum_{k=1}^{m}\sum_{i=1}^{n_o}(1 - y_{i,k}\,\hat{y}_{i,k}) \tag{4.2.5}$$

Since $y_{i,k}^2 = 1$, the definition of E' by (4.2.5) implies that $\phi_1(y_{i,k} - \hat{y}_{i,k}) = y_{i,k}(y_{i,k} - \hat{y}_{i,k})$. Clearly, (4.2.5) attains its minimum value $E' = 0$ if each $\hat{y}_{i,k}$ approaches its target $y_{i,k}$. According to the proposed generalized criterion, the training of the network is initially based on the minimization of the quadratic error criterion (4.2.1). After the initial adaptation cycles, i.e., when λ approaches zero, the training of the network is based on the minimization of the Hamming distance between the entire set of $\mathbf{y}_k$ and $\hat{\mathbf{y}}_k$. The relationship between the objective function E' and the absolute error criterion can be established by observing that if $y_{i,k}$ and $\hat{y}_{i,k}$ take only the values $+1$ or -1, $y_{i,k}^2 = 1$ and $|y_{i,k} - \hat{y}_{i,k}| + y_{i,k}\,\hat{y}_{i,k} = 1 \ \forall\, i = 1, 2, \ldots, n_o$. Thus,

$$\begin{aligned}\sum_{k=1}^{m}\sum_{i=1}^{n_o}(1 - y_{i,k}\,\hat{y}_{i,k}) &= \sum_{k=1}^{m}\sum_{i=1}^{n_o} y_{i,k}(y_{i,k} - \hat{y}_{i,k}) \\ &= \sum_{k=1}^{m}\sum_{i=1}^{n_o} |y_{i,k} - \hat{y}_{i,k}| \end{aligned} \tag{4.2.6}$$

After the initial adaptation cycles, each $\hat{y}_{i,k}$ approaches its target $y_{i,k}$. In this case, the objective function E' defined by (4.2.5) approaches the *absolute error* evaluated over the entire set of expected outputs $\mathbf{y}_k, k = 1, 2, \ldots, m$ and the corresponding estimates $\hat{\mathbf{y}}_k, k = 1, 2, \ldots, m$ provided by the network. Consider also the relationship between E' and the quadratic error criterion (4.2.1) if each $\hat{y}_{i,k}$ approaches its target $y_{i,k}$. Assuming that $y_{i,k}$ and $\hat{y}_{i,k}$ take only the values $+1$ or -1, $\frac{1}{2}(y_{i,k} - \hat{y}_{i,k})^2 = 1 - y_{i,k}\,\hat{y}_{i,k} = y_{i,k}(y_{i,k} - \hat{y}_{i,k}) \ \forall\, i = 1, 2, \ldots, n_o$. This latter equation provides

$$\begin{aligned}\sum_{k=1}^{m}\sum_{i=1}^{n_o}(1 - y_{i,k}\,\hat{y}_{i,k}) &= \sum_{k=1}^{m}\sum_{i=1}^{n_o} y_{i,k}(y_{i,k} - \hat{y}_{i,k}) \\ &= \frac{1}{2}\sum_{k=1}^{m}\sum_{i=1}^{n_o}(y_{i,k} - \hat{y}_{i,k})^2 \end{aligned} \tag{4.2.7}$$

In the case where each $\hat{y}_{i,k}$ approaches its target $y_{i,k}$, the objective function E' approaches the quadratic error criterion (4.2.1). Finally, (4.2.5) indicates that, near the optimum, the minimization of $G(0) = E'$ is equivalent to the maximization of $\sum_{k=1}^{m} c(\mathbf{y}_k, \hat{\mathbf{y}}_k)$, where $c(\mathbf{y}_k, \hat{\mathbf{y}}_k)$ is a measure of similarity of the vectors $\mathbf{y}_k$ and $\hat{\mathbf{y}}_k$, often referred to as the *direction cosine,* defined by

$$c(\mathbf{y}_k, \hat{\mathbf{y}}_k) = \frac{1}{n_o} \sum_{i=1}^{n_o} y_{i,k}\, \hat{y}_{i,k} \tag{4.2.8}$$

If $\phi_2(x) = \frac{1}{2}\, x^2$ and $\phi_1(y_{i,k} - \hat{y}_{i,k}) = y_{i,k}(y_{i,k} - \hat{y}_{i,k})\ \forall\, i = 1, 2, \ldots, n_o$, it can easily be shown that (4.2.2) can equivalently be written as

$$G(\lambda) = \sum_{k=1}^{m} \sum_{i=1}^{n_o} y_{i,k}(y_{i,k} - \hat{y}_{i,k}) + \frac{1}{2} \lambda \sum_{k=1}^{m} \sum_{i=1}^{n_o} (\hat{y}_{i,k}^2 - y_{i,k}^2) \tag{4.2.9}$$

This is the form of the objective function that was proposed for training neural networks with binary output (Karayiannis, 1991a; Karayiannis and Venetsanopoulos, 1991).

The role of the parameter λ during training is studied here in the case where the output of the network is binary. The selection of λ is based on the variation of the objective function caused by the fluctuation of the estimates $\hat{y}_{i,k}, i = 1, 2, \ldots, n_o$, provided by the network. A measure of the variation of the objective function (4.2.9) with respect to a particular estimate, say $\hat{y}_{p,k}$, is provided by $|\partial G(\lambda)/\partial \hat{y}_{p,k}|$, where

$$\frac{\partial G(\lambda)}{\partial \hat{y}_{p,k}} = -y_{p,k} + \lambda\, \hat{y}_{p,k} = -e_{p,k}(\lambda) \tag{4.2.10}$$

For $\lambda = 1$, (4.2.10) reduces to

$$\frac{\partial G(1)}{\partial \hat{y}_{p,k}} = -(y_{p,k} - \hat{y}_{p,k}) \tag{4.2.11}$$

The combination of (4.2.10) and (4.2.11) gives

$$\frac{\partial G(\lambda)}{\partial \hat{y}_{p,k}} = \frac{\partial G(1)}{\partial \hat{y}_{p,k}} \left[1 + (1 - \lambda) \frac{\hat{y}_{p,k}}{y_{p,k} - \hat{y}_{p,k}} \right] \tag{4.2.12}$$

It can be verified that for $y_{p,k} = \pm 1$ and $-1 < \hat{y}_{p,k} < 1$,

$$\frac{\hat{y}_{p,k}}{y_{p,k} - \hat{y}_{p,k}} \geq -\frac{1}{2} \tag{4.2.13}$$

As a result, for $0 \leq \lambda < 1$,

$$1 + (1 - \lambda)\frac{\hat{y}_{p,k}}{y_{p,k} - \hat{y}_{p,k}} \geq \frac{1}{2}(1 + \lambda) > 0 \tag{4.2.14}$$

Therefore, the comparison of $|\partial G(\lambda)/\partial \hat{y}_{p,k}|$ and $|\partial G(1)/\partial \hat{y}_{p,k}|$ depends on the sign of $\hat{y}_{p,k}(y_{p,k} - \hat{y}_{p,k})^{-1}$. Clearly, if $\hat{y}_{p,k}(y_{p,k} - \hat{y}_{p,k})^{-1} < 0$ and $\lambda < 1$, $1 + (1 - \lambda)\hat{y}_{p,k}(y_{p,k} - \hat{y}_{p,k})^{-1} < 1$ and (4.2.12) indicates that

$$\left|\frac{\partial G(\lambda)}{\partial \hat{y}_{p,k}}\right| < \left|\frac{\partial G(1)}{\partial \hat{y}_{p,k}}\right| \ \forall\, \lambda < 1 \tag{4.2.15}$$

On the other hand, if $\hat{y}_{p,k}(y_{p,k} - \hat{y}_{p,k})^{-1} > 0$ and $\lambda < 1$, $1 + (1 - \lambda)\hat{y}_{p,k}(y_{p,k} - \hat{y}_{p,k})^{-1} > 1$. In this case, (4.2.12) implies that

$$\left|\frac{\partial G(\lambda)}{\partial \hat{y}_{p,k}}\right| > \left|\frac{\partial G(1)}{\partial \hat{y}_{p,k}}\right| \ \forall\, \lambda < 1 \tag{4.2.16}$$

The existence of the two possible inequalities (4.2.15) and (4.2.16) suggests further investigation. If $y_{p,k} = +1$, (4.2.15) is valid when $\hat{y}_{p,k} < 0$, while (4.2.16) is valid when $\hat{y}_{p,k} > 0$. If $y_{p,k} = -1$, (4.2.15) is valid if $\hat{y}_{p,k} > 0$, while (4.2.16) is valid when $\hat{y}_{p,k} < 0$. In conclusion, (4.2.15) is valid when the estimate $\hat{y}_{p,k}$ provided by the network is far away from its target $y_{p,k}$, while (4.2.16) is valid when $\hat{y}_{p,k}$ is close to $y_{p,k}$. In fact, when $\hat{y}_{p,k}$ is very close to $y_{p,k}$, (4.2.12) indicates that

$$\left|\frac{\partial G(\lambda)}{\partial \hat{y}_{p,k}}\right| \gg \left|\frac{\partial G(1)}{\partial \hat{y}_{p,k}}\right| \ \forall\, \lambda < 1 \tag{4.2.17}$$

A neural network is trained by adapting its synaptic weights according to some rules provided by the learning algorithm employed, where the training is performed in a sequence of *adaptation cycles*. Each adaptation cycle involves the adaptation of all the synaptic weights of the network with respect to all the associations considered. The development of the alternative training criterion $G(\lambda)$ indicates that λ must change from 1 to 0 in the course of the training, with the value of λ depending on the proximity of the estimate provided by the network to the optimum. At initialization, when the total error E defined by (4.2.1) is large, i.e., the estimate is far away from the optimum, $\lambda = 1$, and as the total error decreases, λ should approach zero. This suggests that the value of λ should be determined in each adaptation cycle from the total error at that point, according to some suitable rule, i.e., $\lambda = \lambda(E)$. The above discussion indicates that $\lambda \approx 1$ when $E \gg 1$. In this case, for any positive integer n, $1/E^n$ approaches zero and, therefore, $\exp(-1/E^n) \approx 1$. On the other hand, when $E \ll 1$, $1/E^n$ becomes very large and, therefore, $\exp(-1/E^n) \approx 0$. As a result, a suitable rule for the reduction of λ from one to zero is the following

$$\lambda = \lambda(E) = \exp(-\mu/E^n) \tag{4.2.18}$$

where μ is a positive real number and n is a positive integer. During the initial adaptation cycles, i.e., when $E \gg 1$, $1/E^n \ll 1$ and $\exp(-\mu/E^n) \approx 1 - \mu/E^n$. The smaller the integer n, the faster the reduction of λ when $E \gg 1$. Obviously, the fastest reduction of λ corresponds to the smallest nonzero integer, that is, $n = 1$. It has been experimentally verified that if λ is much smaller than unity during the initial adaptation cycles the algorithm may be trapped in a local minimum. This is the first indication that n should be greater than 1. Also, when $E \ll 1$, the larger the integer n the faster $\exp(-\mu/E^n)$ converges to zero. Since λ must converge rapidly to zero in this case, this is an additional argument for making $n > 1$. In this chapter, λ is determined during the training of any network according to the following rule

$$\lambda = \lambda(E) = \exp(-\mu/E^2) \tag{4.2.19}$$

The rule proposed here for determining λ during training is not unique. Depending on the training problem under consideration, alternative methods could be employed for computing λ.

The analysis presented in this section admits an interesting interpretation of the proposed generalized criterion. During the initial stage of training, i.e., when $\lambda = 1$, the resulting algorithm is based on the minimization of the error $G(1) = E = \frac{1}{2}\sum_{k=1}^{m}\sum_{i=1}^{n_o} e_{i,k}^2$. The synaptic weights of the network trained on the basis of the generalized criterion are updated by *penalizing error.* During the last stage of training, i.e., when $\lambda = 0$, the resulting algorithm is based on the minimization of (4.2.5), or equivalently the maximization of $C = \sum_{k=1}^{m} \sum_{i=1}^{n_o} y_{i,k}\, \hat{y}_{i,k}$. C is a measure of similarity between the expected response and the response provided by the network. During this stage of training, the synaptic weights of the network trained on the basis of the generalized criterion are updated by *rewarding success.* During the intermediate stage of training, i.e., when $0 < \lambda < 1$, the synaptic weights of the network are updated by partly penalizing error and partly rewarding success. When λ is close to unity, the penalization of the error plays a dominant role in the training. As λ decreases and approaches zero, the rewarding of the success becomes the dominant factor in the training.

The same learning strategy is naturally applied by any adult who teaches a child to speak or to perform some other basic human functions. The training process starts with the selection of a set of simple words which relate to the child's natural environment. During the initial stages of learning, the adult repeats the same simple words after the child in order to help the child to correct its errors. After the initial stages of learning, the adult rewards the child for correctly pronouncing some portions of the target words and also repeats the same words in order that the child will be able to correct the rest of the errors. During the last stages of training, the adult rewards the child when the target words are pronounced correctly. This action provides the child with the sense of what is correct, solidifies the already acquired knowledge, and prepares the child for the formation of meaningful sentences.

4.3 Fast Learning Algorithms for Single-layered Neural Networks

Consider the single-layered neural network shown in Figure 2.7. Assuming that a particular pattern $\mathbf{x}_k$ is the input of the network, the corresponding analog output of the network consists of the elements

$$y^o_{i,k} = \bar{y}_{i,k} = \mathbf{x}^*_k \mathbf{w}_i = w_{i0} + \sum_{j=1}^{n_i} w_{ij} x_{j,k} \; \forall \, i = 1, 2, \ldots, n_o \qquad (4.3.1)$$

where $\mathbf{w}^*_i = [w_{i0}\, w_{i1}\, w_{i2} \ldots w_{in_i}]$, $\mathbf{x}^*_k = [x_{0,k}\, x_{1,k}\, x_{2,k} \ldots x_{n_i,k}]$ and $x_{0,k} = 1 \, \forall \, k = 1, 2, \ldots, m$. Ideally, the binary output of the network is formed by the elements

$$y^o_{i,k} = \text{sgn}(\bar{y}_{i,k}) = \text{sgn}(\mathbf{x}^*_k \mathbf{w}_i) = \text{sgn}(\sum_{j=0}^{n_i} w_{ij} x_{j,k}) \; \forall \, i = 1, 2, \ldots, n_o \qquad (4.3.2)$$

where sgn(.) is the signum function, defined as $\text{sgn}(x) = -1$ if $x < 0$ and $\text{sgn}(x) = +1$ if $x \geq 0$.

The training of single-layered neural networks is based here on the minimization of $G(\lambda)$, given by (4.2.2), where $y_{i,k}, i = 1, 2, \ldots, n_o$ are the elements of the expected output $\mathbf{y}_k$ and $\hat{y}_{i,k}, i = 1, 2, \ldots, n_o$ the corresponding estimates provided by the network (Karayiannis, 1991a). The analytical problems associated with the hard-limiting nonlinearity sgn(.) can be avoided by approximating it by a soft-limiting nonlinearity. In mathematical terms, the estimates of $y_{i,k}, i = 1, 2, \ldots n_o$ provided by the network are considered to be

$$\hat{y}_{i,k} = \sigma(\bar{y}_{i,k}) = \sigma(\mathbf{x}^*_k \mathbf{w}_i) = \sigma(\sum_{j=0}^{n_i} w_{ij} x_{j,k}) \; \forall \, i = 1, 2, \ldots, n_o \qquad (4.3.3)$$

If the output of the network is analog, $\sigma(x) = x$. If the output of the network is binary, $\sigma(.)$ is a continuous, differentiable everywhere function which approximates the signum function. Such a smooth approximation of the hard-limiter has been used extensively in the literature. In this chapter, the two states of the network are chosen to be $\{-1, +1\}$. As described in chapter 2, a good choice for $\sigma(.)$ in this case is the hyperbolic tangent, that is, $\sigma(x) = \tanh(x)$.

4.3.1 Fast Delta Rule

The training of single-layered neural networks is frequently based on the observation that the objective function $G(\lambda)$, defined by (4.2.2), can also be expressed as $G(\lambda) = \sum_{k=1}^{m} G_k(\lambda)$, where

$$G_k(\lambda) = \lambda \sum_{i=1}^{n_o} \phi_2(e_{i,k}) + (1-\lambda) \sum_{i=1}^{n_o} \phi_1(e_{i,k}) \; \forall \; k = 1, 2, \ldots, m \quad (4.3.4)$$

Thus, (4.2.2) is minimized by sequentially minimizing (4.3.4) for $k = 1, 2, \ldots, m$. The gradient descent method is widely used in performing the minimization of (4.3.4). If the output of the network is analog, it is shown in Appendix A4.1 that gradient descent method results in the following update equation for the synaptic weights of the network

$$\mathbf{w}_{p,k} = \mathbf{w}_{p,k-1} + \alpha \, \varepsilon^o_{p,k}(\lambda) \, \mathbf{x}_k \quad (4.3.5)$$

where

$$\varepsilon^o_{p,k}(\lambda) = \lambda(y_{p,k} - \hat{y}_{p,k}) + (1-\lambda) \tanh[\beta(y_{p,k} - \hat{y}_{p,k})] \quad (4.3.6)$$

It is also shown in Appendix A4.1 that if the output of the network is binary the synaptic weights can be updated by equation (4.3.5), where

$$\varepsilon^o_{p,k}(\lambda) = (1 - \hat{y}^2_{p,k})(y_{p,k} - \lambda \, \hat{y}_{p,k}) \quad (4.3.7)$$

The Fast Delta Rule algorithm is summarized in the flowchart presented in Table 4.1.

An alternative formulation of the minimization of (4.2.9) provided the basis for the development of second-order algorithms for feed-forward neural networks with binary output (Karayiannis, 1991a; Karayiannis and Venetsanopoulos, 1991). This formulation is based on the observation that $G(\lambda) = \sum_{i=1}^{n_o} G_{i,m}(\lambda)$, where

$$G_{i,m}(\lambda) = \sum_{k=1}^{m} y_{i,k}(y_{i,k} - \hat{y}_{i,k}) + \frac{1}{2}\lambda \sum_{k=1}^{m} (\hat{y}^2_{i,k} - y^2_{i,k}) \; \forall \; i = 1, 2, \ldots, n_o \quad (4.3.8)$$

(4.2.9) can be minimized by minimizing the objective function defined by (4.3.8) for each $i = 1, 2, \ldots, n_o$, since the estimate of each element $y_{i,k}$, $k = 1, 2, \ldots, m$ depends only on the ith row of the matrix of synaptic weights. This observation suggests that each row $\mathbf{w}^*_i, i = 1, 2, \ldots, n_o$ of the matrix of synaptic weights can be estimated by minimizing the objective function (4.3.8). It was suggested in chapter 3 that (4.3.8) can be minimized using a sequential

Start
Initialize **W** with random values
Select μ
$\lambda = 1$
1 $k = 0$ (k: number of association)
$E = 0$
2 $k \leftarrow k + 1$
$\mathbf{x} = \mathbf{x}_k$
$\mathbf{y} = \mathbf{y}_k$
$\hat{y}_i = \sigma(\sum_{j=1}^{n_i} w_{ij}\, x_j)$
$e_i = y_i - \hat{y}_i$
$\varepsilon_i^o(\lambda) = \lambda e_i + (1 - \lambda) \tanh[\beta e_i]$ (analog output)
$\varepsilon_i^o(\lambda) = (1 - \hat{y}_i^2)(y_i - \lambda \hat{y}_i)$ (binary output)
$\mathbf{w}_i \leftarrow \mathbf{w}_i + \alpha\, \varepsilon_i^o(\lambda)\, \mathbf{x}$
$\hat{y}_i = \sigma(\sum_{j=1} w_{ij}\, x_j)$
$E \leftarrow E + \frac{1}{2} \sum_{i=1}^{n_o} (y_i - \hat{y}_i)^2$
if: $k < m$; then: go to 2
$\lambda = \exp(-\mu / E^2)$
if: $E > E_o$; then: go to 1
Stop

Table 4.1: Flowchart of Fast Delta Rule.

optimization strategy based on the observation that the objective function (4.3.8) is also a function of the number of associations considered. Clearly, $G_{i,m-1}(\lambda)$ depends only on the associations $(\mathbf{y}_k, \mathbf{x}_k), k = 1, 2, \ldots, m-1$. Therefore, the network can be trained with respect to the associations $(\mathbf{y}_k, \mathbf{x}_k), k = 1, 2, \ldots, m-1$ by minimizing $G_{i,m-1}(\lambda)$. The synaptic weights of the network can be updated with respect to the new association $(\mathbf{y}_m, \mathbf{x}_m)$ by minimizing $G_{i,m}(\lambda)$. Assume that $\mathbf{w}_i$ has been updated with respect to the associations $(\mathbf{y}_k, \mathbf{x}_k), k = 1, 2, \ldots, m-1$ by minimizing $G_{i,m-1}(\lambda)$, resulting in the estimate $\mathbf{w}_i = \mathbf{w}_{i,m-1}$. The ith row of the matrix of synaptic weights $\mathbf{W}$ is then updated with respect to the association $(\mathbf{y}_m, \mathbf{x}_m)$ by minimizing $G_{i,m}(\lambda)$, using the update equation

$$\mathbf{w}_{i,m} = \mathbf{w}_{i,m-1} - \alpha\,\mathbf{H}_{i,m}(\lambda)^{-1}|_{\mathbf{w}_i=\mathbf{w}_{i,m-1}} \frac{\partial G_{i,m}(\lambda)}{\partial \mathbf{w}_i}|_{\mathbf{w}_i=\mathbf{w}_{i,m-1}} \quad (4.3.9)$$

where $\partial G_{i,m}(\lambda)/\partial \mathbf{w}_i$ is the gradient of $G_{i,m}(\lambda)$ with respect to $\mathbf{w}_i$, $\mathbf{H}_{i,m}(\lambda)$ the Hessian matrix, and α, the *learning rate,* is a positive real number.

4.3.2 Fast ELEANNE 3

The derivation of this algorithm begins with the evaluation of the gradient of the objective function $G_{i,m}(\lambda)$. From the definition $G_{i,m}(\lambda)$ in (4.3.8),

$$\frac{\partial G_{i,m}(\lambda)}{\partial w_{iq}} = -\sum_{k=1}^{m}(y_{i,k} - \lambda\,\hat{y}_{i,k})\frac{\partial \hat{y}_{i,k}}{\partial w_{iq}} \quad (4.3.10)$$

The gradient of $G_{i,m}(\lambda)$ with respect to $\mathbf{w}_i$ is given by

$$\begin{aligned}\frac{\partial G_{i,m}(\lambda)}{\partial \mathbf{w}_i} &= \left[\frac{\partial G_{i,m}(\lambda)}{\partial w_{i0}}\;\frac{\partial G_{i,m}(\lambda)}{\partial w_{i1}}\;\frac{\partial G_{i,m}(\lambda)}{\partial w_{i2}}\cdots\frac{\partial G_{i,m}(\lambda)}{\partial w_{in_i}}\right]^* \\ &= -\sum_{k=1}^{m}(y_{i,k} - \lambda\,\hat{y}_{i,k})\mathbf{b}_{i,k} \qquad (4.3.11)\end{aligned}$$

where $\mathbf{b}_{i,k}$ is defined by

$$\mathbf{b}_{i,k}^* = \left[\frac{\partial \hat{y}_{i,k}}{\partial w_{i0}}\;\frac{\partial \hat{y}_{i,k}}{\partial w_{i1}}\;\frac{\partial \hat{y}_{i,k}}{\partial w_{i2}}\cdots\frac{\partial \hat{y}_{i,k}}{\partial w_{in_i}}\right] \quad (4.3.12)$$

The update equation (4.3.9) can be simplified by approximating the gradient of $G_{i,m}(\lambda)$ with respect to $\mathbf{w}_i$ at $\mathbf{w}_i = \mathbf{w}_{i,m-1}$ as follows (Karayiannis, 1991a)

$$\frac{\partial G_{i,m}(\lambda)}{\partial \mathbf{w}_i}\Big|_{\mathbf{w}_i=\mathbf{w}_{i,m-1}} = -(y_{i,m} - \lambda\, \hat{y}_{i,m})\mathbf{b}_{i,m} \tag{4.3.13}$$

Summarizing the previous analysis, each row of $\mathbf{W}$ can be updated by the equation

$$\mathbf{w}_{i,m} = \mathbf{w}_{i,m-1} + \alpha\, \mathbf{H}_{i,m}(\lambda)^{-1}|_{\mathbf{w}_i=\mathbf{w}_{i,m-1}} \mathbf{b}_{i,m}(y_{i,m} - \lambda\, \hat{y}_{i,m}) \tag{4.3.14}$$

The final form of the update equations for minimizing $G_{i,m}(\lambda)$ are derived by evaluating the vector $\mathbf{b}_{i,k}$ and the corresponding Hessian matrix. It is shown in Appendix A4.2 that

$$\mathbf{b}_{i,k} = (1 - \hat{y}_{i,k}^2)\, \mathbf{x}_k \tag{4.3.15}$$

It is also shown in Appendix A4.2 that the Hessian matrix can be approximated for $0 \leq \lambda \leq 1$ by

$$\mathbf{H}_{i,m}(\lambda) = \sum_{k=1}^{m} c_{i,k}(\lambda)\, \mathbf{x}_k\, \mathbf{x}_k^* \tag{4.3.16}$$

where

$$c_{i,k}(\lambda) = [(\lambda/3)\, d_{i,k}(1) + (2/3)\, d_{i,k}(\lambda)]\, d_{i,k}(1) \tag{4.3.17}$$

and

$$d_{i,k}(\lambda) = 1 - \lambda\, \hat{y}_{i,k}^2 \tag{4.3.18}$$

It can easily be verified that, for $|\hat{y}_{i,k}| < 1$, $c_{i,k}(\lambda) \geq 0\ \forall\ 0 \leq \lambda \leq 1$. As a result, (4.3.16) is positive semi-definite and $\mathbf{H}_{i,m}(\lambda)$ can be evaluated recursively by using an approach similar to that developed in chapter 3. According to the matrix inversion lemma, $\mathbf{P}_{i,m}(\lambda) = \mathbf{H}_{i,m}(\lambda)^{-1}$ can be evaluated in terms of $\mathbf{P}_{i,m-1}(\lambda) = \mathbf{H}_{i,m-1}(\lambda)^{-1}$ by

$$\mathbf{P}_{i,m}(\lambda) = \mathbf{P}_{i,m-1}(\lambda) - \delta_{i,m}(\lambda)\, \mathbf{P}_{i,m-1}(\lambda)\, \mathbf{x}_m\, \mathbf{x}_m^*\, \mathbf{P}_{i,m-1}(\lambda) \tag{4.3.19}$$

Start
Initialize $\mathbf{W}$ with random values
Select μ
$\lambda = 1$
1 $E = 0$
$i = 0$(i: index of output unit)
2 $i \leftarrow i + 1$
$\bar{\mathbf{w}} = \mathbf{w}_i$
$\mathbf{P} = \mathbf{I}$
$k = 0$(k: number of association)
3 $k \leftarrow k + 1$
$\mathbf{x} = \mathbf{x}_k$
$y = y_{i,k}$
$\hat{y} = \sigma(\sum_{j=1}^{n_i} \bar{w}_j \, x_j)$
$d(\lambda) = 1 - \lambda \, \hat{y}^2$
$c(\lambda) = [(\lambda/3) \, d(1) + (2/3) \, d(\lambda)] \, d(1)$
$\delta(\lambda) = c(\lambda)(1 + c(\lambda) \, \mathbf{x}^* \, \mathbf{P} \, \mathbf{x})^{-1}$
$\mathbf{P} \leftarrow \mathbf{P} - \delta(\lambda) \, \mathbf{P} \, \mathbf{x} \, \mathbf{x}^* \, \mathbf{P}$
$\varepsilon^o(\lambda) = (1 - \hat{y}^2)(1 - \lambda \, \hat{y})$
$\bar{\mathbf{w}} \leftarrow \bar{\mathbf{w}} + \alpha \, \varepsilon^o(\lambda) \, \mathbf{P} \, \mathbf{x}$
$\hat{y} = \rho(\sum_{j=1} \bar{w}_j \, x_j)$
$E = E + \frac{1}{2} \, (y - \hat{y})^2$
if $k < m$; then go to 3
$\mathbf{w}_i = \bar{\mathbf{w}}$
if: $i < n_o$; then: go to 2
$\lambda = \exp(-\mu / E^2)$
if: $E > E_o$; then: go to 1
Stop

Table 4.2: Flowchart of Fast ELEANNE 3.

where $\delta_{i,m}(\lambda) = c_{i,m}(\lambda)(1 + c_{i,m}(\lambda)\,\mathbf{x}_m^* \mathbf{P}_{i,m-1}(\lambda)\,\mathbf{x}_m)^{-1}$, $c_{i,k}(\lambda)$ is given by (4.3.17) and (4.3.18), and $\mathbf{P}_{i,0}(\lambda) = \mathbf{I}$. The update equation for each row of the matrix of synaptic weights can be obtained by substituting $\mathbf{b}_{i,k}$ from (4.3.15) into (4.3.14) as follows

$$\mathbf{w}_{i,m} = \mathbf{w}_{i,m-1} + \alpha\, \varepsilon_{i,m}^{o}(\lambda)\; \mathbf{P}_{i,m}(\lambda)\,\mathbf{x}_m \tag{4.3.20}$$

where $\varepsilon_{i,m}^{o}(\lambda)$ is given by (4.3.7) and $\mathbf{P}_{i,m}(\lambda)$ is evaluated by (4.3.19). This algorithm, called Fast ELEANNE 3, is summarized in the flowchart presented in Table 4.2.

4.3.3 Fast ELEANNE 4

A computationally less demanding version of the previous algorithm is derived here on the basis of some simplifying, though reasonable, assumptions. According to the previous analysis, the Hessian matrix corresponding to the objective function $G_{i,m}(\lambda)$ can be approximated by (4.3.16), where $c_{i,k}(\lambda)$ is given by (4.3.17) and (4.3.18). The fact that there are n_o different coefficients $c_{i,k}(\lambda), i = 1, 2, \ldots, n_o$ requires that n_o matrices $\mathbf{P}_{i,m}(\lambda) = \mathbf{H}_{i,m}(\lambda)^{-1}, i = 1, 2, \ldots, n_o$ be recursively evaluated. A simplification of the Fast ELEANNE 3 algorithm can be obtained by making the assumption that (Karayiannis, 1991a)

$$c_{i,k}(\lambda) = c_{i_0,k}(\lambda) = c_k(\lambda)\ \forall\, i = 1, 2, \ldots, n_o \tag{4.3.21}$$

where $c_{i_0,k}(\lambda)$ is the coefficient defined in terms of the response of a certain output $\hat{y}_{i_0,k}$ by (4.3.17) and (4.3.18). According to this assumption, then

$$\mathbf{H}_{i,m}(\lambda) = \sum_{k=1}^{m} c_k(\lambda)\,\mathbf{x}_k\,\mathbf{x}_k^* = \mathbf{H}_m(\lambda)\ \forall\, i = 1, 2, \ldots, n_o \tag{4.3.22}$$

The replacement of each of the n_o matrices given by (4.3.16) by the 'average' matrix (4.3.22) allows all rows of the matrix of synaptic weights to be simultaneously updated with respect to each of the m associations. The reliable estimation of $c_k(\lambda)$ is the crucial problem resulting from the above assumption.

It is assumed that during the initial adaptation cycles the output of the network is far away from the optimum and, therefore,

Start
Initialize $\mathbf{W}$ with random values
Select μ
$\lambda = 1$
$\nu = 0$ (ν: number of adaptation cycle)
1 $\nu \leftarrow \nu + 1$
$E = 0$
$\mathbf{P} = \mathbf{I}$
$k = 0$ (k: number of association)
2 $k \leftarrow k + 1$
$\mathbf{x} = \mathbf{x}_k$
$d(\lambda) = \exp(-(\lambda + 1)\beta_k^2(\nu - 1)^2)$
$c(\lambda) = (\lambda/3)\, d(1) + (2/3)\, d(\lambda)$
$\delta(\lambda) = c(\lambda)(1 + c(\lambda)\, \mathbf{x}^*\, \mathbf{P}\, \mathbf{x})^{-1}$
$\mathbf{P} \leftarrow \mathbf{P} - \delta(\lambda)\, \mathbf{P}\, \mathbf{x}\, \mathbf{x}^*\, \mathbf{P}$
$i = 0$ (i: index of output unit)
3 $i \leftarrow i + 1$
$\bar{\mathbf{w}} = \mathbf{w}_i$
$y = y_{i,k}$
$\hat{y} = \sigma(\sum_{j=1}^{n_i} \bar{w}_j x_j)$
$\varepsilon^o(\lambda) = (1 - \hat{y}^2)(y - \lambda\, \hat{y})$
$\bar{\mathbf{w}} \leftarrow \bar{\mathbf{w}} + \alpha\, \varepsilon^o(\lambda)\, \mathbf{P}\, \mathbf{x}$
$\hat{y} = \sigma(\sum_{j=1}^{n_i} \bar{w}_j x_j)$
$E \leftarrow E + \frac{1}{2}\, (y - \hat{y})^2$
$\mathbf{w}_i = \bar{\mathbf{w}}$
if $i < n_o$; then go to 3
if $k < m$; then go to 2
$\lambda = \exp(-\mu/E^2)$
if: $E > E_o$; then go to 1
Stop

Table 4.3: Flowchart of Fast ELEANNE 4.

$\hat{y}_{i_o,k} \approx 0 \, \forall \, k = 1, 2, \ldots, m$ (Karayiannis, 1991a; Karayiannis and Venetsanopoulos, 1991). After a certain number of adaptation cycles, the output of the network approaches the optimum and, therefore, $\hat{y}_{i_o,k} \approx y_{i_o,k} \, \forall \, k = 1, 2, \ldots, m$. Finally, it is assumed that $\hat{y}_{i_o,k}$ is a linear function of the number ν of adaptation cycles and also $\hat{y}_{i_o,k} = 0$ when $\nu = 1$ (Karayiannis, 1991a; Karayiannis and Venetsanopoulos, 1991). In mathematical terms,

$$\hat{y}^2_{i_o,k} = \beta_k^2 (\nu - 1)^2 \; \forall \; k = 1, 2, \ldots, m \qquad (4.3.23)$$

where $\beta_k > 0$ determines the speed of convergence of each $\hat{y}_{i_o,k}$ to its target. Since $|\hat{y}_{i_o,k}| < 1$ and $0 \le \lambda \le 1$, the derivation of the estimate of $c_k(\lambda)$ can be further simplified by using the well-known approximation $\exp(-\lambda \hat{y}^2_{i_o,k}) \approx 1 - \lambda \hat{y}^2_{i_o,k}$. Therefore,

$$1 - \lambda \hat{y}^2_{i_o,k} \approx \exp(-\lambda \hat{y}^2_{i_o,k}) = \exp(-\lambda \beta_k^2 (\nu - 1)^2) \qquad (4.3.24)$$

Substituting (4.3.21) and (4.3.24) into (4.3.17) indicates that $c_k(\lambda)$ can be estimated during the νth adaptation cycle by

$$c_k(\lambda) = c_k^\nu(\lambda) = (\lambda/3) d_k^\nu(1) + (2/3) d_k^\nu(\lambda) \qquad (4.3.25)$$

where

$$d_k^\nu(\lambda) = \exp(-(\lambda + 1)\beta_k^2 (\nu - 1)^2) \qquad (4.3.26)$$

and β_k is a positive constant. According to the matrix inversion lemma, $\mathbf{P}_m^\nu(\lambda) = \mathbf{H}_m^\nu(\lambda)^{-1}$ can be evaluated recursively from

$$\mathbf{P}_m^\nu(\lambda) = \mathbf{P}_{m-1}^\nu(\lambda) - \delta_m^\nu(\lambda) \, \mathbf{P}_{m-1}^\nu(\lambda) \, \mathbf{x}_m \, \mathbf{x}_m^* \, \mathbf{P}_{m-1}^\nu(\lambda) \qquad (4.3.27)$$

where $\delta_m^\nu(\lambda) = c_m^\nu(\lambda)(1 + c_m^\nu(\lambda) \, \mathbf{x}_m^* \, \mathbf{P}_{m-1}^\nu(\lambda) \, \mathbf{x}_m)^{-1}$, and $\mathbf{P}_0^\nu(\lambda) = \mathbf{I}$. During the νth adaptation cycle, the rows of the matrix of synaptic weights are updated through the following equations

$$\mathbf{w}_{i,m} = \mathbf{w}_{i,m-1} + \alpha \, \varepsilon_{i,m}^o(\lambda) \, \mathbf{P}_m^\nu(\lambda) \, \mathbf{x}_m \; \forall \; i = 1, 2, \ldots, n_o \qquad (4.3.28)$$

where $\varepsilon_{i,m}^o(\lambda)$ is defined by (4.3.7) and $\mathbf{P}_m^\nu(\lambda)$ is evaluated by (4.3.27). The resulting algorithm, called Fast ELEANNE 4, is summarized in the flowchart presented in Table 4.3.

4.4 Fast Learning Algorithms for Multi-layered Neural Networks

Consider the multi-layered network with one layer of hidden units shown in Figure 2.9. Assuming that the pattern $\mathbf{x}_k$ is the input of the network, the corresponding outputs of the hidden units are

$$\hat{h}_{j,k} = \rho(\bar{h}_{j,k}) = \rho(\mathbf{x}_k^* \mathbf{v}_j) = \rho(\sum_{\ell=0}^{n_i} v_{j\ell} x_{\ell,k}) \qquad (4.4.1)$$

where $\rho(x) = \tanh(x)$, $\mathbf{v}_j^* = [v_{j0}\, v_{j1}\, \dots\, v_{jn_i}]$, $\mathbf{x}_k^* = [x_{0,k}\, x_{1,k}\, \dots\, x_{n_i,k}]$ and $x_{0,k} = 1\, \forall\, k = 1, 2, \dots, m$. Let $\hat{\mathbf{h}}_k$ be the $(n_h + 1) \times 1$ vector formed by the outputs of the hidden units (4.4.1) as follows

$$\hat{\mathbf{h}}_k^* = [\hat{h}_{0,k}\, \hat{h}_{1,k}\, \hat{h}_{2,k} \dots \hat{h}_{n_h,k}] \qquad (4.4.2)$$

where $\hat{h}_{0,k} = 1\, \forall\, k = 1, 2, \dots, m$. Clearly, $\hat{\mathbf{h}}_k$ is a function of the corresponding input pattern $\mathbf{x}_k$ as well as the synaptic weights of the lower network v_{pq}. The output of the network is formed by the elements

$$\hat{y}_{i,k} = \sigma(\bar{y}_{i,k}) = \sigma(\hat{\mathbf{h}}_k^* \mathbf{w}_i) = \sigma(\sum_{j=0}^{n_h} w_{ij} \hat{h}_{j,k}) \qquad (4.4.3)$$

where $\mathbf{w}_i^* = [w_{i0}\, w_{i1}\, w_{i2} \dots w_{in_h}]$. If the output of the network is analog, $\sigma(x) = x$. If the output of the network is formed by binary elements, $\sigma(x) = \tanh(x)$.

Consider also the multi-layered neural network with L layers of hidden units, shown in Figure 2.10. The output of this network is formed by the elements

$$\hat{y}_{i,k} = \sigma(\sum_{j=0}^{n_1} v_{ij}^{(0)} \hat{h}_{j,k}^{(1)})\ \forall\, i = 1, 2, \dots, n_o \qquad (4.4.4)$$

where $\sigma(x) = x$ if the output of the network is analog and $\sigma(x) = \tanh(x)$ if the output of the network is binary. The outputs of the hidden units are given by

$$\hat{h}_{j,k}^{(r)} = \rho(\sum_{\ell=0}^{n_{r+1}} v_{j\ell}^{(r)} \hat{h}_{\ell,k}^{(r+1)})\ ,\ r = 1, 2, \dots, L \qquad (4.4.5)$$

where $h_{\ell,k}^{(L+1)} = \mathbf{x}_{\ell,k} \, \forall \, \ell = 0, 1, \ldots, n_i$.

A multi-layered neural network with one or more layers of hidden units can be trained using the well known Error Back Propagation algorithm (Rumelhart et al., 1986). According to this algorithm, the synaptic weights of the network are updated during the training by propagating back the error between the expected output and the output provided by the network. The same optimization strategy used for the development of the Error Back Propagation algorithm is employed here in conjunction with the generalized criterion proposed in this chapter for the training of multi-layered neural networks.

4.4.1 Fast Back Propagation

This algorithm is derived in Appendix A4.3 by sequentially minimizing the objective function $G_k(\lambda)$, defined by (4.3.4), for $k = 1, 2, \ldots, m$. The update equation for the synaptic weights w_{pq} is obtained as follows

$$\mathbf{w}_{p,k} = \mathbf{w}_{p,k-1} + \alpha \, \varepsilon_{p,k}^{o}(\lambda) \, \hat{\mathbf{h}}_k \tag{4.4.6}$$

If the output of the network is analog,

$$\varepsilon_{p,k}^{o}(\lambda) = \lambda(y_{p,k} - \hat{y}_{p,k}) + (1 - \lambda) \tanh[\beta(y_{p,k} - \hat{y}_{p,k})] \tag{4.4.7}$$

On the other hand, if the network has binary outputs,

$$\varepsilon_{p,k}^{o}(\lambda) = (1 - \hat{y}_{p,k}^{2})(y_{p,k} - \lambda \, \hat{y}_{p,k}) \tag{4.4.8}$$

It is also shown in Appendix A4.3 that the synaptic weights v_{pq} can be updated through the equation

$$\mathbf{v}_{p,k} = \mathbf{v}_{p,k-1} + \alpha \, \varepsilon_{p,k}^{h}(\lambda) \, \mathbf{x}_k \tag{4.4.9}$$

where

$$\varepsilon_{p,k}^{h}(\lambda) = (1 - \hat{h}_{p,k}^{2}) \sum_{i=1}^{n_o} \varepsilon_{i,k}^{o}(\lambda) \, w_{ip} \tag{4.4.10}$$

The output error $\varepsilon_{i,k}^{o}(\lambda)$ is given by (4.4.7) if the output of the network is analog, and (4.4.8) if the output of the network is formed

Start
Initialize **W** and **V** with random values
Select μ
$\lambda = 1$
1 $k = 0$ (k: number of association)
$E = 0$
2 $k \leftarrow k + 1$
$\mathbf{x} = \mathbf{x}_k$
$\mathbf{y} = \mathbf{y}_k$
$\hat{h}_j = \rho(\sum_{\ell=1}^{n_i} v_{j\ell}\, x_\ell)$
$\hat{y}_i = \sigma(\sum_{j=1}^{n_h} w_{ij}\, \hat{h}_j)$
$e_i = y_i - \hat{y}_i$
$\varepsilon_i^o(\lambda) = \lambda e_i + (1 - \lambda)\tanh[\beta e_i]$ (analog output)
$\varepsilon_i^o(\lambda) = (1 - \hat{y}_i^2)(y_i - \lambda\, \hat{y}_i)$ (binary output)
$\mathbf{w}_i \leftarrow \mathbf{w}_i + \alpha\, \varepsilon_i^o(\lambda)\, \hat{\mathbf{h}}$
$\varepsilon_j^h(\lambda) = (1 - \hat{h}_j^2) \sum_{i=1}^{n_o} \varepsilon_i^o(\lambda)\, w_{ij}$
$\mathbf{v}_j \leftarrow \mathbf{v}_j + \alpha\, \varepsilon_j^h(\lambda)\, \mathbf{x}$
$\hat{h}_j = \rho(\sum_{\ell=1}^{n_i} v_{j\ell}\, x_\ell)$
$\hat{y}_i = \sigma(\sum_{j=1}^{n_h} w_{ij}\, \hat{h}_j)$
$E \leftarrow E + \frac{1}{2} \sum_{i=1}^{n_o} (y_i - \hat{y}_i)^2$
if: $k < m$; then: go to 2
$\lambda = \exp(-\mu / E^2)$
if: $E > E_o$; then: go to 1
Stop

Table 4.4: Flowchart of Fast Back Propagation.

by binary elements. The Fast Back Propagation algorithm is summarized in the flowchart presented in Table 4.4.

Because of its simplicity, this algorithm provides an ideal basis for investigating the role of λ during training. Consider a network with analog output. During the initial adaptation cycles, $\lambda \approx 1$. In this case, the first term of (4.4.7) is dominant. As $e_{p,k} = y_{p,k} - \hat{y}_{p,k}$ decreases during the training, λ also decreases and the second term of (4.4.7) becomes dominant. In this case, the behavior of the nonlinearity $\tanh(\beta e_{p,k})$ depends on both β and $e_{p,k} = y_{p,k} - \hat{y}_{p,k}$. If $\beta e_{p,k}$ is sufficiently small, $\tanh(\beta e_{p,k}) \approx \beta e_{p,k}$ and, therefore, $\varepsilon^o_{p,k}(\lambda) \approx [\lambda + (1-\lambda)\beta]e_{p,k}$. In the particular case where $\beta = 1$, $\varepsilon^o_{p,k}(\lambda) \approx e_{p,k} = y_{p,k} - \hat{y}_{p,k}$. If $\beta e_{p,k}$ is sufficiently large, $\tanh(\beta e_{p,k})$ approaches $+1$ or -1, depending on the sign of $e_{p,k} = y_{p,k} - \hat{y}_{p,k}$. It is apparent that the convergence of the resulting algorithm is affected by the value of β, which determines the behavior of the nonlinearity $\tanh(\beta e_{p,k})$. If $\beta \approx 1$, $\tanh(\beta e_{p,k})$ behaves like a linear function and the convergence of the resulting algorithm is expected to be similar to that of the Error Back Propagation algorithm. If the value of β is sufficiently larger than unity, the convergence of the algorithm is expected to improve. As β approaches infinity, $\tanh(\beta e_{p,k})$ asymptotically approaches $\text{sgn}(e_{p,k})$. In this case, the convergence of the resulting algorithm is not guaranteed.

The investigation of the role of λ is even more informative in the case where the output of the network is binary. When $\lambda = 1$, (4.4.8) gives

$$\varepsilon^o_{p,k}(1) = (1 - \hat{y}^2_{p,k})(y_{p,k} - \hat{y}_{p,k}) \tag{4.4.11}$$

Clearly, if $\lambda = 1$ this algorithm coincides with the Error Back Propagation algorithm. In this case, the adaptation of the synaptic weights w_{pq} is determined by the error $e_{p,k} = y_{p,k} - \hat{y}_{p,k}$. It can easily be verified that

$$\varepsilon^o_{p,k}(1) = \begin{cases} (1 + \hat{y}_{p,k})(y_{p,k} - \hat{y}_{p,k})^2 & \text{if } y_{p,k} = +1 \\ -(1 - \hat{y}_{p,k})(y_{p,k} - \hat{y}_{p,k})^2 & \text{if } y_{p,k} = -1 \end{cases} \tag{4.4.12}$$

When the estimate $\hat{y}_{p,k}$ is close to its target $y_{p,k}$, the adaptation of the synaptic weights is dominated by the term $e^2_{p,k} = (y_{p,k} - \hat{y}_{p,k})^2$. This term makes the convergence of the algorithm very slow after the initial adaptation cycles. It can easily be verified that if $\lambda = 0$,

$$\varepsilon^o_{p,k}(0) = y_{p,k}(1-\hat{y}^2_{p,k}) = \begin{cases} y_{p,k}(1+\hat{y}_{p,k})(y_{p,k}-\hat{y}_{p,k}) & \text{if } y_{p,k} = +1 \\ -y_{p,k}(1-\hat{y}_{p,k})(y_{p,k}-\hat{y}_{p,k}) & \text{if } y_{p,k} = -1 \end{cases} \tag{4.4.13}$$

This latter equation indicates that, regardless of the value of λ, the adaptation of the synaptic weights is determined by the error $e_{p,k} = y_{p,k} - \hat{y}_{p,k}$. However, the effect of the error $e_{p,k} = y_{p,k} - \hat{y}_{p,k}$ on the adaptation of the synaptic weights is affected by a bias term which changes during the training of the network.

The above analysis provides the basis for investigating the relationship between the conventional quadratic error criterion (4.2.1), the relative entropy criterion presented in chapter 2, and the generalized criterion proposed in this chapter. If the training of a feed-forward neural network with one layer of hidden units is based on the relative entropy criterion, the synaptic weights of the network are updated by propagating back the output error

$$\eta^o_{p,k} = y_{p,k} - \hat{y}_{p,k} \tag{4.4.14}$$

During the initial adaptation cycles, each $\hat{y}_{p,k}$ is far away from its target $y_{p,k}$. In this case, the output nonlinearity $\sigma(x) = \tanh(x)$ does not approach the saturation level. As a result, the output nonlinearity behaves more like a linear function, that is, $\sigma(x) = \tanh(x) \approx x$. Under this assumption, the binary output of the network is very close to its analog output, i.e., $\hat{y}_{i,k} \approx \bar{y}_{i,k}$. During the initial adaptation cycles, the update equations provided by the relative entropy criterion are similar to those provided by the minimization of $E = G(1) = \frac{1}{2}\sum_{k=1}^{m}\sum_{i=1}^{n_o}(y_{i,k} - \hat{y}_{i,k})^2$ in the case where $\hat{y}_{i,k} = \bar{y}_{i,k}$. In other words, the maximization of the relative entropy criterion during the initial adaptation cycles is almost equivalent to training a network with binary output with respect to its analog output by minimizing the objective function E. It was experimentally verified that training networks with binary output in this fashion substantially degrades their generalization ability (Karayiannis, 1991a). Hence, it is not surprising that neural networks trained by the algorithm based on the relative entropy criterion achieve lower generalization ability than those trained by the conventional Error Back Propagation algorithm (Solla et al., 1988).

Consider the Fast Back Propagation algorithm proposed in this chapter and the algorithm based on the maximization of the relative entropy after the initial adaptation cycles, i.e., when each $\hat{y}_{p,k}$ approaches its target $y_{p,k}$. If $y_{p,k} = 1$ and $\hat{y}_{p,k} \in [0,1)$, then $1 \leq y_{p,k}(1 + \hat{y}_{p,k}) < 2$. The definition of $\varepsilon^o_{p,k}(0)$ in (4.4.13) indicates that $\eta^o_{p,k} \leq \varepsilon^o_{p,k}(0) < 2\,\eta^o_{p,k}$. If $y_{p,k} = -1$ and $\hat{y}_{p,k} \in (-1,0]$, it can similarly be shown that $\eta^o_{p,k} \leq \varepsilon^o_{p,k}(0) < 2\,\eta^o_{p,k}$. This observation reveals a similarity between the Fast Back Propagation algorithm and the algorithm based on the relative entropy criterion after the initial adaptation cycles, and also explains the fast convergence of both algorithms. The crucial difference between these two algorithms is that during the initial adaptation cycles the Fast Back Propagation algorithm is based on the minimization of the quadratic function (4.2.1), which guarantees that the Fast Back Propagation and the Error Back Propagation algorithms provide neural networks with almost identical generalization ability.

The previous discussion revealed a significant similarity between the update equations provided by the maximization of the relative entropy and the Fast Back Propagation algorithm after the initial adaptation cycles i.e., when λ approaches zero. This suggests that the relative entropy criterion could also be incorporated into the formation of the generalized objective function (4.2.2). In fact, a feed-forward neural network can be trained by minimizing the objective function $G(\lambda) = \lambda E + (1-\lambda)E'$, where E is the quadratic error function given by (4.2.1) and $E' = -H$. It can easily be shown that in this case the synaptic weights w_{pq} can be updated by (4.4.7), where

$$\begin{aligned} \varepsilon^o_{p,k}(\lambda) &= [\lambda(1-\hat{y}^2_{p,k}) + (1-\lambda)](y_{p,k} - \hat{y}_{p,k}) \\ &= (1-\lambda\,\hat{y}^2_{p,k})(y_{p,k} - \hat{y}_{p,k}) \end{aligned} \tag{4.4.15}$$

Similarly, the synaptic weights v_{pq} can be updated by (4.4.9), where $\varepsilon^h_{p,k}(\lambda)$ can be determined by (4.4.10) with the output error $\varepsilon^o_{p,k}(\lambda)$ given by (4.4.15).

The direct generalization of the previous analysis leads to the Fast Back Propagation algorithm for multi-layered neural networks with more than one layer of hidden units, such as the network shown in Figure 2.10. The update equations for the synaptic weights

$v_{pq}^{(r)}, r = 0, 1, \ldots, L$ are obtained in Appendix A4.3 as follows

$$\mathbf{v}_{p,k}^{(r)} = \mathbf{v}_{p,k-1}^{(r)} + \alpha\, \varepsilon_{p,k}^{(r)}(\lambda)\, \hat{\mathbf{h}}_k^{(r+1)} \ \forall\, r = 0, 1, \ldots, L \tag{4.4.16}$$

where $\hat{\mathbf{h}}_k^{(L+1)} = \mathbf{x}_k$, the output error $\varepsilon_{p,k}^{(0)}(\lambda) = \varepsilon_{p,k}^{o}(\lambda)$ is given by (4.4.7) if the output of the network is analog and (4.4.8) if the output of the network is formed by binary elements, and

$$\varepsilon_{p,k}^{(r)}(\lambda) = (1 - \hat{h}_{p,k}^{(r)\,2}) \sum_{\ell=1}^{n_{r-1}} \varepsilon_{\ell,k}^{(r-1)}(\lambda)\, v_{\ell p}^{(r-1)} \ \forall\, r = 1, 2, \ldots, L \tag{4.4.17}$$

The remainder of this section presents the derivation of alternative learning algorithms for multi-layered neural networks with binary outputs, which achieve better convergence than the Fast Back Propagation algorithm. This derivation is strongly influenced by the analysis presented in section 4.3, which led to the Fast ELEANNE 3 and the Fast ELEANNE 4 algorithms.

4.4.2 Fast ELEANNE 5

Consider the neural network with binary output and one layer of hidden units, shown in Figure 2.9. In order to simplify the analysis which follows, assume that the multi-layered neural network consists of two single-layered neural networks; the upper network with synaptic weights w_{pq}, and the lower network with synaptic weights v_{pq}. Clearly, the input pattern of the upper network is formed by the outputs of the hidden units. This network can be trained using Fast ELEANNE 5, an algorithm for multi-layered neural networks based on Fast ELEANNE 3. Each adaptation cycle consists of n_o stages, each one corresponding to the rows $\mathbf{w}_i, i = 1, 2, \ldots, n_o$ of the upper network (Karayiannis and Venetsanopoulos, 1991). Each stage consists of the adaptation of the corresponding row of the matrix of synaptic weights $\mathbf{W}$ of the upper network and all the synaptic weights v_{pq} of the lower network by sequentially considering the m associations. Within a certain adaptation stage, when the corresponding row of $\mathbf{W}$ and the synaptic weights v_{pq} have been updated with respect to the associations $(\mathbf{y}_1, \mathbf{x}_1)$, $(\mathbf{y}_2, \mathbf{x}_2), \ldots, (\mathbf{y}_{k-1}, \mathbf{x}_{k-1})$, the same synaptic weights can be adjusted with respect to the association $(\mathbf{y}_k, \mathbf{x}_k)$ by:

- updating the same row of the matrix $\mathbf{W}$ using Fast ELEANNE 3, and

- updating all the synaptic weights v_{pq} of the lower network on the basis of the Delta Rule.

Consider the ith stage of an adaptation cycle and assume that the synaptic weights of the network have been updated with respect to the associations $(\mathbf{y}_1, \mathbf{x}_1), (\mathbf{y}_2, \mathbf{x}_2), \ldots, (\mathbf{y}_{k-1}, \mathbf{x}_{k-1})$. Also, assume that the training with respect to these associations resulted in the estimate $\mathbf{w}_i = \mathbf{w}_{i,k-1}$ of the ith row of the matrix of synaptic weights $\mathbf{W}$ of the upper network and the estimate $\mathbf{V} = \mathbf{V}_{k-1}$ of the matrix of synaptic weights of the lower network. Consider here the adaptation of the ith row of the matrix of synaptic weights $\mathbf{W}$ with respect to the association $(\mathbf{y}_k, \mathbf{x}_k)$. Assuming that $\mathbf{x}_k$ is the input of the lower network, the input of the upper network is the *temporary input pattern* $\hat{\mathbf{h}}_k$. Therefore, $\mathbf{w}_i$ can be updated with respect to the *temporary association* $(\mathbf{y}_k, \hat{\mathbf{h}}_k)$ using Fast ELEANNE 3. It should be emphasized here that the temporary input pattern $\hat{\mathbf{h}}_k$ depends on the constant input pattern $\mathbf{x}_k$ but varies with the synaptic weights v_{pq} of the lower network. According to section 4.3, $\mathbf{w}_i$ can be updated through the equation

$$\mathbf{w}_{i,k} = \mathbf{w}_{i,k-1} + \alpha\, \varepsilon_{i,k}^{o}(\lambda)\, \mathbf{P}_{i,k}(\lambda)\, \hat{\mathbf{h}}_k \tag{4.4.18}$$

where $\hat{\mathbf{h}}_k$ is the temporary input pattern defined by (4.4.1) and (4.4.2), and $\varepsilon_{i,k}^{o}(\lambda)$ is defined by (4.4.8). The matrix $\mathbf{P}_{i,k}(\lambda)$ is evaluated recursively by

$$\mathbf{P}_{i,k}(\lambda) = \mathbf{P}_{i,k-1}(\lambda) - \delta_{i,k}(\lambda)\, \mathbf{P}_{i,k-1}(\lambda)\, \hat{\mathbf{h}}_k\, \hat{\mathbf{h}}_k^*\, \mathbf{P}_{i,k-1}(\lambda) \tag{4.4.19}$$

where $\delta_{i,k}(\lambda) = c_{i,k}(\lambda)(1 + c_{i,k}(\lambda)\, \hat{\mathbf{h}}_k^*\, \mathbf{P}_{i,k-1}(\lambda)\, \hat{\mathbf{h}}_k)^{-1}$, $c_{i,k}(\lambda)$ is given by (4.3.17) and (4.3.18), and $\mathbf{P}_{i,0}(\lambda) = \mathbf{I}$.

The adaptation of the synaptic weights v_{pq} of the lower network with respect to the association $(\mathbf{y}_k, \mathbf{x}_k)$ can be carried out by minimizing the objective function $G_k(\lambda)$, defined by (4.3.4), using the gradient descent method. This function can also be written as $G_k(\lambda) = \bar{G}_k^i(\lambda) + G_k^i(\lambda)$, where $\bar{G}_k^i(\lambda) = \sum_{j \neq i}^{n_o} y_{j,k}(y_{j,k} - \hat{y}_{j,k}) +$

$\frac{1}{2}\lambda \sum_{j \neq i}^{n_o}(\hat{y}_{j,k}^2 - y_{j,k}^2)$, and

$$G_k^i(\lambda) = y_{i,k}(y_{i,k} - \hat{y}_{i,k}) + \frac{1}{2}\lambda(\hat{y}_{i,k}^2 - y_{i,k}^2) \qquad (4.4.20)$$

Obviously, $\bar{G}_k^i(\lambda) = \bar{G}_k^i(\mathbf{w}_j \;\forall\; j \neq i, \lambda)$ and $G_k^i(\lambda) = G_k^i(\mathbf{w}_i, \lambda)$. Since only $\mathbf{w}_i$ was updated, the adaptation of the synaptic weights v_{pq} can be achieved by simply minimizing $G_k^i(\lambda)$. It is shown in Appendix A4.4 that this minimization results in the following update equation for the synaptic weights v_{pq}

$$\mathbf{v}_{p,k} = \mathbf{v}_{p,k-1} + \alpha\, \varepsilon_{pi,k}^h(\lambda)\, \mathbf{x}_k \qquad (4.4.21)$$

where

$$\varepsilon_{pi,k}^h(\lambda) = (1 - \hat{h}_{p,k}^2)\, \varepsilon_{i,k}^o(\lambda)\, w_{ip} \qquad (4.4.22)$$

and $\varepsilon_{i,k}^o(\lambda)$ is defined by (4.4.8). The resulting algorithm, called Fast ELEANNE 5, is summarized in the flowchart presented in Table 4.5.

The algorithm can be generalized for multi-layered networks with more than one hidden layers. Assuming that $\mathbf{x}_k$ is the input of the network with L hidden layers, shown in Figure 2.10, the input of the upper network is the temporary input pattern $\hat{\mathbf{h}}_k^{(1)}$, whose elements can be obtained from (4.4.5) for $r = 1$. According to the analysis presented in section 4.3, the ith row of the matrix of synaptic weights $\mathbf{V}^{(0)}$ can be updated by the equation

$$\mathbf{v}_{i,k}^{(0)} = \mathbf{v}_{i,k-1}^{(0)} + \alpha\, \varepsilon_{i,k}^{(0)}(\lambda)\, \mathbf{P}_{i,k}(\lambda)\, \hat{\mathbf{h}}_k^{(1)} \qquad (4.4.23)$$

where $\varepsilon_{i,k}^{(0)}(\lambda) = \varepsilon_{i,k}^o(\lambda)$ is defined by (4.4.8). The matrix $\mathbf{P}_{i,k}(\lambda)$ is evaluated recursively by

$$\mathbf{P}_{i,k}(\lambda) = \mathbf{P}_{i,k-1}(\lambda) - \delta_{i,k}(\lambda)\, \mathbf{P}_{i,k-1}(\lambda)\, \hat{\mathbf{h}}_k^{(1)}\, \hat{\mathbf{h}}_k^{(1)*}\, \mathbf{P}_{i,k-1}(\lambda) \qquad (4.4.24)$$

where $\delta_{i,k}(\lambda) = c_{i,k}(\lambda)(1 + c_{i,k}(\lambda)\, \hat{\mathbf{h}}_k^{(1)*}\, \mathbf{P}_{i,k-1}(\lambda)\, \hat{\mathbf{h}}_k^{(1)})^{-1}$, $c_{i,k}(\lambda)$ is given by (4.3.17) and (4.3.18), and $\mathbf{P}_{i,0}(\lambda) = \mathbf{I}$. The synaptic weights $v_{pq}^{(r)}, r = 1, 2, \ldots, L$ can be updated by minimizing the objective function $G_k^i(\lambda)$, defined by (4.4.20). It is shown in Appendix A4.4 that each set of synaptic weights $v_{pq}^{(r)}, r = 1, 2, \ldots, L$ can be updated by

$$\mathbf{v}_{p,k}^{(r)} = \mathbf{v}_{p,k-1}^{(r)} + \alpha\, \varepsilon_{pi,k}^{(r)}(\lambda)\, \hat{\mathbf{h}}_k^{(r+1)} \;\forall\; r = 1, 2, \ldots, L \qquad (4.4.25)$$

Start
Initialize **W** and **V** with random values
Select μ
$\lambda = 1$
1 $E = 0$
$i = 0$ (i: index of output unit)
2 $i \leftarrow i + 1$
$\bar{\mathbf{w}} = \mathbf{w}_i$
$\mathbf{P} = \mathbf{I}$
$k = 0$ (k: number of association)
3 $k \leftarrow k + 1$
$\mathbf{x} = \mathbf{x}_k$
$y_i = y_{i,k}$
$\hat{h}_j = \rho(\sum_{\ell=1}^{n_i} v_{j\ell}\, x_\ell)$
$\hat{y} = \sigma(\sum_{j=1}^{n_h} \bar{w}_j\, \hat{h}_j)$
$d(\lambda) = 1 - \lambda\, \hat{y}^2$
$c(\lambda) = [(\lambda/3)\, d(1) + (2/3)\, d(\lambda)]\, d(1)$
$\varepsilon^o(\lambda) = (1 - \hat{y}^2)(y - \lambda\, \hat{y})$
$\delta(\lambda) = c(\lambda)(1 + c(\lambda)\, \hat{\mathbf{h}}^* \mathbf{P}\, \hat{\mathbf{h}})^{-1}$
$\mathbf{P} \leftarrow \mathbf{P} - \delta(\lambda)\, \mathbf{P}\, \hat{\mathbf{h}}\, \hat{\mathbf{h}}^* \mathbf{P}$
$\bar{\mathbf{w}} \leftarrow \bar{\mathbf{w}} + \alpha\, \varepsilon^o(\lambda)\, \mathbf{P}\, \hat{\mathbf{h}}$
$\varepsilon_j^h(\lambda) = (1 - \hat{h}_j^2)\varepsilon^o(\lambda)\, \bar{w}_j$
$\mathbf{v}_j \leftarrow \mathbf{v}_j + \alpha\, \varepsilon^h(\lambda)\, \mathbf{x}$
$\hat{h}_j = \rho(\sum_{\ell=1}^{n_i} v_{j\ell}\, x_\ell)$
$\hat{y} = \sigma(\sum_{j=1}^{n_h} \bar{w}_j\, \hat{h}_j)$
$E \leftarrow E + \frac{1}{2}\,(y - \hat{y})^2$
if: $k < m$; then: go to 2
$\mathbf{w}_i = \bar{\mathbf{w}}$
if: $k < m$; then: go to 2
$\lambda = \exp(-\mu/E^2)$
if: $E > E_o$; then: go to 1
Stop

Table 4.5: Flowchart of Fast ELEANNE 5.

where $\hat{\mathbf{h}}_k^{(L+1)} = \mathbf{x}_k$,

$$\varepsilon_{pi,k}^{(1)}(\lambda) = (1 - \hat{h}_{p,k}^{(1)\,2})\, \varepsilon_{i,k}^{o}(\lambda)\, v_{ip}^{(0)} \tag{4.4.26}$$

and

$$\varepsilon_{pi,k}^{(r)}(\lambda) = (1 - \hat{h}_{p,k}^{(r)\,2}) \sum_{\ell=1}^{n_{r-1}} \varepsilon_{\ell i,k}^{(r-1)}(\lambda)\, v_{\ell p}^{(r-1)} \;\forall\, r = 2, 3, \ldots, L \tag{4.4.27}$$

4.4.3 Fast ELEANNE 6

Consider the multi-layered neural network with binary output and one layer of hidden units, shown in Figure 2.9. The training of this network can also be performed in adaptation cycles, each one consisting of m steps. Assume that during the νth adaptation cycle the synaptic weights w_{pq} of the upper network and v_{pq} of the lower network have been updated with respect to the associations $(\mathbf{y}_1, \mathbf{x}_1), (\mathbf{y}_2, \mathbf{x}_2), \ldots, (\mathbf{y}_{k-1}, \mathbf{x}_{k-1})$. The adaptation of the same synaptic weights with respect to the association $(\mathbf{y}_k, \mathbf{x}_k)$ can be achieved by:

- updating all the rows of the matrix $\mathbf{W}$ using Fast ELEANNE 4, and

- updating the synaptic weights v_{pq} of the lower network on the basis of the Delta Rule.

Consider here the adaptation of the matrix of synaptic weights $\mathbf{W}$ of the upper network with respect to the association $(\mathbf{y}_k, \mathbf{x}_k)$, during the νth adaptation cycle. Assuming that $\mathbf{x}_k$ is the input of the lower network, the input of the upper network is the temporary input pattern $\hat{\mathbf{h}}_k$, defined by the combination of (4.4.1) and (4.4.2). The matrix of synaptic weights $\mathbf{W}$ of the upper network can be updated with respect to the temporary association $(\mathbf{y}_k, \hat{\mathbf{h}}_k)$ using Fast ELEANNE 4. According to the analysis presented in section 4.3, the n_o rows of $\mathbf{W}$ are updated with respect to the temporary association $(\mathbf{y}_k, \hat{\mathbf{h}}_k)$ through the equations

$$\mathbf{w}_{i,k} = \mathbf{w}_{i,k-1} + \alpha\, \varepsilon_{i,k}^{o}(\lambda)\, \mathbf{P}_k^{\nu}(\lambda)\, \hat{\mathbf{h}}_k \;\forall\, i = 1, 2, \ldots, n_o \tag{4.4.28}$$

Start
Initialize **W** and **V** with random values
Select μ
$\lambda = 1$
$\nu = 0$ (ν: number of adaptation cycle)
1 $\nu \leftarrow \nu + 1$
$E = 0$
$\mathbf{P} = \mathbf{I}$
$k = 0$ (k: number of association)
2 $k \leftarrow k + 1$
$\mathbf{x} = \mathbf{x}_k$
$\mathbf{y} = \mathbf{y}_k$
$\hat{h}_j = \rho(\sum_{\ell=1}^{n_i} v_{j\ell}\, x_\ell)$
$d(\lambda) = \exp(-(\lambda+1)\beta_k^2(\nu-1)^2)$
$c(\lambda) = (\lambda/3)\, d(1) + (2/3)\, d(\lambda)$
$\delta(\lambda) = c(\lambda)(1 + c(\lambda)\, \hat{\mathbf{h}}^* \mathbf{P}\, \hat{\mathbf{h}})^{-1}$
$\mathbf{P} \leftarrow \mathbf{P} - \delta(\lambda)\, \mathbf{P}\, \hat{\mathbf{h}}\, \hat{\mathbf{h}}^* \mathbf{P}$
$\hat{y}_i = \sigma(\sum_{j=1}^{n_h} w_{ij}\, \hat{h}_j)$
$\varepsilon_i^o(\lambda) = (1 - \hat{y}_i^2)(y_i - \lambda\, \hat{y}_i)$
$\mathbf{w}_i \leftarrow \mathbf{w}_i + \alpha\, \varepsilon_i^o(\lambda) \mathbf{P}\, \hat{\mathbf{h}}$
$\varepsilon_j^h(\lambda) = (1 - \hat{h}_j^2) \sum_{i=1}^{n_o} \varepsilon_i^o(\lambda)\, w_{ij}$
$\mathbf{v}_j \leftarrow \mathbf{v}_j + \alpha\, \varepsilon_j^h(\lambda)\, \mathbf{x}$
$\hat{h}_j = \rho(\sum_{\ell=1}^{n_i} v_{j\ell}\, x_\ell)$
$\hat{y}_i = \sigma(\sum_{j=1}^{n_h} w_{ij}\, \hat{h}_j)$
$E \leftarrow E + \frac{1}{2} \sum_{i=1}^{n_o} (y_i - \hat{y}_i)^2$
if: $k < m$; then: go to 2
$\lambda = \exp(-\mu/E^2)$
if: $E > E_o$; then: go to 1
Stop

Table 4.6: Flowchart of Fast ELEANNE 6.

where $\hat{\mathbf{h}}_k$ is the temporary input pattern (4.4.2) whose elements are given by (4.4.1) and $\varepsilon^o_{i,k}(\lambda)$ is given by (4.4.8). $\mathbf{P}_k = \mathbf{P}^\nu_k(\lambda)$ is evaluated in terms of $\mathbf{P}_{k-1} = \mathbf{P}^\nu_{k-1}(\lambda)$ by the following equation

$$\mathbf{P}^\nu_k(\lambda) = \mathbf{P}^\nu_{k-1}(\lambda) - \delta^\nu_k(\lambda)\,\mathbf{P}^\nu_{k-1}(\lambda)\,\hat{\mathbf{h}}_k\,\hat{\mathbf{h}}^*_k\,\mathbf{P}^\nu_{k-1}(\lambda) \qquad (4.4.29)$$

where $\delta^\nu_k(\lambda) = c^\nu_k(\lambda)(1 + c^\nu_k(\lambda)\,\hat{\mathbf{h}}^*_k\,\mathbf{P}^\nu_{k-1}(\lambda)\,\hat{\mathbf{h}}_k)^{-1}$, $c^\nu_k(\lambda)$ is defined in terms of ν by (4.3.25), and $\mathbf{P}^\nu_0(\lambda) = \mathbf{I}$.

The simplified algorithm used for the adaptation of the synaptic weights w_{pq} of the upper network allows the simultaneous evaluation of all the rows of the matrix of synaptic weights $\mathbf{W}$. As a result, the synaptic weights v_{pq} of the lower network can be updated with respect to the association $(\mathbf{y}_k, \mathbf{x}_k)$ by minimizing the objective function $G_k(\lambda)$, defined by (4.3.4). It can easily be seen that the resulting update equation is identical to that of the Fast Back Propagation algorithm derived earlier; that is, v_{pq} can be updated with respect to $(\mathbf{y}_k, \mathbf{x}_k)$ by the combination of (4.4.9) and (4.4.10). The resulting algorithm, called Fast ELEANNE 6, is summarized in the flowchart presented in Table 4.6. The generalization of this algorithm to neural networks with more than one layer of hidden units results in the same update equations derived above for the generalization of the Fast Back Propagation algorithm.

4.4.4 Fast ELEANNE 7

This algorithm is derived by replacing the n_o matrices $\mathbf{H}_{i,m}(\lambda), i = 1, 2, \ldots, n_o$ by their average and using the strategy which provided the basis for the development of the Fast ELEANNE 6. Consider the feed-forward neural network with binary output and one layer of hidden units shown in Figure 2.9. The matrices $\mathbf{H}_{i,m}(\lambda), i = 1, 2, \ldots, n_o$ can be evaluated in terms of the temporary input patterns $\hat{\mathbf{h}}_k$ on the basis of the analysis presented in section 4.3. If the output of the network is binary, i.e., $\sigma(x) = \tanh(x)$, these matrices can be approximated by

$$\mathbf{H}_{i,m}(\lambda) = \sum_{k=1}^{m} c_{i,k}(\lambda)\,\hat{\mathbf{h}}_k\,\hat{\mathbf{h}}^*_k \;\forall\, i = 1, 2, \ldots, n_o \qquad (4.4.30)$$

where $c_{i,k}(\lambda)$ is defined by (4.3.17) and (4.3.18). The computational burden imposed by the existence of n_o distinct matrices can be significantly moderated if each of these matrices is replaced by their average, that is,

$$\bar{\mathbf{H}}_m(\lambda) = \frac{1}{n_o} \sum_{i=1}^{n_o} \mathbf{H}_{i,m}(\lambda) = \sum_{k=1}^{m} \bar{c}_k(\lambda)\, \hat{\mathbf{h}}_k\, \hat{\mathbf{h}}_k^* \tag{4.4.31}$$

where

$$\bar{c}_k(\lambda) = \frac{1}{n_o} \sum_{i=1}^{n_o} c_{i,k}(\lambda) = \frac{1}{n_o} \sum_{i=1}^{n_o} [(\lambda/3)\, d_{i,k}(1) + (2/3)\, d_{i,k}(\lambda)]\, d_{i,k}(1) \tag{4.4.32}$$

and $d_{i,k}(\lambda) = 1 - \lambda\, \hat{y}_{i,k}^2$. Under this simplifying assumption, the efficient adaptation of the synaptic weights requires the recursive evaluation of $\bar{\mathbf{P}}_m(\lambda) = \bar{\mathbf{H}}_m(\lambda)^{-1}$.

Since this algorithm only requires,the recursive evaluation of a single matrix inverse, it can be formulated on the basis of the same strategy that provided the Fast ELEANNE 6. Suppose that during a certain adaptation cycle the synaptic weights of the network have been updated with respect to the associations $(\mathbf{y}_1, \mathbf{x}_1)$, $(\mathbf{y}_2, \mathbf{x}_2)$, ..., $(\mathbf{y}_{k-1}, \mathbf{x}_{k-1})$, resulting to the estimates $\mathbf{W} = \mathbf{W}_{k-1}$ and $\mathbf{V} = \mathbf{V}_{k-1}$. The rows $\mathbf{w}_i, i = 1, 2, \ldots, n_o$ of $\mathbf{W}$ can be updated with respect to the association $(\mathbf{y}_k, \mathbf{x}_k)$ by

$$\mathbf{w}_{i,k} = \mathbf{w}_{i,k-1} + \alpha\, \varepsilon_{i,k}^o(\lambda)\, \bar{\mathbf{P}}_k(\lambda)\, \hat{\mathbf{h}}_k \;\; \forall\, i = 1, 2, \ldots, n_o \tag{4.4.33}$$

where $\hat{\mathbf{h}}_k$ is the temporary input pattern and $\varepsilon_{i,k}^o(\lambda)$ is given by (4.4.8). $\bar{\mathbf{P}}_k(\lambda)$ can be evaluated recursively in terms of $\bar{\mathbf{P}}_{k-1}(\lambda)$ by

$$\bar{\mathbf{P}}_k(\lambda) = \bar{\mathbf{P}}_{k-1}(\lambda) - \bar{\delta}_k(\lambda)\, \bar{\mathbf{P}}_{k-1}(\lambda)\, \hat{\mathbf{h}}_k\, \hat{\mathbf{h}}_k^*\, \bar{\mathbf{P}}_{k-1}(\lambda) \tag{4.4.34}$$

where $\bar{\delta}_k(\lambda) = \bar{c}_k(\lambda)(1 + \bar{c}_k(\lambda)\, \hat{\mathbf{h}}_k^*\, \bar{\mathbf{P}}_{k-1}(\lambda)\, \hat{\mathbf{h}}_k)^{-1}$, $\bar{c}_k(\lambda)$ is given by (4.4.32), and $\bar{\mathbf{P}}_0(\lambda) = \mathbf{I}$.

After the adaptation of all the synaptic weights w_{pq} by (4.4.33), the synaptic weights v_{pq} of the lower network can be updated with respect to the same association $(\mathbf{y}_k, \mathbf{x}_k)$ by the combination of (4.4.9) and (4.4.10). The resulting algorithm, called Fast ELEANNE 7, is summarized in the flowchart presented in Table 4.7. The algorithm can be generalized to networks with more than one layer of hidden units as described in section 4.4.3.

Start
Initialize **W** and **V** with random values
Select μ
$\lambda = 1$
$\mathbf{P} = \mathbf{I}$
1 $\quad k = 0$ (k: number of association)
$\quad E = 0$
2 $\quad k \leftarrow k + 1$

$$\mathbf{x} = \mathbf{x}_k$$
$$\mathbf{y} = \mathbf{y}_k$$
$$\hat{h}_j = \rho(\textstyle\sum_{\ell=1}^{n_i} v_{j\ell}\, x_\ell)$$
$$d_i(\lambda) = 1 - \lambda \hat{y}_i^2$$
$$c_i(\lambda) = [(\lambda/3)\, d_i(1) + (2/3)\, d_i(\lambda)] d_i(1)$$
$$\bar{c}(\lambda) = (1/n_o) \textstyle\sum_{i=1}^{n_o} c_i(\lambda)$$
$$\bar{\delta}(\lambda) = \bar{c}(\lambda)(1 + \bar{c}(\lambda)\, \hat{\mathbf{h}}^*\, \bar{\mathbf{P}}\, \hat{\mathbf{h}})^{-1}$$
$$\bar{\mathbf{P}} \leftarrow \bar{\mathbf{P}} - \bar{\delta}(\lambda)\, \bar{\mathbf{P}}\, \hat{\mathbf{h}}\, \hat{\mathbf{h}}^*\, \bar{\mathbf{P}}$$
$$\hat{y}_i = \sigma(\textstyle\sum_{j=1}^{n_h} w_{ij}\, \hat{h}_j)$$
$$\varepsilon_i^o(\lambda) = (1 - \hat{y}_i^2)(y_i - \lambda\, \hat{y}_i)$$
$$\mathbf{w}_i \leftarrow \mathbf{w}_i + \alpha\, \varepsilon_i^o(\lambda) \bar{\mathbf{P}}\, \hat{\mathbf{h}}$$
$$\varepsilon_j^h(\lambda) = (1 - \hat{h}_j^2) \textstyle\sum_{i=1}^{n_o} \varepsilon_i^o(\lambda)\, w_{ij}$$
$$\mathbf{v}_j \leftarrow \mathbf{v}_j + \alpha\, \varepsilon_j^h(\lambda)\, \mathbf{x}$$
$$\hat{h}_j = \rho(\textstyle\sum_{\ell=1}^{n_i} v_{j\ell}\, x_\ell)$$
$$\hat{y}_i = \sigma(\textstyle\sum_{j=1}^{n_h} w_{ij}\, \hat{h}_j)$$
$$E \leftarrow E + \tfrac{1}{2} \textstyle\sum_{i=1}^{n_o} (y_i - \hat{y}_i)^2$$

if: $k < m$; then: go to 2
$\lambda = \exp(-\mu/E^2)$
if: $E > E_o$; then: go to 1
Stop

Table 4.7: Flowchart of Fast ELEANNE 7.

4.5 Experimental Results

In this section, the fast learning algorithms developed in this chapter are compared to existing algorithms on the basis of the number of adaptation cycles required by each algorithm to achieve a prescribed small value of the total error E, defined by (4.2.1). The analysis presented in this chapter indicates that each learning algorithm and the corresponding fast algorithm require approximately the same number of arithmetic operations per adaptation cycle. Therefore, the relationship between the number of adaptation cycles required by each algorithm and its corresponding fast algorithm to reach the same value of the total error also indicates the relationship of the time required by the two algorithms to carry out the same training task. An experimental evaluation of the fast learning algorithms for single-layered neural networks was presented in (Karayiannis, 1991a; Karayiannis and Venetsanopoulos, 1991). This section focuses on the experimental evaluation of the fast algorithms presented in this chapter for training multi-layered neural networks.

The first set of experiments compares the Error Back Propagation and the Fast Back Propagation algorithms in the case where they are used to train a multi-layered neural network with analog output. The two algorithms were used to train an auto-associator network, i.e., one which maps its analog input to its output. In this set of experiments, a discrete signal of length 800 was normalized between -1 and $+1$ and then divided into 25 blocks of length 32. The training set formed by these blocks was used to train a neural network with 32 inputs, 24 hidden units, and 32 output units. The initial set of synaptic weights was provided by a random number generator producing numbers between -0.5 and $+0.5$. Figure 4.1 shows the total error in dB as a function of the number of adaptation cycles when the network was trained by the Error Back Propagation and the Fast Back Propagation algorithms. The learning rate in these experiments was $\alpha = 0.03$. The parameter β which determines the shape of the nonlinearity $\phi_1(.)$, defined by (4.2.3), varied from $\beta = 1$ to $\beta = 6$. The error curve corresponding to $\beta = 1$ was identical to that provided by the Error Back Propagation algorithm. This is not surprising, since in this case $\phi_1(x) \approx \frac{1}{2}\, x^2$ and

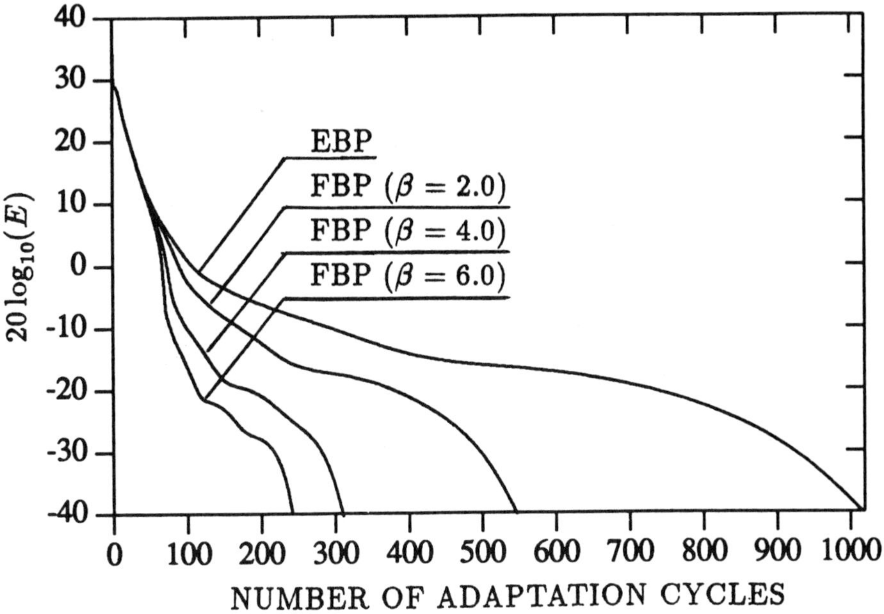

Figure 4.1: Training a multi-layered neural network with analog output using the Error Back Propagation (EBP) and Fast Back Propagation (FBP) algorithms. $20\log_{10}(E)$ as a function of the number of adaptation cycles.

the generalized objective function (4.2.2) is almost identical to the quadratic error function (4.2.1). As the value of β increases, the Fast Back Propagation algorithm converges faster to the target error value than the Error Back Propagation algorithm. However, if β exceeds a certain value, the error starts to oscillate. It was found that in this particular experiment the Fast Back Propagation algorithm oscillates if $\beta > 6$. If the value of β is very large, the Fast Back Propagation algorithm does not converge. In fact, there is a tradeoff between the learning rate and the maximum value of β such that as the learning rate is decreased, the maximum value of β which does not exhibit oscillations increases. The tradeoff between the the learning rate and the parameter β is clearly exhibited in Table 4.8, which shows the number of adaptation cycles required by the Error Back Propagation and Fast Back Propagation algorithms to reduce the total error to $E = 0.01$. It should be emphasized here that the

Algorithm	Error $E = 0.01$
EBP ($\alpha = 0.07$)	410
FBP ($\alpha = 0.07$; $\beta = 2.5$)	176
EBP ($\alpha = 0.05$)	672
FBP ($\alpha = 0.05$; $\beta = 3.0$)	236
EBP ($\alpha = 0.03$)	1017
FBP ($\alpha = 0.03$; $\beta = 6.0$)	242
EBP ($\alpha = 0.01$)	2565
FBP ($\alpha = 0.01$; $\beta = 20.0$)	388

Table 4.8: Number of adaptation cycles required for training a multi-layered neural network with analog output by the Error Back Propagation and Fast Back Propagation algorithms.

selection of a very small target value of the total error is particularly important in the case of networks with analog outputs. The learning rate varied in these experiments from $\alpha = 0.01$ to $\alpha = 0.07$, while the value of β shown in Table 4.8 was the maximum allowable in each case. The Fast Back Propagation converges faster than the Error Back Propagation algorithm in all cases. As the learning rate is decreased and the maximum allowable value of β increases, the difference in convergence becomes even more significant.

The remainder of this section presents an evaluation of the fast algorithms proposed in this chapter for feed-forward neural networks with binary output. In the first experiment, the Error Back Propagation and the Fast Back Propagation algorithms were used to train a neural network consisting of two inputs, two hidden units and one output unit to perform the Exclusive OR operation. The training set consisted of the four associations $(\mathbf{y}_k, \mathbf{x}_k), k = 1,2,3,4$, where $\mathbf{x}_1^* = (-1,-1), y_1 = -1$, $\mathbf{x}_2^* = (-1,+1), y_2 = +1$, $\mathbf{x}_3^* = (+1,-1), y_3 = +1$ and $\mathbf{x}_4^* = (+1,+1), y_4 = -1$. In each adaptation cycle, 20 combinations of the associations $(y_k, \mathbf{x}_k), k = 1,2,3,4$ were presented to the network. The initial set of synaptic weights was provided by a random number generator producing numbers between -0.5 and $+0.5$. The learning rate used in these experiments was $\alpha = 0.01$. The parameter λ was calculated during the training

Algorithm	Error $E = 0.1$	Error $E = 0.01$
EBP ($\alpha = 0.01$)	690	2603
FBP ($\alpha = 0.01; \mu = 0.1$)	513	610
FBP ($\alpha = 0.01; \mu = 1.0$)	488	584
FBP ($\alpha = 0.01; \mu = 10.0$)	479	575

Table 4.9: Number of adaptation cycles required for training a multi-layered neural network to perform the Exclusive OR operation by the Error Back Propagation and Fast Back Propagation algorithms.

by the Fast Back Propagation algorithm from (4.2.19), with $\mu = 0.1$, $\mu = 1$ and $\mu = 10$. Table 4.9 shows the number of adaptation cycles required by the Error Back Propagation and the Fast Back Propagation algorithms to reach the total error value of $E = 0.1$ and $E = 0.01$. According to Table 4.9, the Fast Back Propagation algorithm reaches both target error values faster than the Error Back Propagation algorithm, and the increase of μ results in a slight reduction in the number of adaptation cycles required by the Fast Back Propagation algorithm.

In the next experiment, a multi-layered neural network was trained to function as an n to n encoder. The network consisted of 10 inputs, 5 hidden units and 10 output units. The initial set of synaptic weights was provided by a random number generator producing numbers between -0.5 and $+0.5$. Figure 4.2 shows the error in dB as a function of the number of adaptation cycles when this network was trained by the Error Back Propagation and the Fast Back Propagation algorithms. The learning rate used was $\alpha = 0.03$. The parameter λ was calculated during the training in terms of the total error E by (4.2.19) with $\mu = 1$. Figure 4.2 indicates that the Fast Back Propagation algorithm converges to $E = 0.01$ ($20 \log_{10}(E) = -40$) significanty faster than the Error Back Propagation algorithm. In fact, the Fast Back Propagation also converges faster than the Error Back Propagation when the target error value is $E = 0.1$ ($20 \log_{10}(E) = -20$). This experiment is also useful for illustrating the relationship between the rate at which the parame-

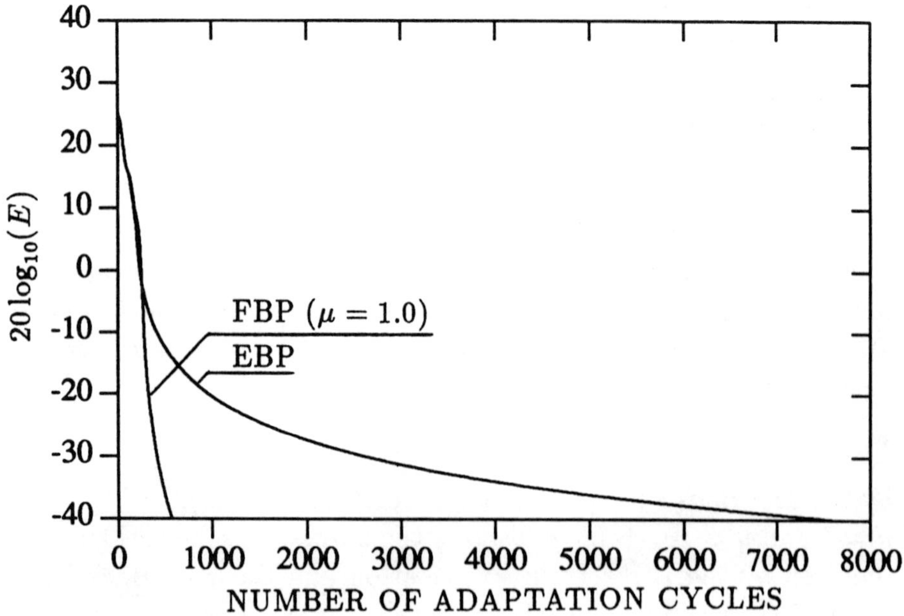

Figure 4.2: Training a multi-layered neural network with binary output to function as an n to n encoder using the Error Back Propagation (EBP) and Fast Back Propagation (FBP) algorithms. $20\log_{10}(E)$ as a function of the number of adaptation cycles.

ter λ is reduced and the learning rate. Figure 4.3 shows the error in dB as a function of the number of adaptation cycles when the same network was trained using the Fast Back Propagation algorithm with a learning rate $\alpha = 0.06$. The two curves show the behavior of the algorithm when λ is computed from (4.2.19), with $\mu = 1$ and $\mu = 0.1$. Because of the larger learning rate, the algorithm is sensitive to the rate at which the parameter λ is reduced during training. In the case where $\mu = 1$, the error oscillates after a certain adaptation cycle. It should be noted that these oscillations are magnified in this Figure by the logarithmic error scale. Figure 4.3 also indicates that the oscillations disappear when the value of μ is reduced to from $\mu = 1$ to $\mu = 0.1$, suggesting that the undesirable oscillations of the error during training can be eliminated by reducing the value of μ. Figure 4.2 indicates that the oscillations can also be eliminated by slightly reducing the learning rate. The performance

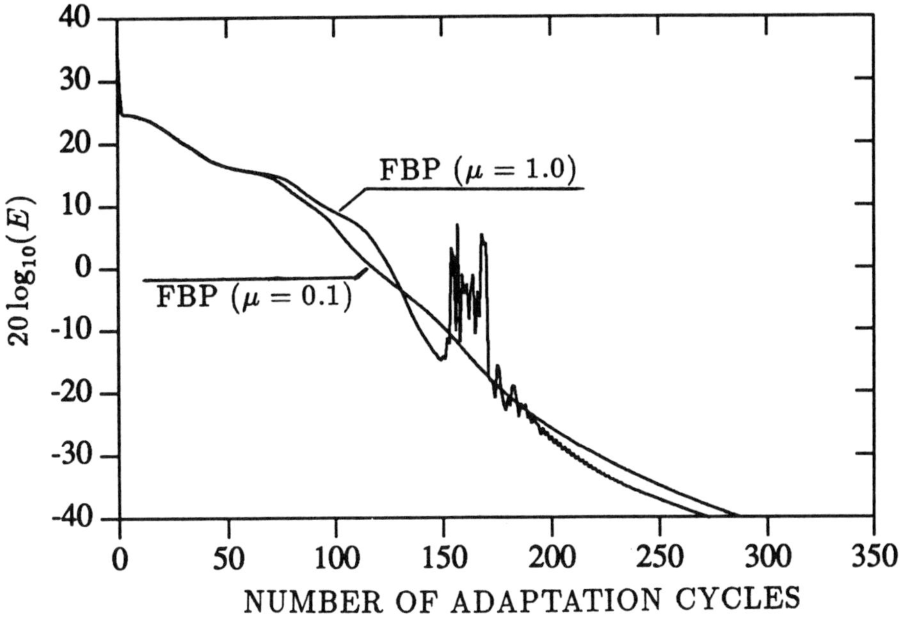

Figure 4.3: Training a multi-layered neural network with binary output to function as an n to n encoder using the Fast Back Propagation (FBP) algorithm with $\mu = 0.1$ and $\mu = 1.0$. $20\log_{10}(E)$ as a function of the number of adaptation cycles.

of the two algorithms has also been tested on a network required to perform an even more difficult mapping. A neural network with 15 inputs, 5 hidden units and 15 output units was trained by the Error Back Propagation and the Fast Back Propagation algorithms to function as an 15 to 15 encoder. Table 4.10 shows the number of adaptation cycles required by the two algorithms to reach the total error values $E = 0.1$ and $E = 0.01$ when the learning rate was varied from $\alpha = 0.01$ to $\alpha = 0.05$. Table 4.10 also shows the value of μ used to compute λ during the training of the network by the Fast Back Propagation algorithm. According to the table, the Fast Back Propagation algorithm converges to both $E = 0.1$ and $E = 0.01$ faster than the Error Back Propagation algorithm. The difference in convergence is significantly greater when the target error value is $E = 0.01$. As the learning rate is decreased, the Fast Back Propagation algorithm represents a very significant reduction

Algorithm	Error $E = 0.1$	Error $E = 0.01$
EBP ($\alpha = 0.05$)	1685	14486
FBP ($\alpha = 0.05; \mu = 1.0$)	401	778
EBP ($\alpha = 0.03$)	2783	24367
FBP ($\alpha = 0.03; \mu = 1.0$)	647	1280
EBP ($\alpha = 0.01$)	8316	72823
FBP ($\alpha = 0.01; \mu = 1.0$)	1965	3858

Table 4.10: Number of adaptation cycles required for training a multi-layered neural network to function as an n to n encoder by the Error Back Propagation and Fast Back Propagation algorithms.

in total time required to train the network.

The same training task was used to compare the efficient learning algorithms for multi-layered neural networks and their fast versions presented in this chapter. The initial set of synaptic weights was provided by a random number generator producing numbers between -0.5 and $+0.5$. Figure 4.4 shows the error in dB as a function of the number of adaptation cycles when the network was trained using the ELEANNE 5 algorithm with $\eta = 1$ and the Fast ELEANNE 5 algorithm. The learning rate used in this experiment was $\alpha = 0.08$. According to Figure 4.4, the Fast ELEANNE 5 algorithm converges much faster than the ELEANNE 5 after the initial adaptation cycles. On the other hand, the ELEANNE 5 achieves faster convergence than the Fast Back Propagation algorithm. However, Fast ELEANNE 5 is the most demanding of the fast algorithms proposed in this chapter for multi-layered neural networks in terms of the number of arithmetic operations required per adaptation cycle. Figure 4.5 shows the error in dB as a function of the number of adaptation cycles when the same multi-layered neural network was trained using ELEANNE 6 with $\eta = 1$ and Fast ELEANNE 6. The learning rate was $\alpha = 0.06$, and $\beta_k = \beta = 0.01 \, \forall \, k = 1, 2, \ldots, m$. The comparison of Figures 4.4 and 4.5 indicates that the Fast ELEANNE 5 algorithm converges faster than the Fast ELEANNE 6 algorithm. However, there is a tradeoff between the number of arithmetic operations per adaptation cycle

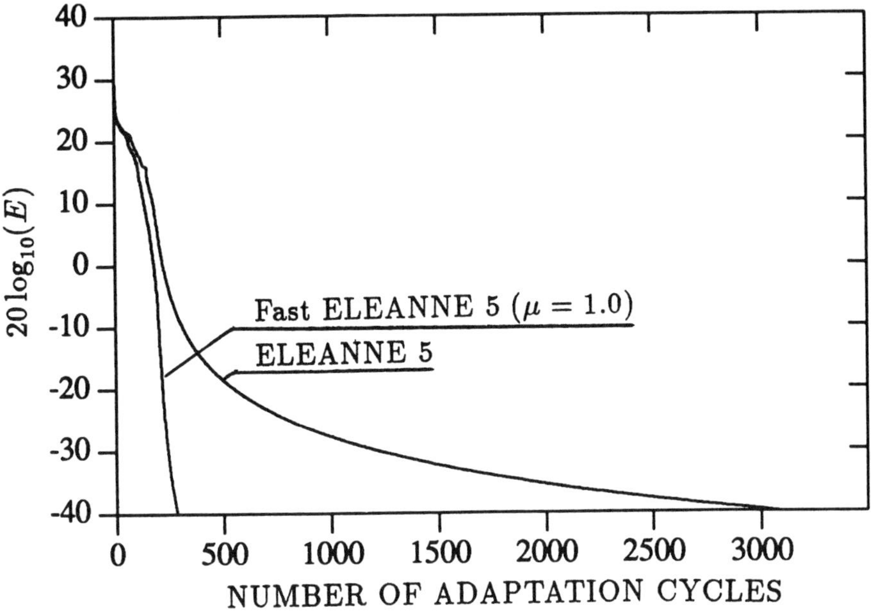

Figure 4.4: Training a multi-layered neural network with binary output to function as an n to n encoder using the ELEANNE 5 and Fast ELEANNE 5 algorithms. $20\log_{10}(E)$ as a function of the number of adaptation cycles.

required by each algorithm and the corresponding rate of convergence. The comparison of the ELEANNE 5 and the ELEANNE 6 algorithms in chapter 3 indicated that ELEANNE 5 performs almost n_o times the number of arithmetic operations per adaptation cycle required by ELEANNE 6. Finally, Figure 4.6 shows the error in dB as a function of the number of adaptation cycles when the same training task was performed using the ELEANNE 7 algorithm with $\eta = 1$ and the Fast ELEANNE 7 algorithm. The learning rate in this case was $\alpha = 0.09$. Fast ELEANNE 5 converges faster than Fast ELEANNE 7, but Fast ELEANNE 7 is computationally less demanding than Fast ELEANNE 5. In fact, Fast ELEANNE 7 requires the same number of arithmetic operations per adaptation cycle as Fast ELEANNE 6. The error curves shown in Figure 4.6 are qualitatively very similar to those generated by the ELEANNE 6 and the Fast ELEANNE 6 algorithms. This suggests that the Fast

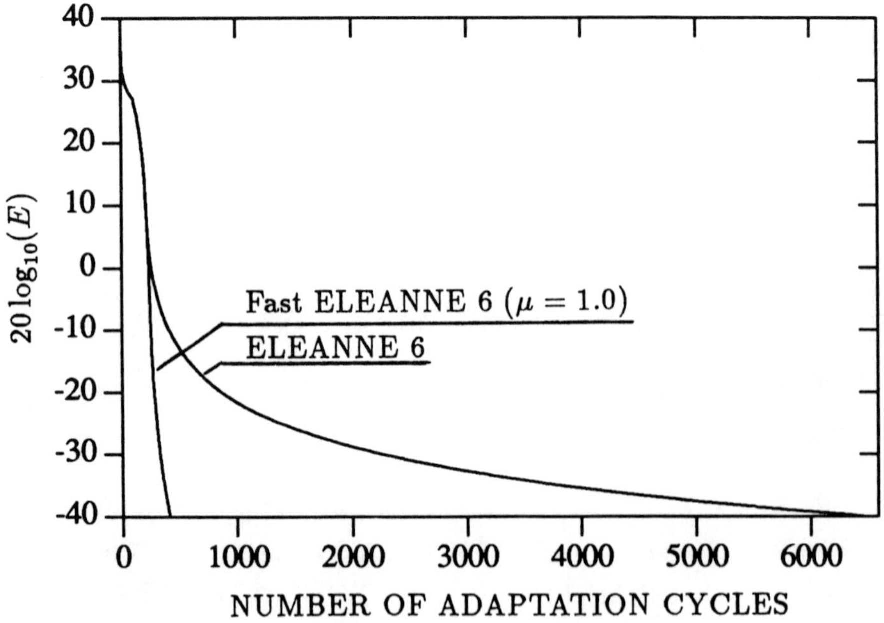

Figure 4.5: Training a multi-layered neural network with binary output to function as an n to n encoder using the ELEANNE 6 and Fast ELEANNE 6 algorithms. $20\log_{10}(E)$ as a function of the number of adaptation cycles.

ELEANNE 7 algorithm is a better alternative than Fast ELEANNE 6, since the application of the Fast ELEANNE 6 algorithm requires the determination of the parameters $\beta_k \, \forall \, k = 1, 2, \ldots, m$, in addition to the learning rate.

4.6 Discussion

This chapter presented a new generalized criterion for training feed-forward neural networks. This criterion provided the basis for the development of several fast learning algorithms for both single-layered and multi-layered neural networks. It was experimentally verified that the proposed algorithms train neural networks much faster than the existing algorithms. The importance of this contribution becomes apparent when one considers the excessive com-

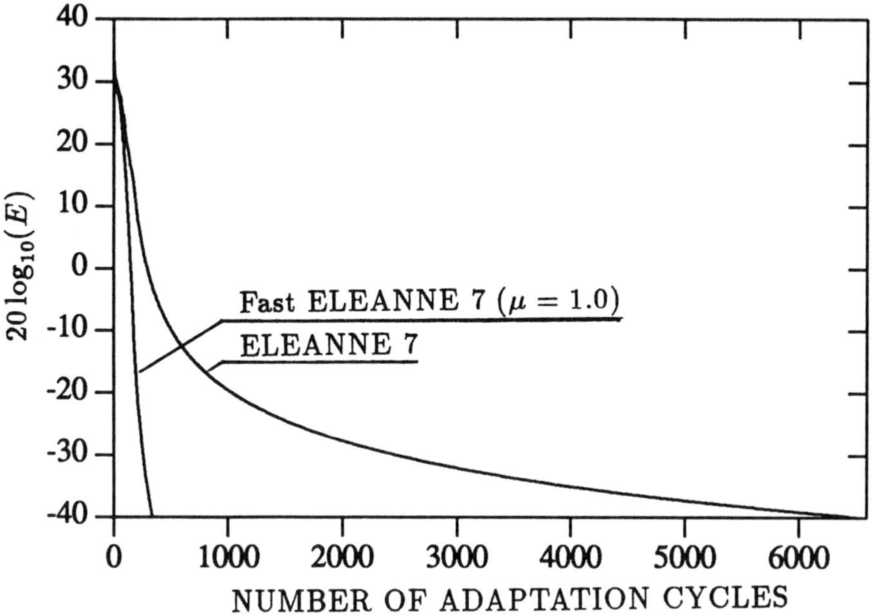

Figure 4.6: Training a multi-layered neural network with binary output to function as an n to n encoder using the ELEANNE 7 and Fast ELEANNE 7 algorithms. $20\log_{10}(E)$ as a function of the number of adaptation cycles.

puter time required by existing algorithms to train the relatively large neural networks found in practical applications. These algorithms will have a strong impact on the development of systems based on feed-forward neural networks. However, the generality of the proposed criterion suggests that its application is not restricted to the training of feed-forward neural networks. Its potential application to the training of alternative neural network architectures is an interesting problem for future research.

Appendix A4.1

The Fast Delta Rule learning algorithm is derived here by minimizing the objective function $G_k(\lambda)$, defined by (4.3.4), using the gradient descent method. The update equation for the synaptic weights w_{pq} of the network is given by

$$\begin{aligned}\mathbf{w}_{p,k} - \mathbf{w}_{p,k-1} &= -\alpha \frac{\partial G_k(\lambda)}{\partial \mathbf{w}_p} \\ &= -\alpha \sum_{i=1}^{n_o} [\lambda \phi_2'(e_{i,k}) + (1-\lambda)\phi_1'(e_{i,k})] \frac{\partial e_{i,k}}{\partial \mathbf{w}_p}\end{aligned} \tag{A4.1.1}$$

According to the definition of $\hat{y}_{i,k}$ in (4.3.3),

$$\frac{\partial e_{i,k}}{\partial \mathbf{w}_p} = -\frac{\partial \hat{y}_{i,k}}{\partial \mathbf{w}_p} = -\frac{d\sigma(\bar{y}_{i,k})}{d\bar{y}_{i,k}} \frac{\partial}{\partial \mathbf{w}_p} (\sum_{j=1}^{n_i} w_{ij} x_{j,k}) = -\sigma'(\bar{y}_{i,k})\, \mathbf{x}_k\, \delta_{ip} \tag{A4.1.2}$$

where δ_{ip} is defined as $\delta_{ip} = 1$ if $i = p$ and $\delta_{ip} = 0$ if $i \neq p$. The combination of (A4.1.1) and (A4.1.2) gives

$$\mathbf{w}_{p,k} = \mathbf{w}_{p,k-1} + \alpha\, \varepsilon_{p,k}^o(\lambda)\, \mathbf{x}_k \tag{A4.1.3}$$

where

$$\varepsilon_{p,k}^o(\lambda) = \sigma'(\bar{y}_{p,k})\, [\lambda \phi_2'(e_{p,k}) + (1-\lambda)\phi_1'(e_{p,k})] \tag{A4.1.4}$$

If the output of the network is analog, $\sigma(x) = x$ and $\sigma'(x) = 1$. In this case, $\phi_1(.)$ is defined by $\phi_1(x) = (1/\beta)\ln(\cosh \beta x)$. It can easily be verified that $\phi_1'(x) = \tanh(\beta x)$. Also, $\phi_2(x) = \frac{1}{2}x^2$ and $\phi_2'(x) = x$. In this case, the synaptic weights w_{pq} can be updated by (A4.1.3), where

$$\varepsilon_{p,k}^o(\lambda) = \lambda(y_{p,k} - \hat{y}_{p,k}) + (1-\lambda)\tanh[\beta(y_{p,k} - \hat{y}_{p,k})] \tag{A4.1.5}$$

If the output of the network is binary, $\sigma(x) = \tanh(x)$. It can easily be verified that $\sigma'(\bar{y}_{p,k}) = 1 - \sigma(\bar{y}_{p,k})^2 = 1 - \hat{y}_{p,k}^2$. According to (4.2.5), the function $\phi_1(.)$ corresponding to the pth unit is given by $\phi_1(e_{p,k}) = y_{p,k}\, e_{p,k}$. Clearly, $\phi_1'(e_{p,k}) = y_{p,k}$. Also, $\phi_2(e_{p,k}) = \frac{1}{2}e_{p,k}^2$ and $\phi_2'(e_{p,k}) = e_{p,k}$. In this case, the synaptic weights w_{pq} can be updated by (A4.1.3), where

$$\begin{aligned}\varepsilon_{p,k}^o(\lambda) &= (1 - \hat{y}_{p,k}^2)[\lambda(y_{p,k} - \hat{y}_{p,k}) + (1-\lambda)y_{p,k}] \\ &= (1 - \hat{y}_{p,k}^2)(y_{p,k} - \lambda\, \hat{y}_{p,k})\end{aligned} \tag{A4.1.6}$$

Appendix A4.2

The Fast ELEANNE 3 algorithm is derived here by evaluating the vector $\mathbf{b}_{i,k}$ which results from the minimization of $G_{i,m}(\lambda)$ and the corresponding Hessian matrix. According to (4.3.3),

$$\frac{\partial \hat{y}_{i,k}}{\partial w_{iq}} = (1 - \hat{y}_{i,k}^2)\, x_{q,k} \tag{A4.2.1}$$

Substituting (A4.2.1) into the definition of $\mathbf{b}_{i,k}$ in (4.3.12) gives

$$\mathbf{b}_{i,k} = (1 - \hat{y}_{i,k}^2)\, \mathbf{x}_k \tag{A4.2.2}$$

The elements of the Hessian matrix which corresponds to $G_{i,m}(\lambda)$ are given by

$$(\mathbf{H}_{i,m}(\lambda))_{pq} = \frac{\partial^2 G_{i,m}(\lambda)}{\partial w_{ip} \partial w_{iq}} = \lambda \sum_{k=1}^{m} \frac{\partial \hat{y}_{i,k}}{\partial w_{ip}} \frac{\partial \hat{y}_{i,k}}{\partial w_{iq}} - \sum_{k=1}^{m} e_{i,k}(\lambda) \frac{\partial^2 \hat{y}_{i,k}}{\partial w_{ip} \partial w_{iq}} \tag{A4.2.3}$$

where $e_{i,k}(\lambda) = y_{i,k} - \lambda \hat{y}_{i,k}$. The second-order partial derivative can be evaluated from (A4.2.1) and (4.3.3) as

$$\frac{\partial^2 \hat{y}_{i,k}}{\partial w_{ip} \partial w_{iq}} = -2\, x_{q,k}\, \hat{y}_{i,k} \frac{\partial \hat{y}_{i,k}}{\partial w_{ip}} = -2\, x_{p,k}\, x_{q,k}\, \hat{y}_{i,k}\, (1 - \hat{y}_{i,k}^2) \tag{A4.2.4}$$

The elements of $\mathbf{H}_{i,m}(\lambda)$ can be obtained by substituting (A4.2.1) and (A4.2.4) into (A4.2.3) as follows

$$(\mathbf{H}_{i,m}(\lambda))_{pq} = \sum_{k=1}^{m} c_{i,k}(\lambda)\, x_{p,k}\, x_{q,k} \tag{A4.2.5}$$

where

$$c_{i,k}(\lambda) = [\lambda\, (1 - \hat{y}_{i,k}^2) + 2\, \hat{y}_{i,k} (y_{i,k} - \lambda \hat{y}_{i,k})](1 - \hat{y}_{i,k}^2) \tag{A4.2.6}$$

It was suggested in section 4.2 that the training process be initialized by setting $\lambda = 1$. It can also be assumed that when the Hessian matrix is evaluated far from the optimum weights, $\hat{y}_{i,k} \approx 0$. Under this assumption,

$$\begin{aligned} \mathbf{H}_{i,m}(\lambda) &= \sum_{k=1}^{m} [1 + 2\hat{y}_{i,k}(y_{i,k} - \lambda \hat{y}_{i,k})(1 - \hat{y}_{i,k}^2)^{-1}](1 - \hat{y}_{i,k}^2)^2 \mathbf{x}_k \mathbf{x}_k^* \\ &\approx \sum_{k=1}^{m} (1 - \hat{y}_{i,k}^2)^2 \mathbf{x}_k \mathbf{x}_k^* \end{aligned} \tag{A4.2.7}$$

When the Hessian matrix is evaluated near to the optimum, it can be assumed that $y_{i,k}\,\hat{y}_{i,k} \approx 1$. Under this assumption, the Hessian matrix can be approximated by

$$\mathbf{H}_{i,m}(\lambda) = \sum_{k=1}^{m} c_{i,k}(\lambda)\,\mathbf{x}_k\,\mathbf{x}_k^* \tag{A4.2.8}$$

where

$$c_{i,k}(\lambda) = [\lambda\, d_{i,k}(1) + 2\, d_{i,k}(\lambda)]\, d_{i,k}(1) \tag{A4.2.9}$$

and

$$d_{i,k}(\lambda) = 1 - \lambda\, \hat{y}_{i,k}^2 \tag{A4.2.10}$$

The development of an easily realizable algorithm requires the derivation of the same set of update equations for any $0 \leq \lambda \leq 1$. This requirement implies the search for a $c_{i,k} = c_{i,k}(\lambda)$, which guarantees that the approximation of the Hessian matrix by (A4.2.8) is valid for any $0 \leq \lambda \leq 1$. In the case where $\lambda = 0$, (A4.2.9) amounts to $c_{i,k}(0) = 2\,(1 - \hat{y}_{i,k}^2)$. During the initial adaptation cycles, i.e., when $\lambda \approx 1$, (A4.2.9) gives

$$c_{i,k}(1) = 3\,(1 - \hat{y}_{i,k}^2)^2 \tag{A4.2.11}$$

The approximation of the Hessian matrix provided by the combination of (A4.2.8) and (A4.2.11) is of the same form as (A4.2.7) with the exception of the multiplicative constant 3. A reliable approximation of the Hessian matrix is very crucial during the initial adaptation cycles, i.e., when $\lambda = 1$. In order for (A4.2.8) to be consistent with (A4.2.7) in the case where $\lambda = 1$, the Hessian matrix is approximated for $0 \leq \lambda \leq 1$ by (A4.2.8), where

$$c_{i,k}(\lambda) = [(\lambda/3)\, d_{i,k}(1) + (2/3)\, d_{i,k}(\lambda)]\, d_{i,k}(1) \tag{A4.2.12}$$

and $d_{i,k}(\lambda)$ is defined by (A4.2.10).

Appendix A4.3

The Fast Back Propagation algorithm is derived here by minimizing the objective function $G_k(\lambda)$, defined by (4.3.4). The update equation for the synaptic weights w_{pq} is given by

$$\begin{aligned}\mathbf{w}_{p,k} - \mathbf{w}_{p,k-1} &= -\alpha \frac{\partial G_k(\lambda)}{\partial \mathbf{w}_p} \\ &= -\alpha \sum_{i=1}^{n_o} [\lambda \phi_2'(e_{i,k}) + (1-\lambda)\phi_1'(e_{i,k})] \frac{\partial e_{i,k}}{\partial \mathbf{w}_p}\end{aligned} \tag{A4.3.1}$$

From the definition of $\hat{y}_{i,k}$ in (4.4.3),

$$\begin{aligned}\frac{\partial e_{i,k}}{\partial \mathbf{w}_p} &= -\frac{\partial \hat{y}_{i,k}}{\partial \mathbf{w}_p} \\ &= -\frac{d\sigma(\bar{y}_{i,k})}{d\bar{y}_{i,k}} \frac{\partial}{\partial \mathbf{w}_p} (\sum_{j=1}^{n_h} w_{ij} \hat{h}_{j,k}) = -\sigma'(\bar{y}_{i,k})\, \hat{\mathbf{h}}_k \,\delta_{ip}\end{aligned} \tag{A4.3.2}$$

Substituting (A4.3.2) into (A4.3.1) gives

$$\mathbf{w}_{p,k} = \mathbf{w}_{p,k-1} + \alpha\, \varepsilon^o_{p,k}(\lambda)\, \hat{\mathbf{h}}_k \tag{A4.3.3}$$

where

$$\varepsilon^o_{p,k}(\lambda) = \sigma'(\bar{y}_{p,k})\, [\lambda \phi_2'(e_{p,k}) + (1-\lambda)\phi_1'(e_{p,k})] \tag{A4.3.4}$$

The update equation for the synaptic weights v_{pq} can be similarly obtained. From the definition of $G_k(\lambda)$,

$$\begin{aligned}\mathbf{v}_{p,k} - \mathbf{v}_{p,k-1} &= -\alpha \frac{\partial G_k(\lambda)}{\partial \mathbf{v}_p} \\ &= -\alpha \sum_{i=1}^{n_o} [\lambda \phi_2'(e_{i,k}) + (1-\lambda)\phi_1'(e_{i,k})] \frac{\partial e_{i,k}}{\partial \mathbf{v}_p}\end{aligned} \tag{A4.3.5}$$

From the definition of $\hat{y}_{i,k}$ in (4.4.3),

$$\frac{\partial e_{i,k}}{\partial \mathbf{v}_p} = -\frac{\partial \hat{y}_{i,k}}{\partial \mathbf{v}_p} = -\frac{d\sigma(\bar{y}_{i,k})}{d\bar{y}_{i,k}} \frac{\partial}{\partial \mathbf{v}_p} (\sum_{j=1}^{n_h} w_{ij} \hat{h}_{j,k}) = -\sigma'(\bar{y}_{i,k}) \sum_{j=1}^{n_h} w_{ij} \frac{\partial \hat{h}_{j,k}}{\partial \mathbf{v}_p} \tag{A4.3.6}$$

Finally, the definition of $\hat{h}_{j,k}$ in (4.4.1) gives

$$\frac{\partial \hat{h}_{j,k}}{\partial \mathbf{v}_p} = \frac{d\rho(\bar{h}_{j,k})}{d\bar{h}_{j,k}} \frac{\partial}{\partial \mathbf{v}_p} (\sum_{\ell=1}^{n_i} v_{j\ell} x_{\ell,k}) = \rho'(\bar{h}_{j,k})\, \mathbf{x}_k\, \delta_{jp} \tag{A4.3.7}$$

The combination of (A4.3.6) and (A4.3.7) with (A4.3.5) gives the following update equation for the synaptic weights v_{pq}

$$\mathbf{v}_{p,k} = \mathbf{v}_{p,k-1} + \alpha\, \varepsilon^h_{p,k}(\lambda)\, \mathbf{x}_k \tag{A4.3.8}$$

where

$$\varepsilon^h_{p,k}(\lambda) = \rho'(\bar{h}_{p,k}) \sum_{i=1}^{n_o} \sigma'(\bar{y}_{i,k})\, [\lambda \phi_2'(e_{i,k}) + (1-\lambda)\phi_1'(e_{i,k})]\, w_{ip} \tag{A4.3.9}$$

If the output of the network is analog, $\sigma(x) = x$ and $\sigma'(x) = 1$, while $\rho(x) = \tanh(x)$ and $\rho'(x) = 1 - \rho(x)^2$. In this case, $\phi_1(.)$ is given by $\phi_1(x) = (1/\beta)\ln(\cosh \beta x)$. It can easily be verified that $\phi_1'(x) = \tanh(\beta x)$. Also, $\phi_2(x) = \frac{1}{2}x^2$ and $\phi_2'(x) = x$. In this case, the synaptic weights w_{pq} can be updated by (A4.3.3), where

$$\varepsilon^o_{p,k}(\lambda) = \lambda(y_{p,k} - \hat{y}_{p,k}) + (1-\lambda)\tanh[\beta(y_{p,k} - \hat{y}_{p,k})] \tag{A4.3.10}$$

The synaptic weights v_{pq} can be updated through the equation (A4.3.8), where

$$\varepsilon^h_{p,k}(\lambda) = (1 - \hat{h}^2_{p,k}) \sum_{i=1}^{n_o} \varepsilon^o_{i,k}(\lambda)\, w_{ip} \tag{A4.3.11}$$

and $\varepsilon^o_{p,k}(\lambda)$ is given by (A4.3.10).

If the output of the network is binary, $\sigma(x) = \tanh(x)$ and $\sigma'(x) = 1 - \sigma(x)^2$. In addition, $\rho(x) = \tanh(x)$ and $\rho'(x) = 1 - \rho(x)^2$. According to (4.2.5), the function $\phi_1(.)$ corresponding to the pth unit is given by $\phi_1(e_{p,k}) = y_{p,k}\, e_{p,k}$. Clearly, $\phi_1'(e_{p,k}) = y_{p,k}$. Also, $\phi_2(e_{p,k}) = \frac{1}{2}e^2_{p,k}$ and $\phi_2'(e_{p,k}) = e_{p,k}$. In this case, the synaptic weights w_{pq} can be updated by (A4.3.3), and (A4.3.4) simplifies to

$$\varepsilon^o_{p,k}(\lambda) = (1 - \hat{y}^2_{p,k})(y_{p,k} - \lambda\, \hat{y}_{p,k}) \tag{A4.3.12}$$

Finally, the update equation for the synaptic weights v_{pq} is given by (A4.3.8), where

$$\varepsilon^h_{p,k}(\lambda) = (1 - \hat{h}^2_{p,k}) \sum_{i=1}^{n_o} \varepsilon^o_{i,k}(\lambda)\, w_{ip} \tag{A4.3.13}$$

and $\varepsilon^o_{i,k}(\lambda)$ is given by (A4.3.12).

A generalization of the above analysis leads to the Fast Back Propagation algorithm for training neural networks with more than one layer of hidden units, such as the network shown in Figure 2.10. The estimates provided by this network are given by (4.4.3), which is identical to (4.4.4) for $\hat{\mathbf{h}}^{(1)}_k = \hat{\mathbf{h}}_k$. As a result, the update equation for the synaptic weights w_{pq} can be obtained from (A4.3.3) for $\hat{\mathbf{h}}_k = \hat{\mathbf{h}}^{(1)}_k$ as follows

$$\mathbf{w}_{p,k} = \mathbf{w}_{p,k-1} + \alpha\, \varepsilon^o_{p,k}(\lambda)\, \hat{\mathbf{h}}^{(1)}_k \tag{A4.3.14}$$

where $\varepsilon^{(0)}_{p,k}(\lambda) = \varepsilon^o_{p,k}(\lambda)$ is given by (A4.3.10) if the output of the network is analog and (A4.3.12) if the output is binary. The synaptic weights $v^{(r)}_{pq}, r = 1, 2, \ldots, L$ can be updated by minimizing the objective function $G_k(\lambda)$, defined by (4.3.4), through the equation

$$\begin{aligned} \mathbf{v}^{(r)}_{p,k} - \mathbf{v}^{(r)}_{p,k-1} &= -\alpha \frac{\partial G_k(\lambda)}{\partial \mathbf{v}^{(r)}_p} \\ &= -\alpha \sum_{i=1}^{n_o} [\lambda \phi_2'(e_{i,k}) + (1-\lambda)\phi_1'(e_{i,k})] \frac{\partial e_{i,k}}{\partial \mathbf{v}^{(r)}_p} \end{aligned} \tag{A4.3.15}$$

It can be shown that (A4.3.15) results in the following update equation for the synaptic weights $v^{(1)}_{pq}$

$$\mathbf{v}^{(1)}_{p,k} = \mathbf{v}^{(1)}_{p,k-1} + \alpha\, \varepsilon^{(1)}_{p,k}(\lambda)\, \hat{\mathbf{h}}^{(2)}_k \tag{A4.3.16}$$

where

$$\varepsilon^{(1)}_{p,k}(\lambda) = (1 - \hat{h}^{(1)\,2}_{p,k}) \sum_{i=1}^{n_o} \varepsilon^o_{i,k}(\lambda)\, w_{ip} \tag{A4.3.17}$$

and $\varepsilon^o_{i,k}(\lambda)$ is defined by (A4.3.10) or (A4.3.12) if the output of the network is analog or binary, respectively. The direct generalization of the above results provides the following update equation for the synaptic weights $v_{pq}^{(r)}, r = 2, 3, \ldots, L$

$$\mathbf{v}_{p,k}^{(r)} = \mathbf{v}_{p,k-1}^{(r)} + \alpha\, \varepsilon_{p,k}^{(r)}(\lambda)\, \hat{\mathbf{h}}_k^{(r+1)} \tag{A4.3.18}$$

where $\hat{\mathbf{h}}_k^{(L+1)} = \mathbf{x}_k$ and

$$\varepsilon_{p,k}^{(r)}(\lambda) = (1 - \hat{h}_{p,k}^{(r)\,2}) \sum_{\ell=1}^{n_{r-1}} \varepsilon_{\ell,k}^{(r-1)}(\lambda)\, v_{\ell p}^{(r-1)} \;\forall\, r = 2, 3, \ldots, L \tag{A4.3.19}$$

Appendix A4.4

The derivation of the Fast ELEANNE 5 algorithm is completed here by considering the adaptation of the synaptic weights v_{pq} of the lower network with respect to the association $(\mathbf{y}_k, \mathbf{x}_k)$. This can be achieved by simply minimizing $G_k^i(\lambda)$. According to the gradient descent method, v_{pq} can be updated by the following equation

$$\mathbf{v}_{p,k} - \mathbf{v}_{p,k-1} = -\alpha \frac{\partial G_k^i(\lambda)}{\partial \mathbf{v}_p} = \alpha\, e_{i,k}(\lambda) \frac{\partial \hat{y}_{i,k}}{\partial \mathbf{v}_p} \tag{A4.4.1}$$

The combination of (A4.3.6) and (A4.3.7) with (A4.4.1) gives the following update equation for the synaptic weights v_{pq} of the lower network

$$\mathbf{v}_{p,k} = \mathbf{v}_{p,k-1} + \alpha\, \varepsilon_{pi,k}^{h}(\lambda)\, \mathbf{x}_k \tag{A4.4.2}$$

where

$$\varepsilon_{pi,k}^{h}(\lambda) = (1 - \hat{h}_{p,k}^2) \varepsilon_{i,k}^{o}(\lambda)\, w_{ip} \tag{A4.4.3}$$

and $\varepsilon_{i,k}^{o}(\lambda)$ is defined by (4.4.8).

A similar analysis provides the Fast ELEANNE 5 algorithm for the multi-layered neural network with L layers of hidden units. The synaptic weights $v_{pq}^{(r)}, r = 1, 2, \ldots, L$ can be updated by minimizing the objective function $G_k^i(\lambda)$, defined by (4.4.20). According to the gradient descent method, the synaptic weights $v_{pq}^{(r)}$ can be updated by the equation

$$\mathbf{v}_{p,k}^{(r)} - \mathbf{v}_{p,k-1}^{(r)} = -\alpha \frac{\partial G_k^i(\lambda)}{\partial \mathbf{v}_p^{(r)}} = \alpha\, e_{i,k}(\lambda) \frac{\partial \hat{y}_{i,k}}{\partial \mathbf{v}_p^{(r)}} \tag{A4.4.4}$$

It can be shown that (A4.4.4) results in the following update equation for the synaptic weights $v_{pq}^{(1)}$

$$\mathbf{v}_{p,k}^{(1)} = \mathbf{v}_{p,k-1}^{(1)} + \alpha\, \varepsilon_{pi,k}^{(1)}(\lambda)\, \hat{\mathbf{h}}_k^{(2)} \tag{A4.4.5}$$

where

$$\varepsilon_{pi,k}^{(1)}(\lambda) = (1 - \hat{h}_{p,k}^{(1)\,2})\, \varepsilon_{i,k}^{o}(\lambda)\, w_{ip} \tag{A4.4.6}$$

and $\varepsilon_{i,k}^{o}(\lambda)$ is defined by (4.4.8). The direct generalization of (A4.4.5) and (A4.4.6) leads to the following update equation for the synaptic weights $v_{pq}^{(r)}, r = 2, 3, \ldots, L$

$$\mathbf{v}_{p,k}^{(r)} = \mathbf{v}_{p,k-1}^{(r)} + \alpha\, \varepsilon_{pi,k}^{(r)}(\lambda)\, \hat{\mathbf{h}}_k^{(r+1)} \tag{A4.4.7}$$

where $\hat{\mathbf{h}}_k^{(L+1)} = \mathbf{x}_k$ and

$$\varepsilon_{pi,k}^{(r)}(\lambda) = (1 - \hat{h}_{p,k}^{(r)\,2}) \sum_{\ell=1}^{n_{r-1}} \varepsilon_{\ell i,k}^{(r-1)}(\lambda)\, v_{\ell p}^{(r-1)} \;\forall\, r = 2, 3, \ldots, L \tag{A4.4.8}$$

Chapter 5

ALADIN: Algorithms for Learning and Architecture DetermINation

5.1 Introduction

There are many multi-layered neural networks which can be used in an application requiring a feed-forward neural network with a fixed number of input and output units. For a given training set, the generalization ability of a multi-layered neural network depends on the number of hidden layers as well as the number of hidden units per layer. There seems to be an agreement in the literature that a neural network with at most two layers of hidden units can approximate any given set of functions, provided that there are enough units per hidden layer (Lapedes and Farber, 1988; Hertz et al., 1991). In addition, it has been shown that a neural network with one layer of enough hidden units can approximate any set of continuous functions (Cybenko, 1989). However, the number of units per hidden layer which guarantees satisfactory performance of the network after training remains an open problem. Some theoretical lower and upper bounds on the sample size versus the size of a network with one layer of hidden units needed for valid generalization have been reported in (Baum and Haussler, 1989). However, there are no reliable criteria for the selection of the number of hidden units per layer required by neural networks with multiple hidden layers. At this stage of research, the selection of the number of hidden layers and the number of hidden units per layer is more an art than a science. In practice, the selection of multi-layered neural networks is usually performed by experimenting with networks of various sizes until the desired network performance is reached. The lack of a procedure for determining the multi-layered network which achieves the best performance in a given application is considered to be one of the major limitations of feed-forward neural networks. This criticism of feed-forward neural networks is fair, since selection of an inadequate network can considerably degrade the performance of the resulting system.

In recent years, there has been considerable research effort towards the development of procedures and algorithms that determine the size and architecture of feed-forward neural networks which guarantee their reliable performance. The development of feed-forward networks of minimum size was attempted in the literature by network construction algorithms (Marchand et al., 1990; Frean, 1990; Mezard and Nadal, 1990) and various pruning or weight decay approaches. These latter approaches involve schemes for eliminating connections or hidden units after training as well as algorithms which eliminate them during the training of the network. Sietsma and Dow attempted to minimize the size of neural networks by selectively pruning hidden units (Sietsma and Dow, 1988; Sietsma and Dow, 1991). According to this approach, a feed-forward neural network much larger than minimum size is initially trained using the Error Back Propagation algorithm. The size of the trained network is subsequently minimized by pruning the "unimportant" hidden units on the basis of various heuristics. Kung and Hu recently proposed a Fobenius approximation method for the reduction of the number of hidden units of a trained multi-layered neural networks (Kung and Hu, 1991). This method minimizes the Fobenius norm of the approximation error by deleting the hidden units contributing the least to the net function of the output layer. The optimal brain damage technique is a method for reducing the size of feed-forward neural networks by selectively deleting synaptic weights (le Cun et al., 1990). This technique uses second-order derivative information to make a tradeoff between the network complexity and the training error. The optimal brain damage technique produces networks with better generalization, reduces the number of examples required for training, and improves the learning speed (le Cun, et al., 1990). The elimination of some "unnecessary" connections during training has been used earlier for reducing the time required for training feed-forward neural networks (Hinton, 1986; Scalettar and Zee, 1988). This can be achieved by forcing the synaptic weights to decay to zero, unless they are reinforced during the training. It was suggested that this strategy can be implemented by replacing the objective function $E = E(\mathbf{w})$, which is conventionally minimized during training, with an alternative objective function of the form $E'(\mathbf{w}) = E(\mathbf{w}) + \gamma\,||\mathbf{w}||^2$, where $||\mathbf{w}||^2$ is the Euclidean norm of the vector $\mathbf{w}$ formed by the synaptic weights of the network (Kramer and Sangiovanni-Vincentelli, 1989). However, the effect of this strategy on the synaptic weights is not uniform, but depends on their relative size (Hertz et al., 1991). This observation motivated the development of alternative rules which guarantee a more uniform decay of the synaptic weights during training (Hertz et al., 1991). These rules result from the minimization of an objective function of the form $E'(\mathbf{w}) = E(\mathbf{w}) + B(\mathbf{w})$, where $B(.)$ is a bias term which imposes additional constraints on the search for the synaptic weights (Hanson and Pratt, 1989). In general, the approaches based on decay of the synaptic weights eliminate connections but not hidden units. It was suggested that the elimination of hidden units can be achieved by making the decay rates larger for units with relatively small outputs or relatively small incoming weights (Chauvin, 1989; Hanson and Pratt, 1989). The constrained search imposed by these approaches does not seem to minimize the number of hidden units required and causes an increase in local minima (Hanson and Pratt, 1989). According to these latter approaches, the hidden units are indirectly eliminated as a result of the elimination of certain connections. Skeletonization is a technique that could be used to simplify neural networks by directly eliminating certain input or hidden units that convey redundant information (Mozer and Smolensky, 1989). This is achieved by iteratively training a network to a certain performance criterion, computing a measure of relevance that indicates which input or hidden units are most critical to performance, and trimming the least relevant units. Hirose et al. proposed a strategy which varies the number of hidden units, in an attempt to force the Error Back Propagation algorithm to escape local minima (Hirose et al., 1991). The addition or deletion of hidden units during the training is regulated by heuristics. In addition to the rather complicated strategy for the addition or removal of hidden units, another disadvantage of this algorithm is that the selection of the final network architecture requires the training of several neural networks which differ in the number of hidden units. This chapter presents the development of *well-defined, computationally efficient,* and *easily reproducible* learning algorithms which are able to determine the architecture of multi-layered neural networks while performing their training (Karayiannis, 1991b; Karayiannis, 1992b).

This chapter is organized as follows: Section 5.2 describes a generalized criterion for training feed-forward neural networks. Section 5.3 presents the development of the ALADIN and Fast ALADIN algorithms for feed-forward neural networks with one layer of hidden units. These algorithms are generalized to feed-forward neural networks with more than one layer of hidden units in section 5.4. Section 5.5 presents an experimental evaluation of the proposed algorithms. Section 5.6 contains concluding remarks.

5.2 Training Criteria

Consider a feed-forward neural network trained with respect to the associations $(\mathbf{y}_k, \mathbf{x}_k), k = 1, 2, \ldots, m$, where $\mathbf{x}_k^* = [x_{1,k}\, x_{2,k} \ldots x_{n_i,k}]$ are the inputs of the network and $\mathbf{y}_k^* = [y_{1,k}\, y_{2,k} \ldots y_{n_o,k}]$ are the corresponding outputs. Neural network training is frequently based on the minimization of the following objective function

$$E = \sum_{k=1}^{m} E_k = \frac{1}{2} \sum_{k=1}^{m} \sum_{i=1}^{n_o} (y_{i,k} - \hat{y}_{i,k})^2 = \frac{1}{2} \sum_{k=1}^{m} \sum_{i=1}^{n_o} e_{i,k}^2 \qquad (5.2.1)$$

where $y_{i,k}$ is the expected output and $\hat{y}_{i,k}$ the corresponding estimate provided by the network. If the output of the network is binary, each $y_{i,k}$ can take the values $+1$ and -1. If the output of the network is analog, the elements $y_{i,k}, i = 1,2,\ldots,n_o$, which form the output patterns $\mathbf{y}_k, k = 1,2,\ldots,m$, are normalized between -1 and $+1$. The minimization of (5.2.1) is usually performed by sequentially adjusting the synaptic weights of the network with respect to each association $(\mathbf{y}_k, \mathbf{x}_k)$. In other words, the minimization of $E = \sum_{k=1}^{m} E_k$ is performed by minimizing $E_k \,\forall\, k = 1,2,\ldots,m$. The training of a feed-forward neural network is performed in *adaptation cycles.* An adaptation cycle consists of the adaptation of all the synaptic weights of the network with respect to all the m associations considered.

The development of fast learning algorithms for feed-forward neural networks, presented in chapter 4, was based on the minimization of an alternative criterion after the initial adaptation cycles. Consider that a feed-forward neural network is trained by minimizing the generalized objective function (Karayiannis, 1992a)

$$\begin{aligned} G(\lambda) &= \lambda E + (1-\lambda)E' \\ &= \lambda \sum_{k=1}^{m}\sum_{i=1}^{n_o} \phi_2(e_{i,k}) + (1-\lambda)\sum_{k=1}^{m}\sum_{i=1}^{n_o} \phi_1(e_{i,k}) \end{aligned} \tag{5.2.2}$$

where $\phi_2(x) = \frac{1}{2}x^2$, and λ decreases from 1 to 0 during the training. If the output of the network is binary,

$$\phi_1(e_{i,k}) = y_{i,k}(y_{i,k} - \hat{y}_{i,k}) \tag{5.2.3}$$

In this case, (5.2.2) can equivalently be written as

$$G(\lambda) = \sum_{k=1}^{m}\sum_{i=1}^{n_o} y_{i,k}(y_{i,k} - \hat{y}_{i,k}) + \frac{1}{2}\lambda \sum_{k=1}^{m}\sum_{i=1}^{n_o}(\hat{y}_{i,k}^2 - y_{i,k}^2) \tag{5.2.4}$$

This is the form of the objective function that was proposed in chapter 4 for training neural networks with binary output (Karayiannis, 1991a; Karayiannis and Venetsanopoulos, 1991).

5.3 Neural Networks with one Hidden Layer

Consider the feed-forward neural network with one layer of n_h hidden units shown in Figure 2.9. Assuming that $\mathbf{x}_k$ is the input to the network, the corresponding outputs of the hidden units are $\hat{h}_{j,k} = \rho(\sum_{\ell=1}^{n_i} v_{j\ell} x_{\ell,k})$, where $\rho(.)$ is a smooth and differentiable everywhere function which approximates the hard-limiting nonlinearity that determines the output of each unit. In this chapter, the hard-limiting nonlinearity is the well-known signum function, defined as $\text{sgn}(x) = +1$ if $x \geq 0$ and $\text{sgn}(x) = -1$ if $x < 0$, and $\rho(.)$ is the hyperbolic tangent, i.e., $\rho(x) = \tanh(x)$. The output of the network is formed by the elements $\hat{y}_{i,k} = \sigma(\sum_{j=1}^{n_h} w_{ij} \hat{h}_{j,k})$, where $\sigma(x) = x$ if the output of the network is analog and $\sigma(x) = \tanh(x)$ if the output of the network is formed by binary elements.

The algorithms presented here determine the architecture of a feed-forward neural network with one layer of hidden units by selectively deactivating redundant hidden units during the training (Karayiannis, 1991b; Karayiannis, 1992b). The development of these algorithms is based on the assumption that the number of hidden units is initially much larger than required for this particular training set. This assumption can always be justified in practice by simply selecting the initial number of hidden units to be very large. The crucial problem here is the derivation of reliable criteria for determining which of the hidden units may be deactivated during each adaptation cycle without significantly affecting the training process and the performance of the trained network. Suppose that the training begins with $n(0) = n_h$ hidden units, where 0 indicates the number of adaptation cycle. Consider the $(\nu + 1)$th adaptation cycle of the learning algorithm, and let $n(\nu)$ be the number of hidden units which are active after the νth adaptation cycle. In addition, let $\mathcal{I}_\nu$ be the set formed by the indices of the hidden units which are active after the νth adaptation cycle. Clearly, $\mathcal{I}_\nu$ consists of positive integers from the interval $[1, n_h]$. During the $(\nu + 1)$th adaptation cycle, the output of the network is formed by the elements

$$\hat{y}_{i,k} = \sigma(\sum_{j \in \mathcal{I}_\nu} w_{ij} \hat{h}_{j,k}) \tag{5.3.1}$$

where

$$\hat{h}_{j,k} = \rho(\sum_{\ell=1}^{n_i} v_{j\ell} x_{\ell,k}) \; \forall j \in \mathcal{I}_\nu \tag{5.3.2}$$

A reliable measure of the effect of the hidden units on the training of the network is the sensitivity of each $G_k(\lambda)$, $k = 1, 2, \ldots, m$ with respect to $\hat{h}_{j,k} \forall j \in \mathcal{I}_\nu$, which is defined here as

$$S(G_k(\lambda), \hat{h}_{j,k}) = |\frac{\partial G_k(\lambda)/\partial \hat{h}_{j,k}}{E_k/\hat{h}_{j,k}}| = |\frac{\hat{h}_{j,k}}{E_k}\frac{\partial G_k(\lambda)}{\partial \hat{h}_{j,k}}| \tag{5.3.3}$$

The sensitivity of $G(\lambda) = \sum_{k=1}^{m} G_k(\lambda)$ with respect to the jth hidden unit can be evaluated by

$$S_j = S_j(\lambda) = \sum_{k=1}^{m} S(G_k(\lambda), \hat{h}_{j,k}) \tag{5.3.4}$$

The development of learning algorithms capable of determining the architecture of a feed-forward neural network while performing its training is based on the requirement that the effect of the active hidden units on the training process be as uniform as possible. The same requirement provided the basis for the development of a preliminary version of the algorithms presented in this chapter (Karayiannis, 1991b; Karayiannis, 1992b). Since the range of $S_j, j \in \mathcal{I}_\nu$ is unknown, the set of $S_j, j \in \mathcal{I}_\nu$ was not directly used for selecting the hidden units that should be deactivated during the $(\nu+1)$th adaptation cycle. Instead, each hidden unit was characterized by a relative sensitivity measure defined by $\bar{S}_j = S_j\,(\sum_{j\in\mathcal{I}_\nu} S_j)^{-1}$. The advantage of using $\bar{S}_j$ instead of S_j as a sensitivity measure for the jth hidden unit is obvious, since the range of each $\bar{S}_j$ is known, i.e., $0 \leq \bar{S}_j \leq 1 \forall j \in \mathcal{I}_\nu$. The assumption that all the active hidden units have the same effect on the training process implies that $\bar{S}_j = \bar{S} \forall j \in \mathcal{I}_\nu$ and, therefore, $\bar{S} = 1/n(\nu)$. This observation motivated the use of the quantity $\bar{S}_t(\gamma) = \gamma\bar{S} = \gamma/n(\nu)$, $0 \leq \gamma \leq 1$, as a threshold in the selection of the hidden units which are deactivated during the $(\nu+1)$th adaptation cycle. In fact, the selection of the hidden units which are deactivated during this adaptation cycle was based on the following test: the jth hidden unit is deactivated if $\bar{S}_j < \bar{S}_t(\gamma) = \gamma/n(\nu) \forall j \in \mathcal{I}_\nu$. If $0 < \gamma < 1$, this test removes the hidden units with the least effect on the training process. If $\gamma = 0$, none of the hidden units are removed and the algorithm trains a neural network of a fixed size and architecture (Karayiannis, 1991b; Karayiannis, 1992b).

According to the deactivation test described above, only the hidden units characterized by low sensitivity measure are deactivated. The deactivation of redundant hidden units depends only on the sample mean of $\bar{S}_j, j \in \mathcal{I}_\nu$, therefore ignoring the variance of the sensitivity measures $S_j, j \in \mathcal{I}_\nu$. The resulting algorithms are very sensitive to the value of γ, especially in difficult training tasks. These observations motivated the development of an alternative test for deactivating of the redundant hidden units, which is presented here.

Let μ_S and σ_S be the sample mean and standard deviation, respectively, of the sensitivity measures $S_j, j \in \mathcal{I}_\nu$ characterizing the active hidden units after the νth adaptation cycle. The requirement that the hidden units have a uniform effect on the training process can be satisfied by deactivating those hidden units whose sensitivity measures lie outside an interval centered on the sample mean μ_S. Assuming an approximate Gaussian distribution, the interval $[\mu_S - \gamma\sigma_S, \mu_S + \gamma\sigma_S]$ is a confidence interval for the mean sensitivity value. For example, the values $\gamma = 1.65$, 1.70, and 2.00 define 90 %, 91 %, and 95 % confidence intervals, respectively. For a given confidence interval, or a value of γ, a hidden unit is deactivated if the corresponding sensitivity measure lies outside the interval $[\mu_S - \gamma\sigma_S, \mu_S + \gamma\sigma_S]$. After the completion of this test, at the $(\nu + 1)$th adaptation cycle the network contains only the $n(\nu)$ active hidden units whose indices belong to the set

$$\mathcal{I}_{\nu+1} = \{j \in \mathcal{I}_\nu : \mu_S - \gamma\sigma_S \leq S_j \leq \mu_S + \gamma\sigma_S\} \qquad (5.3.5)$$

The majority of the redundant hidden units are deactivated quickly during the initial adaptation cycles, but a hidden unit can also be deactivated after the algorithm has reached a small error value. Such a deactivation may slow the convergence of the algorithm or may force it into oscillatory behavior. This possibility can be eliminated by a strategy for preventing the unjustified deactivation of hidden units after the initial adaptation cycles. It was experimentally observed that the unjustified deactivation of a hidden unit is followed by a significant fluctuation of the sample standard deviation of the sensitivity measures. This is intuitively reasonable, since the deactivation of a hidden unit after the initial adaptation cycles significantly changes the role of the remaining active hidden units in the training process. This observation suggests the following deactivation acceptance test: The deactivation of a hidden unit occurring during the νth adaptation cycle is accepted if

$$|\frac{\sigma_S^\nu - \sigma_S^{\nu+1}}{\sigma_S^\nu}| = |1 - \frac{\sigma_S^{\nu+1}}{\sigma_S^\nu}| \leq \delta \tag{5.3.6}$$

where σ_S^ν is the sample standard deviation of the sensitivity measures during the νth adaptation cycle and $\delta \in (0, 1)$ is a *tolerance parameter* which determines the maximum allowable fluctuation of the sample standard deviation. If the condition (5.3.6), or the equivalent condition $\sigma_S^\nu(1-\delta) \leq \sigma_S^{\nu+1} \leq \sigma_S^\nu(1+\delta)$, is not satisfied, the most recently deactivated hidden unit is reactivated, and no further changes are made to the resulting network during subsequent training. The selection of a significantly large value of the parameter δ guarantees that the algorithm provides the smallest network that can be trained to perform the task described by the training set.

5.3.1 ALADIN

Consider training a network by minimizing of $E_k = G_k(1)$. In this case, the update equations can be obtained from the analysis presented in Appendix A5.1. The synaptic weights w_{ij} can be updated with respect to the association $(\mathbf{y}_k, \mathbf{x}_k)$ through the equation obtained in the case where $\lambda = 1$ from (A5.1.1) as

$$w_{ij,k} = w_{ij,k-1} + \alpha\, \varepsilon_{i,k}^o(1)\, \hat{h}_{j,k}\ \forall j \in \mathcal{I}_\nu \tag{5.3.7}$$

where $\varepsilon_{i,k}^o(1) = (1 - \hat{y}_{i,k}^2)(y_{i,k} - \hat{y}_{i,k})$. For $\lambda = 1$, (A5.1.4) provides the corresponding update equation for the synaptic weights $v_{j\ell}$ as

$$v_{j\ell,k} = v_{j\ell,k-1} + \alpha\, \varepsilon_{j,k}^h(1)\, x_{\ell,k}\ \forall j \in I_\nu \tag{5.3.8}$$

where $\varepsilon_{j,k}^h(1) = (1 - \hat{h}_{j,k}^2) \sum_{i=1}^{n_o} \varepsilon_{i,k}^o(1)\, w_{ij}\ \forall j \in \mathcal{I}_\nu$.

The update equations obtained here are essentially identical to those provided by the Error Back Propagation algorithm (Rumelhart et al., 1986). However, ALADIN requires the adaptation of only those synaptic weights which connect the active hidden units with the input and output units. The sensitivity measure $S(E_k, \hat{h}_{j,k})$

Start
Initialize **W** and **V** with random values
Select α, γ, E^o_{tot}
$n_j = 1 \; \forall j = 1, 2, \ldots, n_h$ ($n_j = 0$: jth hidden unit inactive)
$\nu = 0$ (ν: number of adaptation cycles)
1 $\nu \leftarrow \nu + 1$
$\mathcal{I} = \{j \in [1, n_h] : n_j = 1\}$
$S_j = 0 \;\; \forall j \in \mathcal{I}$
$E_{tot} = 0$
$k = 0$ (k: number of association)
2 $k \leftarrow k + 1$
$\mathbf{x} = \mathbf{x}_k$
$\mathbf{y} = \mathbf{y}_k$
$\hat{h}_j = \rho(\sum_{\ell=1}^{n_i} v_{j\ell}\, x_\ell) \;\; \forall j \in \mathcal{I}$
$\hat{y}_i = \sigma(\sum_{j \in \mathcal{I}} w_{ij}\, \hat{h}_j)$
$\varepsilon^o_i = (1 - \hat{y}_i^2)(y_i - \hat{y}_i)$
$w_{ij} \leftarrow w_{ij} + \alpha\, \varepsilon^o_i\, \hat{h}_j \;\; \forall j \in \mathcal{I}$
$\varepsilon^h_j = (1 - \hat{h}_j^2) \sum_{i=1}^{n_o} \varepsilon^o_i\, w_{ij} \;\; \forall j \in \mathcal{I}$
$v_{j\ell} \leftarrow v_{j\ell} + \alpha\, \varepsilon^h_j\, x_\ell \;\; \forall j \in \mathcal{I}$
$\hat{h}_j = \rho(\sum_{\ell=1}^{n_i} v_{j\ell}\, x_\ell) \;\; \forall j \in \mathcal{I}$
$\hat{y}_i = \sigma(\sum_{j \in \mathcal{I}} w_{ij}\, \hat{h}_j)$
$\varepsilon^o_i = (1 - \hat{y}_i^2)(y_i - \hat{y}_i)$
$E = \frac{1}{2} \sum_{i=1}^{n_o} (y_i - \hat{y}_i)^2$
$S_j(E) = |(\hat{h}_j / E) \sum_{i=1}^{n_o} \varepsilon^o_i\, w_{ij}| \;\; \forall j \in \mathcal{I}$
$S_j \leftarrow S_j + S_j(E)$
$E_{tot} \leftarrow E_{tot} + E$
if: $k < m$; then: go to 2
Evaluate mean μ_S and standard deviation σ_S of $S_j, j \in \mathcal{I}$
if: $S_j < \mu_S - \gamma\sigma_S$ or $S_j > \mu_S + \gamma\sigma_S$; then: $n_j = 0 \;\; \forall j \in \mathcal{I}$
if: $E_{tot} > E^o_{tot}$; then: go to 1
Stop

Table 5.1: Flowchart of ALADIN.

which evaluates the effect of each active hidden unit on the objective function $E_k = G_k(1)$ can be obtained in the case where $\lambda = 1$ by combining (5.3.3) and (A5.1.3) as

$$S(E_k, \hat{h}_{j,k}) = S(G_k(1), \hat{h}_{j,k}) = |(\hat{h}_{j,k}/E_k) \sum_{i=1}^{n_o} \varepsilon_{i,k}^o(1)\, w_{ij}| \qquad (5.3.9)$$

The resulting algorithm, called ALADIN, is summarized in the flowchart presented in Table 5.1. If a value of γ corresponding to a confidence interval of 100 % is used, all the hidden units remain active and ALADIN trains a network of a fixed architecture. In this case, ALADIN coincides with the Error Back Propagation algorithm.

5.3.2 Fast ALADIN

The Fast ALADIN algorithm results from the minimization of $E_k = G_k(\lambda)$, where λ is determined during training. The update equation for the synaptic weights w_{ij} can be directly obtained from Appendix A5.1 as follows

$$w_{ij,k} = w_{ij,k-1} + \alpha\, \varepsilon_{i,k}^o(\lambda)\, \hat{h}_{j,k} \;\forall j \in \mathcal{I}_\nu \qquad (5.3.10)$$

where $\varepsilon_{i,k}^o(\lambda) = (1 - \hat{y}_{i,k}^2)(y_{i,k} - \lambda \hat{y}_{i,k})$. The synaptic weights $v_{j\ell}$ can be updated with respect to the association $(\mathbf{y}_k, \mathbf{x}_k)$ through the equation

$$v_{j\ell,k} = v_{j\ell,k-1} + \alpha\, \varepsilon_{j,k}^h(\lambda)\, x_{\ell,k} \;\forall j \in \mathcal{I}_\nu \qquad (5.3.11)$$

where $\varepsilon_{j,k}^h(\lambda) = (1 - \hat{h}_{j,k}^2) \sum_{i=1}^{n_o} \varepsilon_{i,k}^o(\lambda)\, w_{ij} \;\forall j \in \mathcal{I}_\nu$.

The sensitivity of the objective function $G_k(\lambda)$ with respect to each active hidden unit is measured by $S(G_k(\lambda), \hat{h}_{j,k})$, which can be obtained by combining (5.3.3) and (A5.1.3) as

$$S(G_k(\lambda), \hat{h}_{j,k}) = |(\hat{h}_{j,k}/E_k) \sum_{i=1}^{n_o} \varepsilon_{i,k}^o(\lambda)\, w_{ij}| \qquad (5.3.12)$$

The training of the network begins by minimizing $E_k = G_k(1)$, which corresponds to $\lambda = 1$. After the first adaptation cycle, λ

Start
Initialize $\mathbf{W}$ and $\mathbf{V}$ with random values
Select α, γ, μ, E^o_{tot}
$\lambda = 1$
$n_j = 1 \; \forall j = 1, 2, \ldots, n_h$ ($n_j = 0$: jth hidden unit inactive)
$\nu = 0$ (ν: number of adaptation cycles)
1 $\nu \leftarrow \nu + 1$
$\mathcal{I} = \{j \in [1, n_h] : n_j = 1\}$
$S_j = 0 \;\; \forall j \in \mathcal{I}$
$E_{tot} = 0$
$k = 0$ (k: number of association)
2 $k \leftarrow k + 1$
$\mathbf{x} = \mathbf{x}_k$
$\mathbf{y} = \mathbf{y}_k$
$\hat{h}_j = \rho(\sum_{\ell=1}^{n_i} v_{j\ell}\, x_\ell) \;\; \forall j \in \mathcal{I}$
$\hat{y}_i = \sigma(\sum_{j \in \mathcal{I}} w_{ij}\, \hat{h}_j)$
$\varepsilon^o_i = (1 - \hat{y}_i^2)(y_i - \lambda \hat{y}_i)$
$w_{ij} \leftarrow w_{ij} + \alpha\, \varepsilon^o_i\, \hat{h}_j \;\; \forall j \in \mathcal{I}$
$\varepsilon^h_j = (1 - \hat{h}_j^2) \sum_{i=1}^{n_o} \varepsilon^o_i\, w_{ij} \;\; \forall j \in \mathcal{I}$
$v_{j\ell} \leftarrow v_{j\ell} + \alpha\, \varepsilon^h_j\, x_\ell \;\; \forall j \in \mathcal{I}$
$\hat{h}_j = \rho(\sum_{\ell=1}^{n_i} v_{j\ell}\, x_\ell) \;\; \forall j \in \mathcal{I}$
$\hat{y}_i = \sigma(\sum_{j \in \mathcal{I}} w_{ij}\, \hat{h}_j)$
$\varepsilon^o_i = (1 - \hat{y}_i^2)(y_i - \lambda \hat{y}_i)$
$E = \frac{1}{2} \sum_{i=1}^{n_o} (y_i - \hat{y}_i)^2$
$S_j(E) = |(\hat{h}_j / E) \sum_{i=1}^{n_o} \varepsilon^o_i\, w_{ij}| \;\; \forall j \in \mathcal{I}$
$S_j \leftarrow S_j + S_j(E)$
$E_{tot} \leftarrow E_{tot} + E$
if: $k < m$; then: go to 2
Evaluate mean μ_S and standard deviation σ_S of $S_j, j \in \mathcal{I}$
if: $S_j < \mu_S - \gamma\sigma_S$ or $S_j > \mu_S + \gamma\sigma_S$; then: $n_j = 0 \;\; \forall j \in \mathcal{I}$
$\lambda = \exp(-\mu/(E_{tot})^2)$
if: $E_{tot} > E^o_{tot}$; then: go to 1
Stop

Table 5.2: Flowchart of Fast ALADIN.

is determined at every adaptation cycle in terms of the total error achieved by the algorithm during the previous adaptation cycle as $\lambda = \exp(-\mu/E^2)$, where μ is a positive real number (Karayiannis, 1992a). During the initial adaptation cycles, $E \gg 1$ and, therefore, $\lambda \approx 1$. As the training progresses, the total error E decreases and λ also decreases. Finally, when $E \ll 1$, $\lambda \approx 0$. The resulting algorithm, called Fast ALADIN, is summarized in the flowchart presented in Table 5.2. If a value of γ corresponding to a confidence interval of 100 % is used, the architecture of the network is fixed during the training and Fast ALADIN coincides with the Fast Back Propagation algorithm (Karayiannis, 1992a).

5.4 Neural Networks with Multiple Hidden Layers

The algorithms presented in section 5.3 can be generalized for training feed-forward neural networks with multiple layers of hidden units (Karayiannis, 1991b; Karayiannis, 1992b). Consider the multilayered neural network with L layers of hidden units shown in Figure 2.10 and assume that the input of the network is $\mathbf{x}_k$. Suppose that when the training begins each hidden layer consists of $n^{(r)}(0)\,\forall\, r = 1, 2, \ldots, L$ units. Consider the $(\nu + 1)$th adaptation cycle of the algorithm and let $n^{(r)}(\nu)\,\forall\, r = 1, 2, \ldots, L$ be the number of hidden units in each layer which are active after the νth adaptation cycle. Also let $\mathcal{I}_\nu^{(r)}\,\forall\, r = 1, 2, \ldots, L$ be the sets formed by the indices of the units which are active in each hidden layer after the νth adaptation cycle and $\mathcal{I}_0^{(r)} = \{1, 2, \ldots, n_o\}$. During the $(\nu + 1)$th adaptation cycle, the output of the network consists of the elements

$$\hat{y}_{i,k} = \sigma\Big(\sum_{j \in \mathcal{I}_\nu^{(1)}} v_{ij}^{(0)} \hat{h}_{j,k}^{(1)}\Big) \qquad (5.4.1)$$

The outputs of the units which belong to the hidden layers indexed by $r = 1, 2, \ldots, L$ are given by

$$\hat{h}_{j,k}^{(r)} = \rho\Big(\sum_{\ell \in \mathcal{I}_\nu^{(r)}} v_{j\ell}^{(r)} \hat{h}_{\ell,k}^{(r+1)}\Big)\ \forall\, r = 1, 2, \ldots, L \qquad (5.4.2)$$

where $\hat{h}_{\ell,k}^{(L+1)} = x_{\ell,k} \,\forall\, \ell, k$.

The effect of the jth unit of the rth hidden layer on the objective function $G_k(\lambda)$ can be measured by

$$S(G_k(\lambda), \hat{h}_{j,k}^{(r)}) = \left|\frac{\partial G_k(\lambda)/\partial \hat{h}_{j,k}^{(r)}}{E_k/\hat{h}_{j,k}^{(r)}}\right| = \left|\frac{\hat{h}_{j,k}^{(r)}}{E_k}\frac{\partial G_k(\lambda)}{\partial \hat{h}_{j,k}^{(r)}}\right| \tag{5.4.3}$$

The sensitivity of $G(\lambda) = \sum_{k=1}^{m} G_k(\lambda)$ with respect to the jth unit of the rth hidden layer can be evaluated by $S_j^{(r)} = S_j^{(r)}(\lambda) = \sum_{k=1}^{m} S(G_k(\lambda), \hat{h}_{j,k}^{(r)})$. Let $\mu_S^{(r)}$ and $\sigma_S^{(r)}$ be the sample mean and standard deviation of $S_j^{(r)} \in \mathcal{I}_\nu^{(r)}$, respectively. The selection of the units from each hidden layer to be deactivated during the νth adaptation cycle is based on the following test: the jth unit of the rth hidden layer is deactivated if $\bar{S}_j^{(r)} < \mu_S^{(r)} - \gamma_r \sigma_S^{(r)}$ or $\bar{S}_j^{(r)} > \mu_S^{(r)} + \gamma_r \sigma_S^{(r)}$. After this test, each hidden layer contains only the units whose indices belong to the set $\mathcal{I}_{\nu+1}^{(r)} = \{j \in \mathcal{I}_\nu^{(r)} : \mu_S^{(r)} - \gamma_r \sigma_S^{(r)} \leq S_j^{(r)} \leq \mu_S^{(r)} + \gamma_r \sigma_S^{(r)}\}$. The unjustified deactivation of a hidden unit may be avoided by applying a deactivation acceptance test similar to that presented in section 5.3. The parameters $\gamma_r, r = 1, 2, \ldots, L$ are not necessarily equal. However, the algorithm has the same effect on each layer of hidden units if $\gamma_r = \gamma \,\forall\, r = 1, 2, \ldots, L$.

5.4.1 ALADIN

The generalization of the ALADIN algorithm to the neural network with L layers of hidden units shown in Figure 2.10 requires the update equations for the synaptic weights $v_{j\ell}^{(r)} \,\forall\, r = 0, 1, \ldots, L$, which are derived in Appendix A5.2 by minimizing $G_k(\lambda), k = 1, 2, \ldots, m$. For $\lambda = 1$, (A5.2.6) provides the update equations for the synaptic weights $v_{j\ell}^{(r)} \,\forall\, r = 0, 1, \ldots, L$ as

$$v_{j\ell,k}^{(r)} = v_{j\ell,k-1}^{(r)} + \alpha\, \varepsilon_{j,k}^{(r)}(1)\, \hat{h}_{\ell,k}^{(r+1)} \,\forall\, r = 0, 1, \ldots, L \tag{5.4.4}$$

where $\hat{h}_{\ell,k}^{(L+1)} = x_{\ell,k} \,\forall\, \ell, k$, $\varepsilon_{j,k}^{(0)}(1) = (1 - \hat{y}_{j,k}^2)(y_{j,k} - \hat{y}_{j,k})$, and $\varepsilon_{j,k}^{(r)}(1) = (1 - \hat{h}_{j,k}^{(r)2}) \sum_{i \in \mathcal{I}_\nu^{(r-1)}} \varepsilon_{i,k}^{(r-1)}(1) v_{ij}^{(r-1)} \,\forall\, r = 1, 2, \ldots, L$.

Once again, the resulting update equations are identical to those provided by the Error Back Propagation algorithm. The difference is that the update equations apply only to the active hidden units which are selected on the basis of the sensitivity analysis presented above. The sensitivity measure for the jth unit of the rth layer can be obtained in the case where $\lambda = 1$ by combining (5.4.3) and (A5.2.9) as

$$S(E_k, \hat{h}_{j,k}^{(r)}) = |(\hat{h}_{j,k}^{(r)}/E_k) \sum_{i \in \mathcal{I}_\nu^{(r-1)}} \varepsilon_{i,k}^{(r-1)}(1)\, v_{ij}^{(r-1)}| \; \forall r = 1, 2, \ldots, L \tag{5.4.5}$$

5.4.2 Fast ALADIN

The Fast ALADIN algorithm for a multi-layered neural network with L layers of hidden units can be derived by minimizing $E_k = G_k(\lambda)$, where λ is determined during the training. The update equations for the synaptic weights $v_{j\ell}^{(r)} \; \forall r = 0, 1, \ldots, L-1$ can be directly obtained from (A5.2.6) as

$$v_{j\ell,k}^{(r)} = v_{j\ell,k-1}^{(r)} + \alpha\, \varepsilon_{j,k}^{(r)}(\lambda)\, \hat{h}_{\ell,k}^{(r+1)} \; \forall r = 0, 1, \ldots, L \tag{5.4.6}$$

where $\hat{h}_{\ell,k}^{(L+1)} = \mathbf{x}_{\ell,k} \; \forall \ell, k$, $\varepsilon_{j,k}^{(0)}(\lambda) = (1 - \hat{y}_{j,k}^2)(y_{j,k} - \lambda \hat{y}_{j,k})$, and $\varepsilon_{j,k}^{(r)}(\lambda) = (1 - \hat{h}_{j,k}^{(r)\,2}) \sum_{i \in \mathcal{I}_\nu^{(r-1)}} \varepsilon_{i,k}^{(r-1)}(\lambda)\, v_{ij}^{(r-1)} \; \forall r = 1, 2, \ldots, L$.

The sensitivity measure for the jth unit of the rth hidden layer can be obtained by combining (5.4.3) and (A5.2.9) as

$$S(G_k(\lambda), \hat{h}_{j,k}^{(r)}) = |(\hat{h}_{j,k}^{(r)}/E_k) \sum_{i \in \mathcal{I}_\nu^{(r-1)}} \varepsilon_{i,k}^{(r-1)}(\lambda)\, v_{ij}^{(r-1)}| \; \forall r = 1, 2, \ldots, L \tag{5.4.7}$$

The parameter λ is not constant during training. Training begins with $\lambda = 1$, and after the first adaptation cycle it is evaluated at every adaptation cycle in terms of the total error achieved by the algorithm during the previous adaptation cycle as $\lambda = \exp(-\mu/E^2)$.

5.5 Experimental Results

The algorithms presented in this chapter were used for training feed-forward neural networks with one layer of hidden units to function

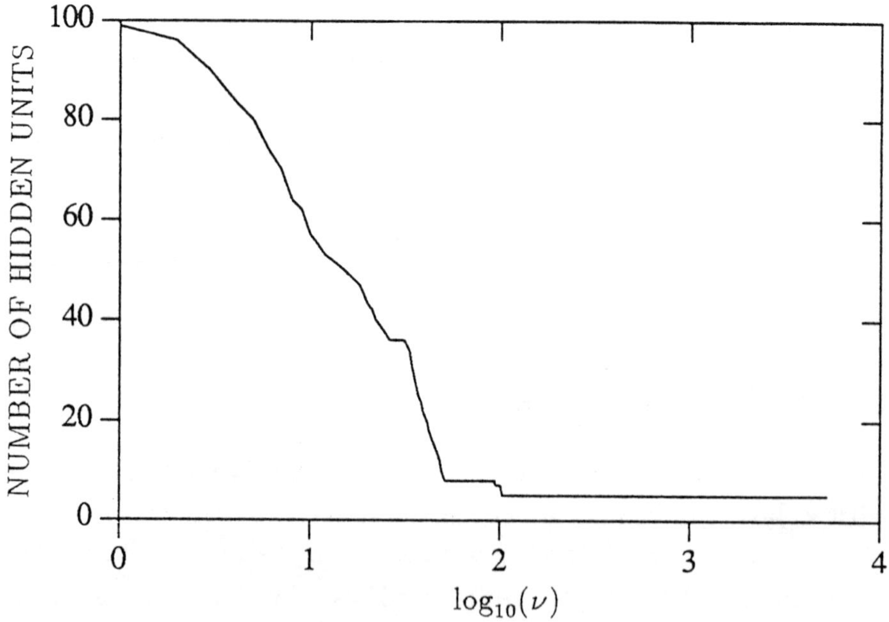

Figure 5.1: Training a neural network by the Fast ALADIN algorithm to function as a 20 to 20 encoder. Number of active hidden units as a function of $\log_{10}(\nu)$, where ν is the number of adaptation cycles.

as n to n encoders. A neural network trained to function as an n to n encoder requires at least $[\log_2(n)]$ hidden units, where $[p]$ denotes the largest integer which is smaller than the real number p, i.e., $[p] < p < [p] + 1$. The fact that the minimum number of hidden units required in this training task is a priori known makes this experiment a useful test for evaluating the ability of the algorithms presented in this chapter to determine the minimal network architecture during training.

In the first experiment, a neural network with 20 inputs, one layer of hidden units, and 20 nonlinear output units was trained to function as a 20 to 20 encoder using ALADIN with $\gamma = 1.65$ and Fast ALADIN with $\mu = 0.1$ and $\gamma = 1.65$. In all the experiments, the initial number of hidden units was $n_h = 100$. Figure 5.1 shows the number of active hidden units as a function of the logarithm of the number of adaptation cycles when the network is trained by

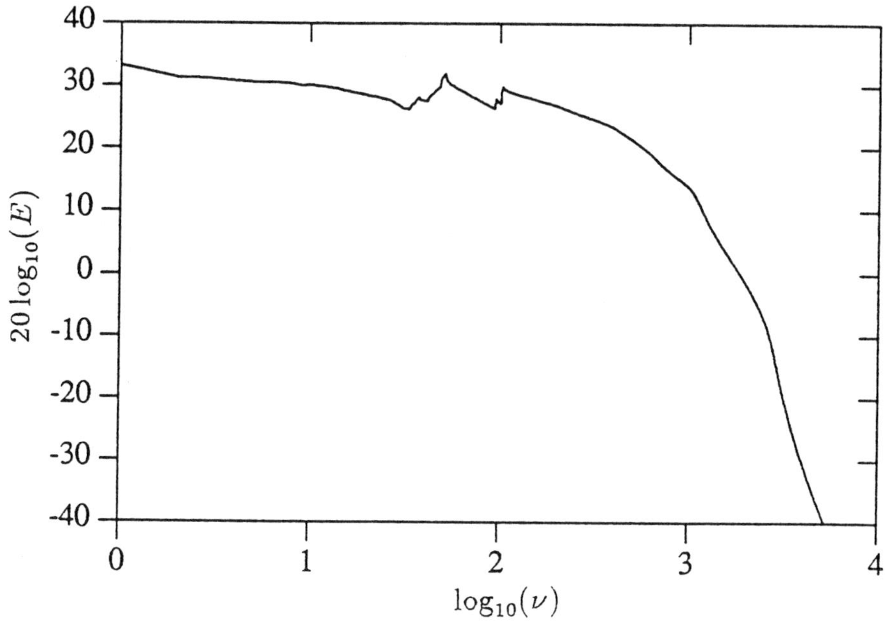

Figure 5.2: Training a neural network by the Fast ALADIN algorithm to function as a 20 to 20 encoder. Total error in dB as a function of $\log_{10}(\nu)$, where ν is the number of adaptation cycles.

the Fast ALADIN algorithm with $\delta = 0.9$. The logarithmic scale was selected in order to emphasize the architectural modifications occurring during the initial adaptation cycles, which are crucial for the training process. Figure 5.2 shows the total error in dB as a function of the logarithm of the number of adaptation cycles. According to Figure 5.1, most of the hidden units are deactivated during the initial adaptation cycles. A comparison of Figures 5.1 and 5.2 indicates that the total error achieved by the algorithm increases temporarily when some of the hidden units are deactivated during training.

The role of the tolerance parameter δ is illustrated in Tables 5.3 and 5.4, which summarize the results of the training by the ALADIN and Fast ALADIN algorithms, respectively. These tables show the number of hidden units corresponding to values of δ between 0.1 and 0.9 and the number of adaptation cycles required by the algorithms to reach an error value of 0.1 and 0.01. Although

Tolerance (δ)	Hidden Units	$E = 0.1$	$E = 0.01$
0.9	5	11949	102276
0.8	5	11956	102284
0.7	5	11956	102284
0.6	5	11956	102284
0.5	5	11968	102327
0.4	7	5682	45727
0.3	32	878	6517
0.2	36	765	6124
0.1	97	483	3612

Table 5.3: Training a neural network by ALADIN to function as a 20 to 20 encoder. Number of active hidden units and number of adaptation cycles for different values of the tolerance parameter δ.

the two algorithms produce the same network architecture for the same value of δ, there is a significant difference in the number of adaptation cycles required for the training which demonstrates the efficiency of the Fast ALADIN algorithm. Any value of the tolerance parameter δ between 0.5 and 0.9 leads to a network with 5 hidden units, which is in fact the minimum number of hidden units required for this training task. As the value of δ decreases below 0.5, the elimination of hidden units stops early in the training process due to the small fluctuations of the variance allowed. A clear advantage of the algorithms presented in this chapter is that the training process is not sensitive to the changes of the tolerance parameter δ, as indicated by the wide range of values of δ which result in a network of minimum size.

The Fast ALADIN algorithm was also used to train a neural network with 30 inputs, one layer of hidden units, and 30 nonlinear output units to function as a 30 to 30 encoder. The initial number of hidden units was $n_h = 100$, while the parameter used in the elimination of the hidden units was $\gamma = 1.70$. Table 5.5 summarizes the network architectures produced by various values of the tolerance parameter δ and the number of adaptation cycles required for the reduction of the total error to 0.1 and 0.01. When the tolerance parameter δ is between 0.5 and 0.9 the algorithm produces a net-

Tolerance (δ)	Hidden Units	$E = 0.1$	$E = 0.01$
0.9	5	3193	5241
0.8	5	3187	5232
0.7	5	3178	5224
0.6	5	3209	5253
0.5	5	3194	5241
0.4	7	1452	2446
0.3	32	226	343
0.2	36	214	338
0.1	97	182	243

Table 5.4: Training a neural network by Fast ALADIN to function as a 20 to 20 encoder. Number of active hidden units and number of adaptation cycles for different values of the tolerance parameter δ.

Tolerance (δ)	Hidden Units	$E = 0.1$	$E = 0.01$
0.9	5	10936	18534
0.8	5	6985	11532
0.7	6	4154	6883
0.6	6	4567	7316
0.5	6	4443	7167
0.4	20	624	852
0.3	95	188	267
0.2	95	188	267
0.1	95	188	267

Table 5.5: Training a neural network by Fast ALADIN to function as a 30 to 30 encoder. Number of active hidden units and number of adaptation cycles for different values of the tolerance parameter δ.

Tolerance (δ)	Hidden Units	Classified Test Samples
0.8	4	93.96 %
0.6	4	93.96 %
0.4	6	92.58 %
0.2	93	88.89 %

Table 5.6: Training a neural network by Fast ALADIN to perform a classification task. Number of active hidden units and percentage of correctly classified test samples for different values of the tolerance parameter δ.

work with 5 hidden units, the minimum number required for this training set, or 6 hidden units.

The last set of experiments evaluates the generalization ability of feed-forward neural networks trained by the proposed algorithms. The problem considered in these experiments is the determination of lithologies from well-logging data. A feed-forward neural network was trained to classify lithologies such as sandstone, shale, and limestone from the resistivity and spontaneous potential in different depths. The data used in these experiments were based on an application of neural networks to lithology presented in (McCormack, 1991). The input data were normalized between -1 and $+1$. The feed-forward neural network used consists of two inputs, corresponding to the resistivity and the spontaneous potential, and three output units. The output patterns $(+1, -1, -1)$, $(-1, +1, -1)$, and $(-1, -1, +1)$ were assigned to sandstone, shale, and limestone, respectively. A feed-forward neural network with $n_h = 100$ initial hidden units was trained using the Fast ALADIN algorithm with $\gamma = 1.65$ and different values of the tolerance parameter δ. It was experimentally found that the neural network requires at least 4 active hidden units in order to perform this classification task. The generalization ability of the trained neural networks was tested using a set of 2600 test samples, formed by various combinations of resistivity and spontaneous potential values. Table 5.6 summarizes the results of these tests. For each value of δ used, Table 5.6 shows the number of active hidden units of the trained network as well as the percentage of the test samples correctly classified by the network.

Regardless of the number of active hidden units, the networks tested in these experiments correctly classified the test samples which were close to the samples in the training set. However, there are some input vectors that cannot be classified by the networks tested. This is a clear consequence of the formation of the training set in these experiments. Since the networks consist of three binary outputs, there are $2^3 = 8$ possible output vectors. According to the formulation of the problem, only three of these patterns are assigned to lithologies. The unclassified test samples correspond to the remaining five output patterns. Table 5.6 indicates that the generalization ability of the networks is affected by the number of hidden units and, therefore, the value of the tolerance parameter δ used in the training. Clearly, the smallest percentage of unclassified test samples corresponds to the network with the smallest number of active hidden units. When the number of hidden units increases to 6, the percentage of unclassified input vectors increases slightly. Finally, there is a significant increase in the percentage of unclassified input vectors when the number of active hidden units increases from 6 to 93.

5.6 Discussion

This chapter presented the development and evaluation of learning algorithms capable of determining the architecture of feed-forward neural networks while performing their training. The criteria for architecture determination during the training were initially established for feed-forward neural networks with one layer of hidden units. The resulting algorithms were subsequently generalized for training and architecture determination in feed-forward neural networks with more than one layer of hidden units. It was experimentally verified that the algorithms presented in this chapter provide the smallest possible network for a given training set. Among the possible network architectures that can be trained to perform a given task, the algorithms presented in this chapter are capable of selecting the one with the highest generalization ability during the training process.

Appendix A5.1

This Appendix derives the update equations for the synaptic weights of a neural network with one layer of hidden units by minimizing $G_k(\lambda), k = 1, 2, \ldots, m$, and also evaluates the corresponding partial derivative $\partial G_k(\lambda)/\partial \hat{h}_{p,k}$. If $\rho(x) = \tanh(x)$, it can easily be shown that $\rho'(x) = (1 - \rho(x)^2)$. In this case,

$$
\begin{aligned}
w_{pq,k} - w_{pq,k-1} &= -\alpha \frac{\partial G_k(\lambda)}{\partial w_{pq}} \\
&= -\alpha \sum_{i=1}^{n_o} [\lambda \phi_2'(e_{i,k}) + (1 - \lambda)\phi_1'(e_{i,k})] \frac{\partial e_{i,k}}{\partial w_{pq}} \\
&= \alpha\, \varepsilon_{p,k}^o(\lambda)\, \hat{h}_{q,k} \qquad \text{(A5.1.1)}
\end{aligned}
$$

where

$$\varepsilon_{p,k}^o(\lambda) = \sigma'(\bar{y}_{p,k})\, [\lambda \phi_2'(e_{p,k}) + (1 - \lambda)\phi_1'(e_{p,k})] \qquad \text{(A5.1.2)}$$

According to the definition of $G_k(\lambda)$ in (5.2.2),

$$\frac{\partial G_k(\lambda)}{\partial \hat{h}_{p,k}} = \sum_{i=1}^{n_o} [\lambda \phi_2'(e_{i,k}) + (1 - \lambda)\phi_1'(e_{i,k})] \frac{\partial e_{i,k}}{\partial \hat{h}_{p,k}} = -\sum_{i=1}^{n_o} \varepsilon_{i,k}^o(\lambda)\, w_{ip} \qquad \text{(A5.1.3)}$$

where $\varepsilon_{i,k}^o(\lambda)$ is given by (A5.1.2). Using (A5.1.3) and the definition of $\hat{h}_{p,k}$,

$$v_{pq,k} - v_{pq,k-1} = -\alpha \frac{\partial G_k(\lambda)}{\partial v_{pq}} = -\alpha \frac{\partial G_k(\lambda)}{\partial \hat{h}_{p,k}} \frac{\partial \hat{h}_{p,k}}{\partial v_{pq}} = \alpha\, \varepsilon_{p,k}^h(\lambda)\, x_{q,k} \qquad \text{(A5.1.4)}$$

where

$$\varepsilon_{p,k}^h(\lambda) = (1 - \hat{h}_{p,k}^2) \sum_{i=1}^{n_o} \varepsilon_{i,k}^o(\lambda)\, w_{ip} \qquad \text{(A5.1.5)}$$

If the output of the network is binary, $\sigma(x) = \tanh(x)$ and $\sigma'(x) = 1 - \sigma(x)^2$. In addition, $\rho(x) = \tanh(x)$ and $\rho'(x) = 1 - \rho(x)^2$. The discrepancy measure $\phi_1(.)$ which corresponds to the pth output unit is given by $\phi_1(e_{p,k}) = y_{p,k} e_{p,k}$. Clearly, $\phi_1'(e_{p,k}) = y_{p,k}$. In this case, (A5.1.2) gives

$$\varepsilon_{p,k}^o(\lambda) = (1 - \hat{y}_{p,k}^2)(y_{p,k} - \lambda \hat{y}_{p,k}) \qquad \text{(A5.1.6)}$$

Appendix A5.2

This Appendix derives the update equations for the synaptic weights of the multi-layered neural network with L layers of hidden units shown in Figure 2.10 by minimizing $G_k(\lambda), k = 1, 2, \ldots, m$, and also evaluates the partial derivatives $\partial G_k(\lambda)/\partial \hat{h}^{(r)}_{p,k} \, \forall \, r = 1, 2, \ldots, L$. The response of this network is identical to that of the network considered in Appendix A5.1, provided that $w_{pq} = v^{(0)}_{pq}$ and $\hat{\mathbf{h}}_k = \hat{\mathbf{h}}^{(1)}_k$. Therefore, $\partial G_k(\lambda)/\partial v^{(0)}_{pq}$ can be obtained from (A5.1.1) by replacing w_{pq} with $v^{(0)}_{pq}$ and $\hat{\mathbf{h}}_k$ with $\hat{\mathbf{h}}^{(1)}_k$ as follows

$$v^{(0)}_{pq,k} - v^{(0)}_{pq,k-1} = -\alpha \frac{\partial G_k(\lambda)}{\partial v^{(0)}_{pq}} = \alpha \, \varepsilon^{(0)}_{p,k}(\lambda) \, \hat{h}^{(1)}_{q,k} \tag{A5.2.1}$$

where

$$\varepsilon^{(0)}_{p,k}(\lambda) = (1 - \hat{y}^2_{p,k})(y_{p,k} - \lambda \hat{y}_{p,k}) \tag{A5.2.2}$$

Similarly, the same substitutions in (A5.1.3) give $\partial G_k(\lambda)/\partial \hat{h}^{(1)}_{p,k}$ as

$$\frac{\partial G_k(\lambda)}{\partial \hat{h}^{(1)}_{p,k}} = -\sum_{i=1}^{n_o} \varepsilon^{(0)}_{i,k}(\lambda) \, v^{(0)}_{ip} \tag{A5.2.3}$$

Using (A5.2.3) and the definition of $\hat{h}^{(1)}_{p,k}$,

$$v^{(1)}_{pq,k} - v^{(1)}_{pq,k-1} = -\alpha \frac{\partial G_k(\lambda)}{\partial v^{(1)}_{pq}} = -\alpha \frac{\partial G_k(\lambda)}{\partial \hat{h}^{(1)}_{p,k}} \frac{\partial \hat{h}^{(1)}_{p,k}}{\partial v^{(1)}_{pq}} = \alpha \, \varepsilon^{(1)}_{p,k}(\lambda) \, \hat{h}^{(2)}_{q,k} \tag{A5.2.4}$$

where

$$\varepsilon^{(1)}_{p,k}(\lambda) = (1 - \hat{h}^{(1)\,2}_{p,k}) \sum_{i=1}^{n_o} \varepsilon^{(0)}_{i,k}(\lambda) \, v^{(0)}_{ip} \tag{A5.2.5}$$

The direct generalization of (A5.2.1) and (A5.2.4) gives

$$v^{(r)}_{pq,k} - v^{(r)}_{pq,k-1} = -\alpha \frac{\partial G_k(\lambda)}{\partial v^{(r)}_{pq}} = \alpha \, \varepsilon^{(r)}_{p,k}(\lambda) \, \hat{h}^{(r+1)}_{q,k} \; \forall \, r = 0, 1, \ldots, L \tag{A5.2.6}$$

where $\hat{h}^{(L+1)}_{q,k} = x_{q,k} \, \forall \, q, k$, $\varepsilon^{(0)}_{p,k}(\lambda)$ is given by (A5.2.2), and

$$\varepsilon^{(r)}_{p,k}(\lambda) = (1 - \hat{h}^{(r)\,2}_{p,k}) \sum_{\ell \in \mathcal{I}^{(r-1)}_{\nu}} \varepsilon^{(r-1)}_{\ell,k}(\lambda) \, v^{(r-1)}_{\ell p} \; \forall \, r = 1, 2, \ldots, L \tag{A5.2.7}$$

In the case where $r = 2$, it can be shown that

$$\frac{\partial G_k(\lambda)}{\partial \hat{h}^{(2)}_{p,k}} = \sum_{i=1}^{n_o} [\lambda \phi_2'(e_{i,k}) + (1-\lambda)\phi_1'(e_{i,k})] \frac{\partial e_{i,k}}{\partial \hat{h}^{(2)}_{p,k}} = - \sum_{j \in \mathcal{I}_\nu^{(1)}} \varepsilon^{(1)}_{j,k}(\lambda)\, v^{(1)}_{jp} \tag{A5.2.8}$$

where $\varepsilon^{(1)}_{j,k}(\lambda)$ is given by (A5.2.5). Finally, the generalization of (A5.2.3) and (A5.2.8) gives

$$\frac{\partial G_k(\lambda)}{\partial \hat{h}^{(r)}_{p,k}} = - \sum_{j \in \mathcal{I}_\nu^{(r-1)}} \varepsilon^{(r-1)}_{j,k}(\lambda)\, v^{(r-1)}_{jp} \;\forall\, r = 1, 2, \ldots, L \tag{A5.2.9}$$

Chapter 6

Performance Evaluation of Single-layered Neural Networks

6.1 Introduction

The performance of neural networks is determined by their *capacity* and *generalization ability* or robustness to noise. Generalization is the property of trained neural networks to classify an input correctly even if it is not a member of the training set. The capacity of a neural network is determined by the amount of information that can be reliably stored in the network. The performance of neural networks depends on their architecture and the learning scheme employed for their training. The performance of single-layered neural networks trained using the outer-product rule has been extensively studied in recent years (Amari and Maginu 1988; McEliece et al. 1987). According to these analytical studies, neural networks trained using the outer-product rule are characterized by low capacity and low generalization ability. The real efficiency of such networks in practical applications is expected to be even lower than predicted, due to the simplifying assumptions made in these analytical studies. The performance evaluation of optimally trained neural networks has not attracted attention proportional to their potential, probably because of the analytical difficulties involved. Nevertheless, the performance evaluation of optimally trained neural networks provides a reliable upper bound on operational efficiency that can be attained by single-layered neural networks.

This chapter studies the performance of single-layered neural networks. It begins with an analysis of the performance of single-layered neural networks trained using the outer-product rule. Extensive analysis reveals the improvement in performance when the network is trained using optimal least-squares methods. The study of the effect of the training scheme on the performance of neural networks is completed by an experimental comparison of the performance of single-layered neural networks with binary output trained with respect to their analog and binary outputs.

This chapter is organized as follows: Section 6.2 focuses on the performance of single-layered neural networks trained with respect to their analog output using optimal least-squares techniques. This section also studies the efficiency of the approximation of optimal least-squares training by the outer-product rule. Section 6.3 evaluates the capacity of optimally trained neural networks. The performance of single-layered neural networks trained with respect to their analog and binary output is experimentally compared in section 6.4. Finally, section 6.5 contains concluding remarks.

6.2 Optimal Least-squares Training of Single-layered Neural Networks

A neural network maps a set of input vectors to a set of output vectors. For a given input $\mathbf{x}$, the response of the network is given by $\hat{\mathbf{y}} = T(\mathbf{x})$, where $T(.)$ is, in general, a nonlinear operator which depends on the architecture of the neural network considered. A neural network can be trained to perform a certain task by appropriately creating a *training set,* that is, a set of output-input vectors $(\mathbf{y}_k, \mathbf{x}_k), k = 1, 2, \ldots, m$. According to a standard notation, $\mathbf{x}_k^* = [x_{1,k}\, x_{2,k} \ldots x_{n_i,k}]$ is the argument, or the *key pattern,* and $\mathbf{y}_k^* = [y_{1,k}\, y_{2,k} \ldots y_{n_o,k}]$ is the function, or the *associated pattern.* In the case where the key and associated patterns are identical, i.e., $\mathbf{x}_k = \mathbf{y}_k, k = 1, 2, \ldots, m$, $\mathbf{x}_k$ and $\mathbf{y}_k$ will be referred to as *stored patterns.* The synaptic weights of the network can be determined during training by minimizing the quadratic error function $E_k = \frac{1}{2}\sum_{i=1}^{n_o}(y_{i,k} - \hat{y}_{i,k})^2 \,\forall\, k = 1, 2, \ldots, m$.

Assuming that $\mathbf{x}_k$ is the input to the single-layered neural network shown in Figure 2.7, the corresponding analog output of the network consists of the elements $\bar{y}_{i,k} = \mathbf{x}_k^* \,\mathbf{w}_i = \mathbf{w}_i^* \,\mathbf{x}_k = \sum_{j=1}^{n_i} w_{ij} x_{j,k}$ $\forall\, i = 1, 2, \ldots, n_o$. The binary output of the network is formed by the elements $y_{i,k}^o = \text{sgn}(\bar{y}_{i,k}) = \text{sgn}(\sum_{j=1}^{n_i} w_{ij} x_{j,k}) \;\forall\; i = 1, 2, \ldots, n_o$, where sgn(.) is the signum function, defined as $\text{sgn}(x) = -1$ if $x < 0$ and $\text{sgn}(x) = +1$ if $x \geq 0$.

Consider a single-layered neural network trained with respect to its analog output, consisting of the elements $\bar{y}_{i,k} = \mathbf{x}_k^* \,\mathbf{w}_i = \mathbf{w}_i^* \,\mathbf{x}_k$. The synaptic weights of the network can be determined on the basis of least-squares techniques, as shown in chapter 2. In this case, the training can also be formulated as an *optimal associative recall* problem. Optimal associative recall is the problem of determining the mapping which optimally associates two sets of patterns. The application of optimal linear associative recall in the training of neural networks was initially proposed in the early seventies (Kohonen and Ruohonen, 1973). According to this formulation, the associated data $\mathbf{y}_k, k = 1, 2, \ldots, m$ can be related to the key data $\mathbf{x}_k, k = 1, 2, \ldots, m$ through the following linear mapping

$$\mathbf{y}_k = \mathbf{W}_m \,\mathbf{x}_k \,\forall\, k = 1, 2, \ldots, m \tag{6.2.1}$$

where $\mathbf{W}_m$ is the $n_o \times n_i$ matrix to be optimally determined. In the context of artificial neural networks, $\mathbf{W}_m$ is the *matrix of synaptic weights.* If $\mathbf{X}_m(\mathbf{Y}_m)$ is the matrix whose column vectors are $\mathbf{x}_k(\mathbf{y}_k), k = 1, 2, \ldots, m$, (6.2.1) is equivalent to

$$\mathbf{Y}_m = \mathbf{W}_m \,\mathbf{X}_m \tag{6.2.2}$$

The optimal solution of (6.2.2) in the least-squares sense is given by (Kohonen and Ruohonen, 1973)

$$\mathbf{W}_m = \mathbf{Y}_m \,\mathbf{X}_m^+ \tag{6.2.3}$$

where $\mathbf{X}_m^+$ is the generalized inverse of $\mathbf{X}_m$. The validity of the above result, derived for $\mathbf{Y}_m \neq \mathbf{X}_m$, can easily be verified in the case where

$\mathbf{Y}_m = \mathbf{X}_m$. Under this assumption, the matrix of synaptic weights $\mathbf{U}_m$ satisfies the following condition

$$\mathbf{X}_m = \mathbf{U}_m \, \mathbf{X}_m \tag{6.2.4}$$

From the properties of the generalized inverse, $\mathbf{X} \, \mathbf{X}^+ \, \mathbf{X} = \mathbf{X}$ (Albert, 1972). Clearly, the condition (6.2.4) is satisfied by the following $n_i \times n_i$ matrix

$$\mathbf{U}_m = \mathbf{X}_m \, \mathbf{X}_m^+ \tag{6.2.5}$$

In fact, (6.2.5) provides the optimal matrix of synaptic weights for the neural network trained with respect to the associations $(\mathbf{x}_k, \mathbf{x}_k), k = 1, 2, \ldots, m$.

The matrix of synaptic weights given by (6.2.3) is unique, as a result of the uniqueness of the generalized inverse. Using well-known properties of the generalized inverse, the optimal matrix of synaptic weights (6.2.3) can be written as (Rao and Mitra, 1971; Albert, 1972)

$$\mathbf{W}_m = \mathbf{Y}_m \, (\mathbf{X}_m^* \, \mathbf{X}_m)^+ \, \mathbf{X}_m^* \tag{6.2.6}$$

In the case where $\mathbf{Y}_m = \mathbf{X}_m$, $\mathbf{U}_m = \mathbf{X}_m \, (\mathbf{X}_m^* \, \mathbf{X}_m)^+ \, \mathbf{X}_m^*$. Assuming that $\mathbf{X}_m^* \, \mathbf{X}_m$ is a nonsingular matrix, the generalized inverse coincides with the conventional matrix inverse. In such a case, $\mathbf{U}_m = \mathbf{X}_m \, (\mathbf{X}_m^* \, \mathbf{X}_m)^{-1} \, \mathbf{X}_m^*$ (Personnaz et al., 1985).

In general, there is no guarantee that the matrix $\mathbf{X}_m^* \, \mathbf{X}_m$ is nonsingular. The necessary and sufficient condition for any matrix of this form to be nonsingular is presented here. Consider the $n \times 1$ real vectors $\mathbf{a}^* = [a_1 \, a_2 \, \ldots \, a_n]$ and $\mathbf{b}^* = [b_1 \, b_2 \, \ldots \, b_n]$ and let the inner product be defined as follows

$$< \mathbf{a}, \mathbf{b} >= \mathbf{a}^* \, \mathbf{b} = \sum_{i=1}^{n} a_i \, b_i \tag{6.2.7}$$

By its definition, $\mathbf{X}_m = [\mathbf{x}_1 \, \mathbf{x}_2 \, \ldots \, \mathbf{x}_m]$. Therefore,

$$\mathbf{X}_m^* \mathbf{X}_m = \begin{bmatrix} < \mathbf{x}_1, \mathbf{x}_1 > & < \mathbf{x}_1, \mathbf{x}_2 > & & < \mathbf{x}_1, \mathbf{x}_m > \\ < \mathbf{x}_2, \mathbf{x}_1 > & < \mathbf{x}_2, \mathbf{x}_2 > & \vdots & < \mathbf{x}_2, \mathbf{x}_m > \\ \vdots & \vdots & & \vdots \\ < \mathbf{x}_m, \mathbf{x}_1 > & < \mathbf{x}_m, \mathbf{x}_2 > & & < \mathbf{x}_m, \mathbf{x}_m > \end{bmatrix} \tag{6.2.8}$$

Clearly, $\det(\mathbf{X}_m^* \mathbf{X}_m) = G(\mathbf{x}_1, \mathbf{x}_2, \ldots, \mathbf{x}_m)$, where $G(\mathbf{x}_1, \mathbf{x}_2, \ldots, \mathbf{x}_m)$ is the Gramian of the vectors $\mathbf{x}_k, k = 1, 2, \ldots, m$ (Gantmacher, 1977). The Gramian is nonzero if and only if the vectors $\mathbf{x}_k, k = 1, 2, \ldots, m$ are linearly independent (Gantmacher, 1977). Therefore, $\mathbf{X}_m^* \mathbf{X}_m$ is nonsingular if and only if the vectors $\mathbf{x}_k, k = 1, 2, \ldots, m$ are linearly independent.

6.2.1 Optimal Least-squares Training and the Outer-product Rule

The outer-product rule is the among the earliest learning schemes for neural networks (Anderson, 1972; Kohonen, 1972; Nakano, 1972). This learning scheme was also associated with the Hopfield neural network (Hopfield, 1984). Although the outer-product rule is a simple and intuitively reasonable scheme, it is not optimal in any sense (Karayiannis, 1991a; Karayiannis and Venetsanopoulos, 1992d). In fact, the performance of first-order neural networks trained using the outer-product rule is very sensitive to the correlation of the key patterns. This interpretation suggests a more detailed study on optimal least-squares training of neural networks. Furthermore, the simplicity of the outer-product rule motivates an investigation of the relationship between the outer-product rule and optimal least-squares training, with particular emphasis on the efficiency of the approximation of optimal least-squares training by the outer-product rule.

Consider the optimal matrix of synaptic weights given by (6.2.6) and assume that the key patterns are uncorrelated, that is, $< \mathbf{x}_k, \mathbf{x}_\ell >= n_i\, \delta_{k\ell}$, where $\delta_{k\ell} = 1$ if $k = \ell$ and $\delta_{k\ell} = 0$ if $k \neq \ell$. In this case, $\mathbf{X}_m^* \mathbf{X}_m = n_i \mathbf{I}$, where $\mathbf{I}$ denotes the $m \times m$ identity matrix. Subsequently, the optimal matrix of synaptic weights becomes

$$\mathbf{W}_m[0] = \frac{1}{n_i} \mathbf{Y}_m \mathbf{X}_m^* \tag{6.2.9}$$

where $\mathbf{W}_m[0]$ denotes the matrix of synaptic weights that corresponds to the outer-product rule. Thus, the outer-product rule is the optimal solution of (6.2.2) in the least-squares sense, given that the key patterns are uncorrelated. In some simulation experiments reported in the literature, the key patterns were randomly generated. That is, each element of the patterns was assigned to one of the two possible states ± 1, with equal probability of $1/2$. However, the use of such artificially created data in simulation experiments may create false impressions regarding the performance of neural networks trained using the outer-product rule. In fact, the orthogonality assumption is very rarely satisfied in practical situations. As an example, consider the situation where the neural network is used in pattern recognition applications, with a set of given patterns being the key patterns.

Now consider the more realistic situation in which the key patterns are correlated. The information describing the correlation of the key patterns is contained in the matrix $\mathbf{X}_m^* \mathbf{X}_m$. If the elements of $\mathbf{X}_m$ can only take the values ± 1, $\mathbf{X}_m^* \mathbf{X}_m$ has a special form. A class of such matrices which are guaranteed to be nonsingular is specified here. The diagonal elements of $\mathbf{X}_m^* \mathbf{X}_m$ are $< \mathbf{x}_k, \mathbf{x}_k >= \sum_{i=1}^{n_i} x_{i,k}\, x_{i,k} = n_i \,\forall\, k = 1, 2, \ldots, m$. Therefore,

$$\mathbf{X}_m^* \mathbf{X}_m = n_i\,(\mathbf{I} + \mathbf{C}) \tag{6.2.10}$$

where $\mathbf{C}$ is an $m \times m$ matrix with elements

$$(\mathbf{C})_{k\ell} = \begin{cases} 0 & \text{if } k = \ell \\ \frac{1}{n_i} \sum_{i=1}^{n_i} x_{i,k}\, x_{i,\ell} & \text{if } k \neq \ell \end{cases} \tag{6.2.11}$$

Assume that $\lambda_k, k = 1, 2, \ldots, m$, are the distinct eigenvalues of $\mathbf{C}$ and let λ_d be its dominant eigenvalue, defined as $\lambda_d = \lambda_i : |\lambda_i| > |\lambda_j| \,\forall\, j \neq i$. The existence of a quantitative measure for the correlation of the key patterns is particularly important in order to reliably evaluate the generalization ability of neural networks. $\mathbf{X}_m^* \mathbf{X}_m$ describes the correlation of the key patterns, but its diagonal elements are all equal to n_i, regardless of the key patterns. Therefore, the useful information regarding the correlation of the key patterns is eventually carried by $\mathbf{C}$. In other words, there is no loss of information if $\mathbf{C}$ is chosen to be studied instead of $\mathbf{X}_m^* \mathbf{X}_m$. On the basis of the above arguments, the magnitude of the dominant eigenvalue of $\mathbf{C}$ is proposed here as the quantitative measure for the correlation of the key patterns. In fact, the key patterns are classified as follows:

- The key patterns are *weakly correlated* if $|\lambda_d| < 1$.

- The key patterns are *strongly correlated* if $|\lambda_d| \geq 1$.

Since $\mathbf{C}$ is a matrix of a special structure, it is possible to determine the range of its eigenvalues. By its definition, $\mathbf{C}$ is a matrix with zero diagonal elements. In addition, $0 \leq |(\mathbf{C})_{k\ell}| \leq 1 \, \forall \, k \neq \ell$. The range of the eigenvalues of $\mathbf{C}$ can be determined by considering the following two extreme, and rather trivial, situations:

- The key patterns are uncorrelated. In this case, $\mathbf{C}$ is an identically zero matrix and $\lambda_d = 0$.

- All the key patterns are identical. In this case, $\mathbf{C} = \mathbf{C}^*$ with elements $(\mathbf{C}^*)_{k\ell} = 1 - \delta_{k\ell}$, where $\delta_{k\ell} = 1$ for $k = \ell$ and $\delta_{k\ell} = 0$ for $k \neq \ell$. It is shown in Appendix A6.1 that $\lambda = -1$ is an eigenvalue of $\mathbf{C}^*$ of multiplicity $m - 1$, while the dominant eigenvalue of $\mathbf{C}^*$ is $\lambda_d = m - 1$.

In conclusion, $|\lambda_d| \in [0, m-1]$. However, $\lambda_d = 0$ and $\lambda_d = m-1$ correspond to very trivial cases. In practical situations, $|\lambda_d| \in (0, m-1)$. It should be remarked here that in the case of only two key patterns, i.e., when $m = 2$, $|\lambda_d| \in [0, 1]$. If, in addition, the two key patterns are distinct, then $|\lambda_d| < 1$. Therefore, $\mathbf{X}_m^* \mathbf{X}_m$ is always nonsingular only in the rather trivial case of two distinct key patterns.

It is shown here that if the key patterns are weakly correlated, i.e., if the dominant eigenvalue of $\mathbf{C}$ satisfies $|\lambda_d| < 1$, the optimal matrix of synaptic weights (6.2.3) can be expanded in the form of an alternating infinite series (Karayiannis, 1991a). This expansion is derived here as a special case of the Neumann-type expansion of the generalized inverse (Rao and Mitra, 1971). Assume that $\mu_k, k = 1, 2, \ldots, m$ are the non-negative, distinct eigenvalues of $\mathbf{X}_m^* \mathbf{X}_m$ and let μ_d be its dominant eigenvalue. According to (6.2.10), the dominant eigenvalue of $\mathbf{X}_m^* \mathbf{X}_m$ is given in terms of the dominant eigenvalue of $\mathbf{C}$ by $\mu_d = n_i (1 + \lambda_d)$. The non-negative definition of μ_d implies that $\lambda_d > -1$. For $0 < \alpha < 2/\mu_d$,

$$\mathbf{X}_m^+ = \alpha \sum_{k=0}^{\infty} (\mathbf{I}_m - \alpha \mathbf{X}_m^* \mathbf{X}_m)^k \mathbf{X}_m^* \qquad (6.2.12)$$

Assuming that $\alpha = 1/n_i$, the definition of $\mathbf{C}$ in (6.2.10) implies that $\mathbf{I}_m - (1/n_i)\,\mathbf{X}_m^*\,\mathbf{X}_m = -\mathbf{C}$. Under these assumptions, (6.2.12) amounts to

$$\mathbf{X}_m^+ = \frac{1}{n_i}\sum_{k=0}^{\infty}(-1)^k\,\mathbf{C}^k\,\mathbf{X}_m^* \tag{6.2.13}$$

The maximum value of λ_d for which the expansion (6.2.13) is valid is determined by the inequality $\alpha = 1/n_i < 2/(n_i(1+\lambda_d)) = 2/\mu_d$. Obviously, this inequality holds if $\lambda_d < 1$. Since $\lambda_d > -1$, the expansion (6.2.13) is valid if $|\lambda_d| < 1$. The combination of (6.2.3) and (6.2.13) indicates that the optimal matrix of synaptic weights $\mathbf{W}_m$ can be expanded as follows

$$\begin{aligned}\mathbf{W}_m &= \frac{1}{n_i}\sum_{k=0}^{\infty}(-1)^k\,\mathbf{Y}_m\,\mathbf{C}^k\,\mathbf{X}_m^* \\ &= \mathbf{W}_m[0] + \frac{1}{n_i}\sum_{k=1}^{\infty}(-1)^k\,\mathbf{Y}_m\,\mathbf{C}^k\,\mathbf{X}_m^* \end{aligned} \tag{6.2.14}$$

where $\mathbf{W}_m[0]$ is given by (6.2.9). A validity test of the above expansion does not require knowledge of all of the eigenvalues of $\mathbf{C}$. In fact, it depends only on the dominant eigenvalue of $\mathbf{C}$, which can easily be approximated using the power method such as the Rayleigh quotient (Noble and Daniel, 1977).

The expansion (6.2.14) indicates that if the dominant eigenvalue of $\mathbf{C}$ satisfies $|\lambda_d| < 1$, the outer-product rule can also be seen as a "rough" approximation of the optimal matrix of synaptic weights. That is, the outer-product rule is the approximation of an infinite series of terms by the first term. Since $|\lambda_d| < 1$, $||\mathbf{C}^k||$ is a decreasing function of k. Therefore, the efficiency of such an approximation depends on the first term which follows, that is, $\mathbf{Y}_m\,\mathbf{C}\,\mathbf{X}_m^*$. The effect of cutting off the infinite series after the first term is now investigated by determining the upper bound of $||\mathbf{C}||$ as a function of the dominant eigenvalue of $\mathbf{C}$. Consider the similarity transformation $\mathbf{P}^{-1}\,\mathbf{C}\,\mathbf{P} = \mathbf{\Lambda}$, where $\mathbf{\Lambda}$ is the diagonal matrix whose diagonal elements are the eigenvalues of $\mathbf{C}$. Obviously, $\mathbf{C}^k = \mathbf{P}\mathbf{\Lambda}^k\mathbf{P}^{-1}$. For any matrix norm,

$$||\mathbf{C}^k|| = ||\mathbf{P}\mathbf{\Lambda}^k\mathbf{P}^{-1}|| \leq ||\mathbf{P}||\,||\mathbf{\Lambda}^k||\,||\mathbf{P}^{-1}|| \tag{6.2.15}$$

Consider here the infinite norm $||.||_\infty$, defined for any $m \times m$ matrix $\mathbf{B}$ as $||\mathbf{B}||_\infty = \max_i \sum_{j=1}^{m} |b_{ij}|$. Since $\mathbf{\Lambda}^k = \mathrm{diag}(\lambda_1^k, \lambda_2^k, \ldots, \lambda_m^k)$, the infinite norm of $\mathbf{\Lambda}^k$ is $||\mathbf{\Lambda}^k||_\infty = \max_i\{|\lambda_i|^k\} = |\lambda_d|^k$. Therefore,

$$||\mathbf{C}^k||_\infty \leq |\lambda_d|^k ||\mathbf{P}||_\infty ||\mathbf{P}^{-1}||_\infty \tag{6.2.16}$$

In the case where $k = 1$, the upper bound of $||\mathbf{C}||_\infty$ is a function of $|\lambda_d|$. Therefore, the efficiency of the outer-product rule as an approximation of the optimal matrix of synaptic weights depends on the magnitude of the dominant eigenvalue of $\mathbf{C}$ relative to unity. Obviously, this approximation is satisfactory only if $|\lambda_d| \ll 1$. In fact, this is the case when the key patterns are randomly generated.

According to the above analysis, the optimal matrix of synaptic weights $\mathbf{W}_m$ can be expressed as the summation of the matrix $\mathbf{W}_m[0]$, which corresponds to the outer-product rule, and an infinite series of additional terms. Consider an approximation of $\mathbf{W}_m$, obtained from the sum of $\mathbf{W}_m[0]$ and a finite number of additional terms. Let $\mathbf{W}_m[L]$ be the approximation of $\mathbf{W}_m$ which corresponds to L additional terms. Then,

$$\begin{aligned} \mathbf{W}_m[L] &= \mathbf{W}_m[0] + \frac{1}{n_i} \sum_{k=1}^{L} (-1)^k \, \mathbf{Y}_m \, \mathbf{C}^k \, \mathbf{X}_m^* \\ &= \mathbf{W}_m[L-1] + \frac{1}{n_i} (-1)^L \, \mathbf{Y}_m \, \mathbf{C}^L \, \mathbf{X}_m^* \end{aligned} \tag{6.2.17}$$

If the elements of $\mathbf{X}_m$ can only take the values ± 1, the evaluation of $\mathbf{W}_m[L]$ can be significantly simplified. In fact, any finite approximation of the optimal matrix of synaptic weights $\mathbf{W}_m$ can be expressed in terms of $\mathbf{W}_m[0]$ and $\mathbf{U}_m[0]$ only. It is shown in Appendix A6.2 that $\mathbf{W}_m[L]$ can be evaluated through the following formula

$$\mathbf{W}_m[L] = \mathbf{W}_m[L-1] + (-1)^L \, \mathbf{W}_m[0] \, (\mathbf{U}_m[0] - \mathbf{I})^L \tag{6.2.18}$$

Any finite approximation of $\mathbf{W}_m$ can also be the basis for training neural networks. Assuming that the key patterns are weakly correlated, a neural network whose matrix of synaptic weights is any finite approximation of $\mathbf{W}_m$ is expected to be more efficient, in terms of capacity and generalization ability, than one trained using the outer-product rule. The capacity of networks whose matrix of synaptic weights is any finite approximation of $\mathbf{W}_m$ is studied in the next section. The effect of additional terms on the generalization ability of the network can be studied through an alternative form of the elements of any finite approximation of $\mathbf{W}_m$, which is derived in Appendix A6.2. According to this analysis, the approximation of $\mathbf{W}_m$, which corresponds to L additional terms, consists of elements of the following form

$$(\mathbf{W}_m[L])_{ij} = \frac{1}{n_i} \sum_{k=1}^{m} y_{i,k}[L]\, x_{j,k} \tag{6.2.19}$$

where $y_{i,k}[L]$ is given in terms of $y_{i,k}[L-1]$ as follows

$$y_{i,k}[L] = y_{i,k}[L-1] + \frac{(-1)^L}{n_i^L} \sum_{q_1,\ldots,q_L=1}^{n_i} \sum_{k_1 \neq k_2}^{m} \cdots \sum_{k_L \neq k}^{m} y_{i,k_1} x_{q_1,k_1} x_{q_1,k_2} x_{q_2,k_2} \ldots x_{q_L,k_L} x_{q_L,k} \tag{6.2.20}$$

and $y_{i,k}[0] = y_{i,k}$ corresponds to the outer-product rule. It is also shown in Appendix A6.2 that the elements of any finite approximation $\mathbf{W}_m[L]$ of $\mathbf{W}_m$ can be equivalently written as follows

$$(\mathbf{W}_m[L])_{ij} = \frac{1}{n_i} \sum_{k=1}^{m} y_{i,k}\, x_{j,k}[L] \tag{6.2.21}$$

where $x_{j,k}[L]$ is given in terms of $x_{j,k}[L-1]$ by

$$x_{j,k}[L] = x_{j,k}[L-1] + \frac{(-1)^L}{n_i^L} \sum_{q_1,\ldots,q_L=1}^{n_i} \sum_{k_L \neq k_{L-1}}^{m} \cdots \sum_{k_1 \neq k}^{m} x_{q_1,k} x_{q_1,k_1} x_{q_2,k_1} x_{q_2,k_2} \ldots x_{q_L,k_L} x_{j,k_L} \tag{6.2.22}$$

with $x_{j,k}[0] = x_{j,k}$. The optimal matrix of synaptic weights $\mathbf{W}_m$ is given in terms of $\mathbf{W}_m[L]$ as follows

$$\mathbf{W}_m = \lim_{L \to \infty} \mathbf{W}_m[L] \tag{6.2.23}$$

Therefore, when L approaches infinity, both (6.2.19) and (6.2.21) provide an outer-product rule like expression for the optimal matrix of synaptic weights. The equivalent representation of the elements of any finite approximation of $\mathbf{W}_m$ by (6.2.19) provides the basis for the evaluation of the capacity of optimally trained neural networks, presented in the next section. The alternative form (6.2.21) can be used to illustrate the effect of considering additional terms in the approximation of $\mathbf{W}_m$, called here *orthogonalization effect*.

Consider the finite approximation of $\mathbf{W}_m$ consisting of L additional terms, and assume that $\mathbf{x}_\ell$ is the input to the network with matrix of synaptic weights $\mathbf{W}_m[L]$. According to (6.2.21), the response of the network to this input is given by

$$y_i = \operatorname{sgn}\Big(\sum_{j=1}^{n_i}(\mathbf{W}_m[L])_{ij}\, x_{j,\ell}\Big) = \operatorname{sgn}\Big(\sum_{k=1}^{m} y_{i,k}\, c(\mathbf{x}_k[L], \mathbf{x}_\ell)\Big) \qquad (6.2.24)$$

where $c(\mathbf{x}, \mathbf{y})$ denotes the direction cosine between the $n \times 1$ vectors $\mathbf{x}$ and $\mathbf{y}$, defined by

$$c(\mathbf{x}, \mathbf{y}) = \frac{1}{n}\sum_{j=1}^{n} x_j\, y_j \qquad (6.2.25)$$

By its definition, the direction cosine $c(\mathbf{x}, \mathbf{y})$ is a measure of the similarity between the patterns $\mathbf{x}$ and $\mathbf{y}$, taking values in the interval $[-1, +1]$. Clearly, $c(\mathbf{x}, \mathbf{y}) = 0$ if the patterns $\mathbf{x}$ and $\mathbf{y}$ are uncorrelated.

When $\mathbf{x}_\ell$ is the input of the network, its expected response is $\mathbf{y}_\ell$. According to the above analysis, the actual response of the network consists of the elements

$$y_i = \operatorname{sgn}(y_{i,\ell} + n_{i,\ell}[L])\, \forall\, i = 1, 2, \ldots, n_o \qquad (6.2.26)$$

and depends on the noise or interference terms

$$n_{i,\ell}[L] = \sum_{k \neq \ell}^{m} y_{i,k}\, c(\mathbf{x}_k[L], \mathbf{x}_\ell)\, \forall\, i = 1, 2, \ldots, n_o \qquad (6.2.27)$$

Assume that the network is trained using the outer-product rule, i.e., $L = 0$. In the ideal situation where the key patterns are uncorrelated, $c(\mathbf{x}_k[0], \mathbf{x}_\ell) = 0$ and, therefore, $n_{i,\ell}[0] = 0$. In general, if the key patterns are correlated, $n_{i,\ell}[0] \neq 0$. In this case, the response of the network depends on the magnitude of the noise term relative to unity. The analysis which follows studies the response of the network whose matrix of synaptic weights is any finite approximation of the optimal matrix of synaptic weights $\mathbf{W}_m$ and, eventually, the matrix $\mathbf{W}_m$ itself.

Consider any two stored patterns $\mathbf{x}_k$ and $\mathbf{x}_\ell$, such that $k \neq \ell$. It is shown in Appendix A6.3 that

$$c(\mathbf{x}_k[L], \mathbf{x}_\ell) = (-1)^L (\mathbf{C}^{L+1})_{k\ell} \, \forall \, k \neq \ell \tag{6.2.28}$$

where $\mathbf{C}$ is defined by (6.2.10). For $|\lambda_d| < 1$, $||\mathbf{C}^L||_\infty$ is a decreasing function of L, that is, $||\mathbf{C}^{L+1}||_\infty < ||\mathbf{C}^L||_\infty \, \forall \, L$. This last inequality is valid if $|(\mathbf{C}^L)_{k\ell}|$ is also a decreasing function of L. Taking into account (6.2.28), $|c(\mathbf{x}_k[L],\ \mathbf{x}_\ell)|$ is also a decreasing function of L, that is,

$$|c(\mathbf{x}_k[L+1], \mathbf{x}_\ell)| < |c(\mathbf{x}_k[L], \mathbf{x}_\ell)| \, \forall \, L \tag{6.2.29}$$

The above analysis exhibits the orthogonalization effect, caused by an increase of the terms of the finite series (6.2.17) which approximates the optimal matrix of synaptic weights $\mathbf{W}_m$. In other words, the correlation between $\mathbf{x}_k[L]$ and $\mathbf{x}_\ell$ decreases when L increases.

The same analysis provides some results of particular importance when L approaches infinity. In this case, $\lim_{L \to \infty} ||\mathbf{C}^L||_\infty = 0$, which implies that

$$\lim_{L \to \infty} |(\mathbf{C}^L)_{k\ell}| = 0 \tag{6.2.30}$$

The combination of (6.2.28) and (6.2.30) indicates that

$$\lim_{L \to \infty} |c(\mathbf{x}_k[L], \mathbf{x}_\ell)| = 0 \tag{6.2.31}$$

Therefore, when L approaches infinity, the correlation between $\mathbf{x}_k[L]$ and $\mathbf{x}_\ell$ vanishes. This latter conclusion, in conjunction with the combination of (6.2.19) and (6.2.21), indicates that the optimal matrix of synaptic weights can be seen as the realization of the outer-product rule between orthogonalized key patterns. The orthogonalization of the key patterns is the direct result of the optimal derivation of $\mathbf{W}_m$.

6.2.2 Experimental Results

The first set of experiments demonstrates the effect of the correlation of the stored patterns on the generalization ability of optimally trained neural networks and networks of the same architecture whose matrices of synaptic weights result from various approximations of the optimal one. The length of the stored patterns was $n = 150$. The number of associations stored in the network was chosen to be $m = 10$, that is, the network was operating below the "optimistic" bound $n/(2\log n) \approx 15$, but above the "pessimistic" one $n/(4\log n) \approx 7$. Several sets of stored patterns were generated in such a way that the magnitude of the dominant eigenvalue λ_d of $\mathbf{C}$ was in the range 0.4–1.5. 1000 test patterns were created for each set of stored patterns in such a way that the normalized Hamming distance between each test pattern and one of the stored patterns was uniformly distributed between 0 and 1/3. This set of experiments compared the optimally trained neural network and three networks of the same size and architecture whose matrix of synaptic weights was an approximation of the optimal matrix of synaptic weights $\mathbf{U}_m$ by $\mathbf{U}_m[0]$, which corresponds to the outer-product rule, $\mathbf{U}_m[1]$ and $\mathbf{U}_m[3]$. Figure 6.1 shows the percentage of perfectly recalled patterns, by each of these networks, as a function of $|\lambda_d|$, or equivalently, the correlation of the stored patterns. According to Figure 6.1, the generalization ability of the network trained using the outer-product rule drops rather quickly with increased correlation of the stored patterns. This is an experimental verification of the analysis presented in this section, which indicated that the outer-product rule is optimal when the stored patterns are uncorrelated and provides a satisfactory approximation of the optimal matrix of synaptic weights when $|\lambda_d| \ll 1$. Figure 6.1 also indicates that the generalization ability of the optimally trained neural network is essentially unaffected by correlation of the stored patterns. The robustness of this network to correlation of the stored patterns is a direct consequence of the optimal derivation of the matrix of synaptic weights. In addition, when the normalized Hamming distance between any of the test patterns and one of the stored patterns is not greater than 1/3, this network achieves perfect recall of the stored patterns. According to Figure 6.1, when $|\lambda_d| \ll 1$,

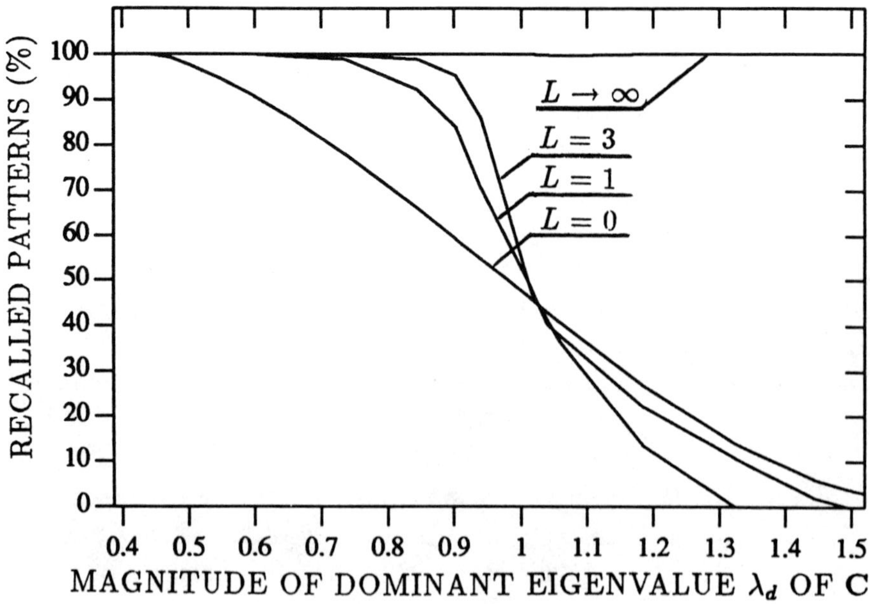

Figure 6.1: Percentage of perfectly recalled patterns by a neural network with matrix of synaptic weights $\mathbf{U}_m[L]$ as a function of the dominant eigenvalue $|\lambda_d|$ of $\mathbf{C}$.

the network whose matrix of synaptic weights is the approximation of the optimal matrix of synaptic weights $\mathbf{U}_m$ by $\mathbf{U}_m[1]$ or $\mathbf{U}_m[3]$ achieves generalization ability which is comparable to that achieved by the optimally trained neural network. However, the generalization efficiency of these networks drops rather quickly when $|\lambda_d| \geq 1$. This is not surprising, since the approximation of $\mathbf{U}_m$ by $\mathbf{U}_m[1]$ or $\mathbf{U}_m[3]$, resulting from the infinite expansion (6.2.14), is valid only when $|\lambda_d| < 1$. The network trained using $\mathbf{U}_m[3]$ is more efficient than the one trained using $\mathbf{U}_m[1]$ when $|\lambda_d| < 1$. However, when $|\lambda_d| \geq 1$, the generalization ability of this network drops faster than that of the network trained using $\mathbf{U}_m[1]$. This is because the expansion (6.2.14) is not valid for $|\lambda_d| \geq 1$ and, in addition, $||\mathbf{C}^k||$ approaches infinity as k increases when $|\lambda_d| > 1$.

The second set of experiments evaluates the effect of the noise level of the input patterns on the generalization ability of single-layered neural networks. Once again, the length of the stored pat-

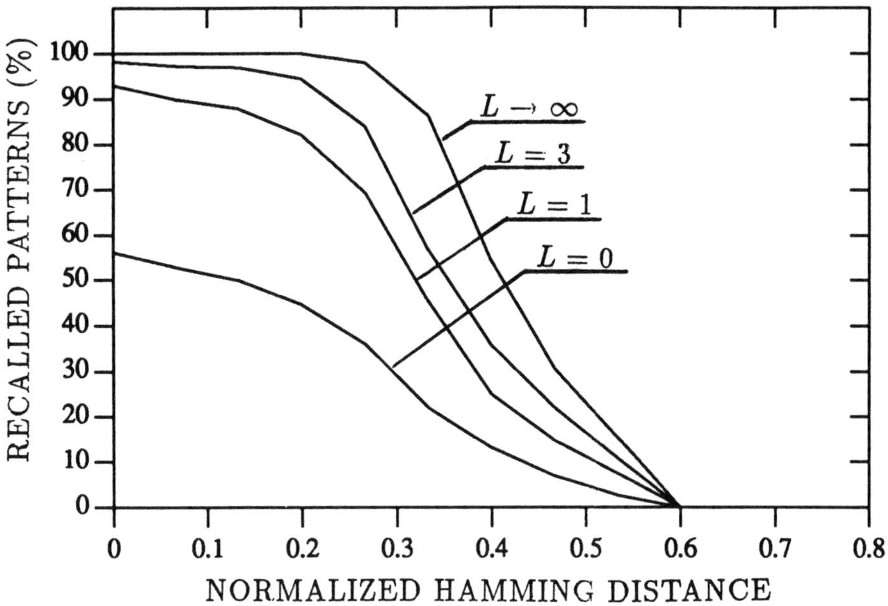

Figure 6.2: Percentage of perfectly recalled patterns by a neural network with matrix of synaptic weights $\mathbf{U}_m[L]$ as a function of the normalized Hamming distance between the input patterns and one of the stored patterns.

terns was $n = 150$, while the number of associations stored in the network was $m = 10$. A set of stored patterns was generated in such a way that $|\lambda_d| \approx 0.9$. This set of stored patterns consists of weakly correlated elements. However, with this value of $|\lambda_d|$, the stored patterns in this set can only marginally be classified as weakly correlated. For each set of stored patterns, 1000 patterns were also generated to play the role of the input patterns, but in this case the normalized Hamming distance between each one of the input patterns and one of the stored patterns was uniformly distributed between 0 and 1. Once again, this set of experiments compared the optimally trained neural network and three neural networks of the same size and architecture whose matrix of synaptic weights was an approximation of the optimal matrix of synaptic weights $\mathbf{U}_m$ by $\mathbf{U}_m[0]$, which corresponds to the outer-product rule, $\mathbf{U}_m[1]$, and $\mathbf{U}_m[3]$. Figure 6.2 shows the percentage of perfectly recalled patterns by each of these networks as a function of the normalized Hamming distance. The curves in Figure 6.2 were obtained as follows: The interval $(0,1)$ was divided into 15 subintervals of length $1/15$. For each interval $i, 1 \leq i \leq 15$, the percentage of perfectly recalled patterns was evaluated among the key patterns which differed from one of the stored patterns in $n \times d_i$ positions, where $(i-1)/15 < d_i \leq i/15$. Consider the curve representing the generalization ability of the optimally trained neural network. The interval $(0,1)$, which represents the normalized Hamming distance, is divided into three subintervals. When the normalized Hamming distance between the input patterns and one of the stored patterns lies in the first, second, or third subinterval, the network achieves perfect recall of all, some, or none of the stored patterns, respectively. Clearly, reliable operation of the network can be guaranteed only if the normalized Hamming distance between the input patterns and one of the stored patterns lies in the first interval. The optimally trained neural network achieves perfect recall of any stored pattern only if the normalized Hamming distance between the input pattern and one of the stored patterns does not exceed a given threshold. Finally, Figure 6.2 indicates that the generalization ability of neural networks whose matrix of synaptic weights is any finite approximation of the one corresponding to the optimally trained neural network is clearly inferior to that of the optimally trained neural network. In fact, for this particular set of stored patterns, only the optimally trained neural network is capable of achieving perfect recall.

6.3 Capacity Considerations

The capacity of a neural network can be evaluated by studying its response, provided that its input is one of the key patterns (Amari and Maginu, 1988). For the sake of simplicity, it is assumed in this analysis that $\mathbf{y}_k = \mathbf{x}_k \, \forall \, k = 1, 2, \ldots, m$ and, in addition, $n_i = n_o = n$. Assume that $\mathbf{U}_m[L]$ is the matrix of synaptic weights and the pattern $\mathbf{x}_\alpha$ is the input of the network. It is shown in Appendix A6.4

that the response of the network is determined by the equation

$$x_i = \mathrm{sgn}(\sum_{j=1}^{n}(\mathbf{U}_m[L])_{ij}\, x_{j,\alpha}) = \mathrm{sgn}(x_{i,\alpha} + n_{i,\alpha}[L]) \qquad (6.3.1)$$

where $x_{i,\alpha}$ is the signal and $n_{i,\alpha}[L]$ is the noise, or interference, term given as follows

$$n_{i,\alpha}[L] = \frac{(-1)^L}{n^{L+1}} \sum_{j,q_1,\ldots,q_L=1}^{n} \sum_{k_1 \neq k_2}^{m} \cdots \sum_{k_L \neq k}^{m} \sum_{k \neq \alpha}^{m} x_{i,k_1} x_{q_1,k_1} x_{q_1,k_2} \cdots x_{q_L,k} x_{j,k} x_{j,\alpha} \qquad (6.3.2)$$

Equation (6.3.1) is general enough that it can determine the response of the network whose matrix of synaptic weights corresponds to any finite approximation of $\mathbf{U}_m$. The performance of a given network is determined by the noise component $n_{i,\alpha}[L]$, which depends on the learning scheme realized by the network. In the analysis which follows, $n_{i,\alpha}[L]$ is considered to be a Gaussian zero-mean random variable with variance $\sigma_1[L]^2$.

Since $x_{i,\alpha}$ can take the values $+1$ and -1 with equal probability, the probability that the ith component of $\mathbf{x}_\alpha$ is perfectly recalled is $1-p$, where $p = \mathrm{Prob}(n_{i,\alpha}[L] < -1) = G(-\rho)$, $\rho = 1/\sigma_1[L]$, and

$$G(x) = \int_{-\infty}^{x} \frac{1}{\sqrt{2\pi}} \exp\{-\frac{t^2}{2}\} dt \qquad (6.3.3)$$

The probability that the pattern $\mathbf{x}_\alpha$ is perfectly recalled is given by $P_n = (1-p)^n = (1-G(-\rho))^n$. Thus,

$$\log P_n = n \log(1 - G(-\rho)) = n \log G(\rho) \qquad (6.3.4)$$

For large values of ρ, (6.3.4) can be approximated as follows

$$\log P_n \approx n \log(1 - \frac{1}{\rho\sqrt{2\pi}} \exp\{-\frac{\rho^2}{2}\}) \qquad (6.3.5)$$

Finally, under the assumption that $\rho \gg 1$, (6.3.5) can be approximated by

$$\begin{aligned}\log P_n &\approx -\frac{1}{\sqrt{2\pi}}\frac{n}{\rho}\exp\{-\frac{\rho^2}{2}\} \\ &= -\frac{1}{\sqrt{2\pi}}\exp\{-\frac{\rho^2}{2}+\log n-\log\rho\} \qquad (6.3.6)\end{aligned}$$

The above analysis can be the basis for evaluating various capacity bounds proposed for a first-order neural network trained using the outer-product rule (Amari and Maginu, 1988; McEliece et al., 1987). Assuming that $\mathbf{x}_\alpha$ is the input of the network and $\mathbf{U}_m[0]$ the matrix of synaptic weights, the response of the network is determined by $x_i = \text{sgn}(x_{i,\alpha}+n_{i,\alpha}[0])$, where $x_{i,\alpha}$ is the ith component of the stored pattern $\mathbf{x}_\alpha$ and $n_{i,\alpha}[0]$ is the noise term, obtained from (6.3.2) for $L=0$ as follows (Amari and Maginu, 1988)

$$n_{i,\alpha}[0] = \frac{1}{n}\sum_{j=1}^{n}\sum_{k\neq\alpha}^{m} x_{i,k}\,x_{j,k}\,x_{j,\alpha} \qquad (6.3.7)$$

Assuming that $x_{i,k}$ are independent random variables taking the values $+1$ and -1 with equal probability of $1/2$, $n_{i,\alpha}[0]$ can be considered a Gaussian zero-mean random variable with variance (Amari and Maginu, 1988)

$$\sigma_1^2 = \sigma_1[0]^2 = \frac{m-1}{n} \approx \frac{m}{n} \qquad (6.3.8)$$

Assume that m_0 is the largest allowable number of associations and let $c_1 = c_1[0] = m_0/n$ be the capacity ratio for this network. Thus,

$$\rho = \frac{1}{\sigma_1} = (\frac{n}{m_0})^{\frac{1}{2}} = (\frac{1}{c_1})^{\frac{1}{2}} \qquad (6.3.9)$$

Since $n \gg m_0$ is a typical situation in first-order neural networks, the approximations given by (6.3.5) and (6.3.6) are valid for $\rho = \sqrt{n/m_0}$. The capacity of this network can be evaluated by determining the largest $c_1 = c_1(n)$ for which $\lim_{n\to\infty} P_n = 1$, where

$$\log P_n \approx -\frac{1}{(2\pi)^{1/2}}\exp\{-\frac{1}{2c_1}+\log n+\frac{1}{2}\log c_1\} \qquad (6.3.10)$$

Capacity ratio c_1	Probability of perfect recall P_n
$(2\log n)^{-1}$	$\exp\left\{-\left(2(\pi\log n)\right)^{-\frac{1}{2}}\right\}$
$(4\log n)^{-1}$	$\exp\left\{-\left(2n(2\pi\log n)\right)^{-\frac{1}{2}}\right\}$
$(2\log n - \log\log n)^{-1}$	$\exp\left\{-(\log n)^{\frac{1}{2}}\left(2\pi\log(n^2/\log n)\right)^{-\frac{1}{2}}\right\}$
$(2\log n + \log\log n)^{-1}$	$\exp\left\{-\left(2\pi\log n\log(n^2\log n)\right)^{-\frac{1}{2}}\right\}$

Table 6.1: Probability of perfect recall P_n as a function of n for various capacity ratios c_1.

Equation (6.3.10) can be the basis for the comparison and evaluation of various capacity bounds proposed for single-layered neural networks trained using the outer-product rule. Ideally, the condition $\lim_{n\to\infty} P_n = 1$ is satisfied when $c_1 \to 0$. In practice, the derivation of a capacity bound for this network is based on the search for the largest c_1 which guarantees that P_n converges to unity, at least asymptotically. The speed of convergence is among the most crucial criteria for the selection of the capacity bound.

Under the requirement that most of the stored patterns can be perfectly recalled, the following capacity bound was proposed for first-order neural networks trained using the outer-product rule (McEliece et al., 1987)

$$c_1 = \frac{1}{2\log n} \tag{6.3.11}$$

Obviously, the same bound can also be derived from (6.3.10) by requiring that $-1/(2c_1) + \log n = 0$. With the additional requirement that every one of the stored patterns can be perfectly recalled, the following "tighter" capacity bound was also proposed by the same authors

$$c_1 = \frac{1}{4\log n} \tag{6.3.12}$$

The most optimistic capacity bound proposed for neural networks trained using the outer-product rule is (Amari and Maginu, 1988)

$$c_1 = \frac{1}{2\log n - \log\log n} \tag{6.3.13}$$

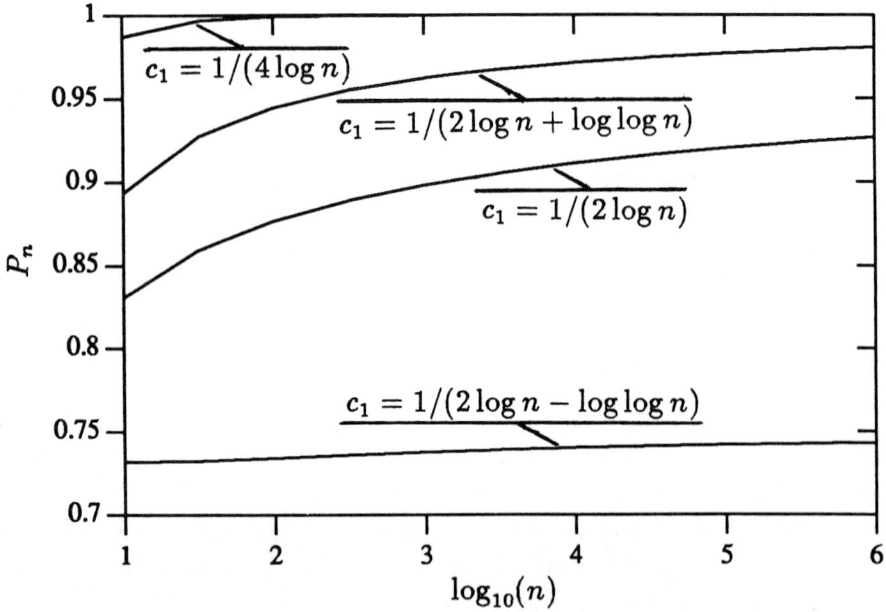

Figure 6.3: Probability of perfect recall P_n for the neural network trained using the outer-product rule as a function of n for various capacity ratios.

A compromise between the "optimistic" capacity ratio (6.3.11) and the "pessimistic" one (6.3.12), results by slightly modifying the capacity ratio given by (6.3.13) as follows (Karayiannis, 1991a)

$$c_1 = \frac{1}{2\log n + \log\log n} \tag{6.3.14}$$

Clearly, the capacity ratio defined by (6.3.14) satisfies the inequality $1/(4\log n) < c_1 < 1/(2\log n)$.

Table 6.1 shows the probability of perfect recall P_n as a function of n for the capacity ratios c_1 mentioned above. In addition, the probability of perfect recall P_n corresponding to different choices of c_1 is plotted as a function of n in Figure 6.3. According to Figure 6.3, the capacity bound given by (6.3.12) is the most reliable one. However, Figure 6.3 shows that the capacity bound defined by (6.3.13) cannot guarantee the asymptotic convergence of P_n to unity. In this particular case, P_n is only slightly affected by increasing n.

Finally, Figure 6.3 shows that the capacity bound given by (6.3.14) is a reasonable compromise between the two extremes proposed in (McEliece et al., 1987).

Consider the capacity of the neural network whose matrix of synaptic weights is the approximation of $\mathbf{U}_m$ by $\mathbf{U}_m[1]$. If $\mathbf{U}_m[1]$ is the matrix of synaptic weights and $\mathbf{x}_\alpha$ is the input of the network, the response of the network is determined by $x_i = \text{sgn}(x_{i,\alpha} + n_{i,\alpha}[1])$, where the noise term can be obtained from (6.3.2) for $L = 1$ as follows

$$n_{i,\alpha}[1] = -\frac{1}{n^2} \sum_{j=1}^{n} \sum_{q_1=1}^{n} \sum_{k_1 \neq k}^{m} \sum_{k \neq \alpha}^{m} x_{i,k_1}\, x_{q_1,k_1}\, x_{q_1,k}\, x_{j,k}\, x_{j,\alpha} \qquad (6.3.15)$$

Under the same assumptions regarding $x_{i,k}$, $E\{n_{i,\alpha}[1]\} = 0$. Therefore, $n_{i,\alpha}[1]$ can again be considered as a Gaussian zero-mean random variable. The variance of $n_{i,\alpha}[1]$ can be evaluated as follows

$$\begin{aligned} \sigma_1[1]^2 &= \frac{1}{n^4} \sum_{j,j',q_1,q_1'=1}^{n} \sum_{k_1 \neq k}^{m} \sum_{k_1' \neq k'}^{m} \sum_{k,k' \neq \alpha}^{m} \\ &\quad E\{x_{i,k_1} x_{q_1,k_1} x_{q_1,k} x_{j,k} x_{j,\alpha} x_{i,k_1'} x_{q_1',k_1'} x_{q_1',k'} x_{j',k'} x_{j',\alpha}\} \end{aligned} \qquad (6.3.16)$$

Since $x_{i,k}$ are independent random variables, $\sigma_1[1]^2$ is equal to the number of terms for which $j = j'$, $q_1 = q_1'$, $k_1 = k_1'$ and $k = k'$. It can easily be verified that

$$\sigma_1[1]^2 = \frac{1}{n^4} n^2 (m-1)^2 = (\frac{m-1}{n})^2 \approx (\frac{m}{n})^2 \qquad (6.3.17)$$

If m_1 is the largest allowable number of associations for this network and $c_1[1] = m_1/n$ the corresponding capacity ratio,

$$\rho = \frac{1}{\sigma_1[1]} = \frac{n}{m_1} = \frac{1}{c_1[1]} \qquad (6.3.18)$$

The capacity ratio of this network can be determined as the largest $c_1[1]$ which guarantees the asymptotic convergence of P_n to unity. A comparison of (6.3.9) and (6.3.18) indicates that $c_1[1]$ is related to c_1 by

$$c_1[1] = c_1^{\frac{1}{2}} \qquad (6.3.19)$$

Since $c_1 < 1$, (6.3.19) implies that the capacity of the network whose matrix of synaptic weights is the approximation of $\mathbf{U}_m$ by $\mathbf{U}_m[1]$ is larger than that of the network trained using the outer-product rule. In addition, specific bounds for this network can be found from the previous evaluation of the capacity bounds proposed for the neural network trained using the outer-product rule. For example, Figure 6.3 indicates that the most reliable capacity bound for the network with matrix of synaptic weights $\mathbf{U}_m[1]$ is $c_1[1] = 1/(4\log n)^{1/2}$.

Now consider the case where $\mathbf{U}_m[2]$ is the matrix of synaptic weights. If $\mathbf{x}_\alpha$ is the input of the network, its response is determined in this case by $x_i = \text{sgn}(x_{i,\alpha} + n_{i,\alpha}[2])$, where the noise component $n_{i,\alpha}[2]$ can be obtained from (6.3.2) as follows

$$n_{i,\alpha}[2] = \frac{1}{n^3}\sum_{j=1}^{n}\sum_{q_1,q_2=1}^{n}\sum_{k_1\neq k_2}^{m}\sum_{k_2\neq k}^{m}\sum_{k\neq\alpha}^{m} x_{i,k_1}x_{q_1,k_1}x_{q_1,k_2}x_{q_2,k_2}x_{q_2,k}x_{j,k}x_{j,\alpha} \tag{6.3.20}$$

Under the same assumptions regarding the stored patterns, $n_{i,\alpha}[2]$ can be considered to be a Gaussian zero-mean random variable. It can be similarly shown that the variance of $n_{i,\alpha}[2]$ is given by

$$\sigma_1[2]^2 = (\frac{m-1}{n})^3 \approx (\frac{m}{n})^3 \tag{6.3.21}$$

If $c_1[2] = m_2/n$ is the capacity ratio of the network with matrix of synaptic weights $\mathbf{U}_m[2]$,

$$\rho = \frac{1}{\sigma_1[2]} = (\frac{n}{m_2})^{\frac{3}{2}} = (\frac{1}{c_1[2]})^{\frac{3}{2}} \tag{6.3.22}$$

A comparison of (6.3.9) and (6.3.22) provides the relationship between $c_1[2]$ and c_1 as follows

$$c_1[2] = c_1^{\frac{1}{3}} \tag{6.3.23}$$

This analysis can be generalized to neural networks whose matrix of synaptic weights is any finite approximation of $\mathbf{U}_m$. Assuming that $\mathbf{U}_m[L]$ is the matrix of synaptic weights and $\mathbf{x}_\alpha$ the input of the network, the output of the network is determined by

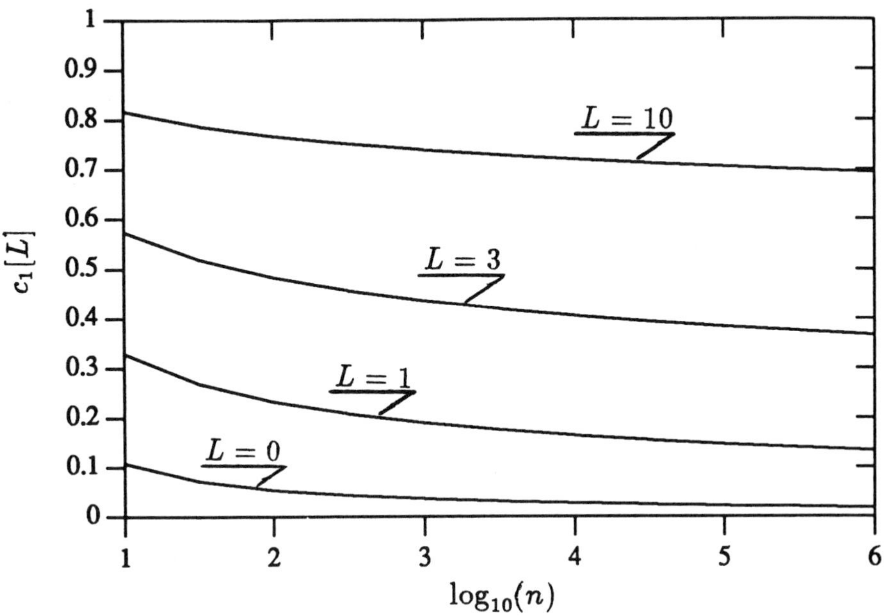

Figure 6.4: Analytically predicted capacity ratios for the neural network trained using various approximations of the optimal matrix of synaptic weights $\mathbf{U}_m$ as a function of n.

the combination of (6.3.1) and (6.3.2). The noise component $n_{i,\alpha}[L]$ is a zero-mean Gaussian random variable with variance

$$\sigma_1[L]^2 = (\frac{m-1}{n})^{L+1} \approx (\frac{m}{n})^{L+1} \tag{6.3.24}$$

Assuming that $c_1[L] = m_L/n$ is the capacity ratio for this network,

$$\rho = \frac{1}{\sigma_1[L]} = (\frac{n}{m_L})^{\frac{L+1}{2}} = (\frac{1}{c_1[L]})^{\frac{L+1}{2}} \tag{6.3.25}$$

A comparison of (6.3.9) and (6.3.25) indicates that

$$c_1[L] = c_1^{\frac{1}{L+1}} \tag{6.3.26}$$

In conclusion, the capacity of the network whose matrix of synaptic weights is the approximation of $\mathbf{U}_m$ by $\mathbf{U}_m[L]$ is given as a function of the capacity of the network trained using the outer-product rule by (6.3.26). Figure 6.4 shows the capacity ratios $c_1[L]$ as a function of the pattern length n for different values of L.

The previous analysis can be the basis for the derivation of the capacity bound of the optimally trained neural network. According to (6.2.23), the optimal matrix of synaptic weights $\mathbf{U}_m$ is the limit of $\mathbf{U}_m[L]$ when L approaches infinity. Therefore, the capacity bound of the optimally trained neural network is given by

$$c_1^{opt} = \lim_{L\to\infty} c_1[L] = \lim_{L\to\infty} c_1{}^{\frac{1}{L+1}} = 1 \tag{6.3.27}$$

Under the assumptions regarding the stored patterns made above, the capacity of the optimally trained neural network converges asymptotically to unity. The asymptotic convergence is imposed by the assumptions upon which this analysis was based. In fact, (6.3.26) indicates that c_1^{opt} cannot be unity, since the approximations (6.3.5) and (6.3.6) are valid only for $\rho \gg 1$. Ideally, the capacity of this network is independent of the length of the stored patterns, n. On the other hand, the derivation of this capacity bound is not affected by the specific capacity bound assumed for the network trained using the outer-product rule, since (6.3.27) is valid for any $c_1 < 1$. The capacity bound for the optimally trained neural network is a rather optimistic one, due to the strong assumptions regarding the correlation of the stored patterns. In fact, in practical situations the real capacity of the network is expected to be lower than the analytically derived one. However, the capacity bound derived for the optimally trained first-order neural network operating under favorable conditions, provides an indication for the maximum capacity which can be achieved by single-layered neural networks.

6.3.1 Experimental Results

The derivation of the capacity bound for the optimally trained neural networks was based on the assumption that the input of the network was one of the stored patterns. When the input of the network is a noisy or incomplete version of one of the stored patterns, it is expected that the capacity of the network will be lower than the analytically predicted one. The capacity of the optimally trained

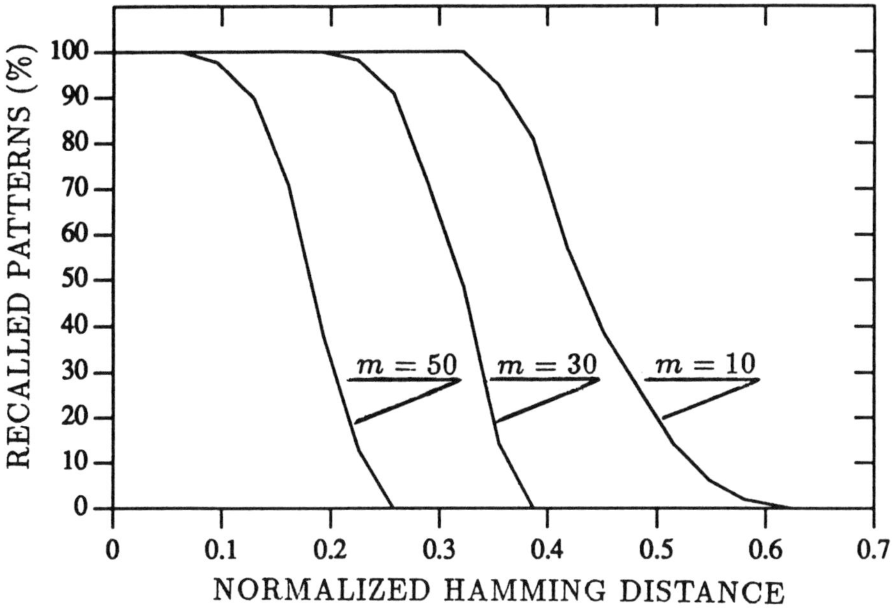

Figure 6.5: Percentage of perfectly recalled patterns by an optimally trained neural network as a function of the normalized Hamming distance between the input patterns and one of the stored patterns for different values of m.

neural network when its inputs are not members of the training set is evaluated by the next two sets of experiments.

The first set of experiments evaluates the generalization ability of the optimally trained network when the number of associations stored in the network increases. The length of the stored patterns was $n = 150$, while the number of associations stored in the network varied from $m = 10$ to $m = 50$. For each m, 1000 noisy input patterns were generated. The normalized Hamming distance between each one of the input patterns and one of the stored patterns was uniformly distributed between 0 and 1. Figure 6.5 shows the percentage of perfectly recalled patterns, for each m, as a function of the normalized Hamming distance between the input patterns and one of the stored patterns. Figure 6.5 indicates that the performance of the network is substantially affected by an increase in the number of stored patterns. However, the curves which correspond to different numbers of stored patterns are qualitatively very similar. In fact, regardless of the number of patterns stored in the network, the interval $(0, 1)$, which represents the normalized Hamming distance, can be divided into three subintervals. When the normalized Hamming distance between the input pattern and one of the stored patterns lies in the first, second, or third subinterval, the network achieves perfect recall of all, some, or none of the stored patterns, respectively. The reliable operation of the network is guaranteed only if the normalized Hamming distance between the input pattern and one of the stored patterns lies in the first interval. Figure 6.5 indicates that the interval which guarantees reliable operation of the network decreases as the number of stored patterns increases. In other words, the price paid for increasing the number of patterns stored in the network is a decrease in the generalization ability of the network.

The second set of experiments evaluates the effect of the noise level of the input patterns on the capacity of the optimally trained neural network. The length of the stored patterns was $n = 150$, while the number of patterns stored in the network varied from $m = 2$ to $m = 200$. For each m, four sets of input patterns of different noise levels were presented to the network. Figure 6.6 shows the percentage of perfectly recalled patterns as a function of the capacity ratio when the noise level of the input patterns was below a given threshold. Each curve in Figure 6.6 corresponds to the performance of the network when the normalized Hamming distance between the input patterns and one of the stored patterns was zero or uniformly distributed between $0 - 1/10$, $0 - 1/4$, and $0 - 1/2$. Figure 6.6 indicates that the network can guarantee perfect recall of the stored patterns only if the capacity ratio does not exceed a certain threshold, regardless of the noise level of the input patterns. The only exception is the case where the noise level of the input patterns is severely high, that is, when the normalized Hamming distance between the input patterns and one of the stored patterns is uniformly distributed between 0 and 1/2. In this case, the network cannot guarantee perfect recall of the stored patterns, even if the number of patterns stored in the network is very small. For input patterns of different noise level, Figure 6.6 provides the

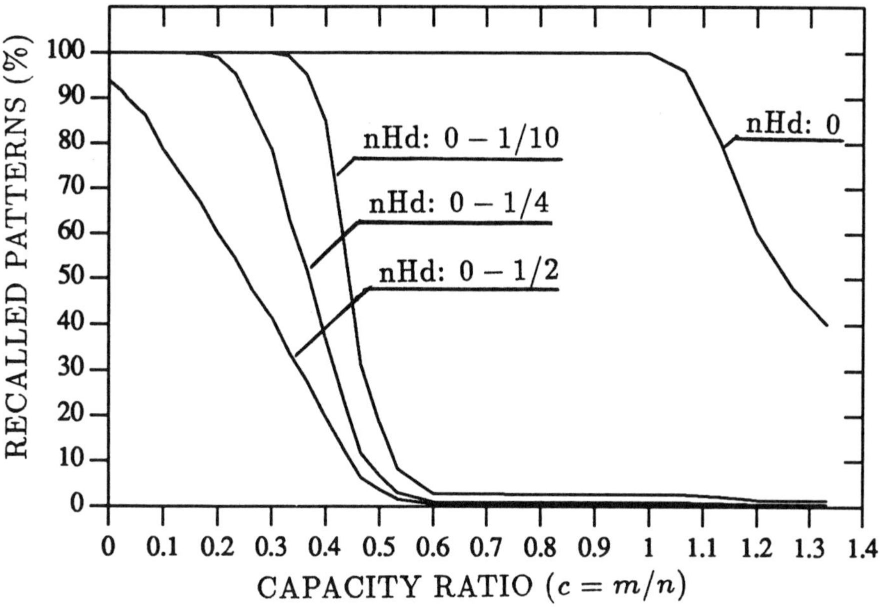

Figure 6.6: Percentage of perfectly recalled patterns by an optimally trained neural network as a function of the capacity ratio. The normalized Hamming distance (nHd) between the input patterns and one of the stored patterns was zero or uniformly distributed between $0 - 1/10$, $0 - 1/4$, and $0 - 1/2$.

maximum allowable number of patterns which can be stored in the optimally trained neural network and still guarantee perfect recall of the stored patterns. When the input patterns are identical to the stored patterns, the maximum allowable number of patterns is approximately equal to the length n of the stored patterns. This is an experimental verification of the analysis presented in this section, which indicated that the capacity ratio of the optimally trained neural network approaches unity. This analysis was based on the assumption that the input of the network was one of the stored patterns. Not surprisingly, the allowable number of stored patterns decreases as the noise level of the input patterns increases. Finally, regardless of the noise level of the input patterns, the capacity ratio which guarantees reliable operation of the optimally trained neural network is much smaller than the analytically predicted one.

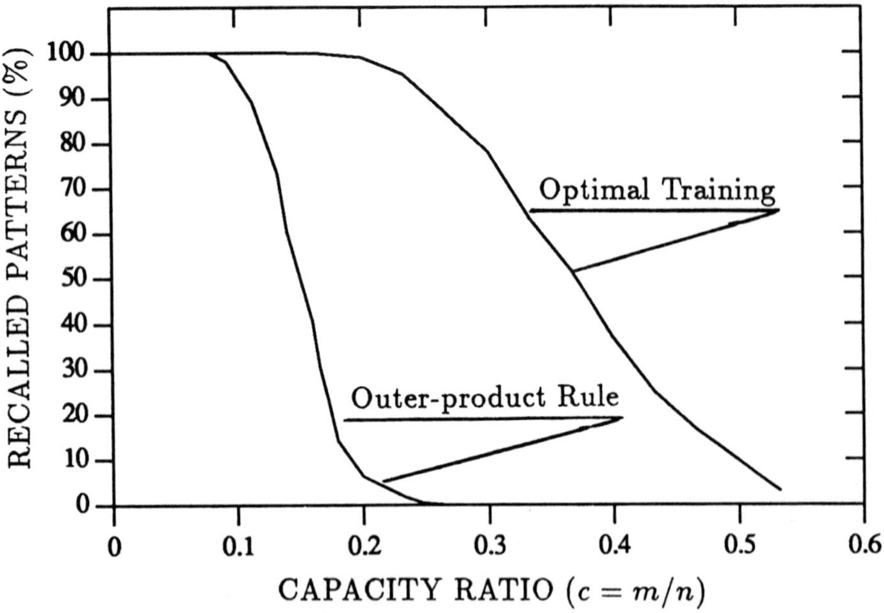

Figure 6.7: Percentage of perfectly recalled patterns by an optimally trained neural network and the same network trained using the outer-product rule as a function of the capacity ratio.

The last set of experiments compares the capacity of the neural network trained using the outer-product rule and the optimally trained neural network when the input patterns are not members of the training set. The number of patterns stored in the network varied from $m = 2$ to $m = 80$. The length of the stored patterns was $n = 150$. For each experiment, 1000 noisy patterns were generated to be used as input patterns. The normalized Hamming distance between the input patterns and one of the stored patterns was uniformly distributed between 0 and 1/4. In each experiment, the inputs of both networks were identical. Figure 6.7 shows the percentage of perfectly recalled patterns by the two networks as a function of the capacity ratio. Figure 6.7 indicates that, regardless of the learning scheme employed, the network can guarantee perfect recall of the stored patterns only if the capacity ratio does not exceed a certain threshold. For each network, this threshold provides the maximum allowable number of stored patterns which guarantee perfect recall. According to Figure 6.7, the optimally trained neural network can reliably store a much larger number of patterns than the network trained using the outer-product rule. When the number of stored patterns exceeds the maximum allowable one, the recall ability of the network trained using the outer-product rule drops much faster than that of the optimally trained neural network.

6.4 Output Nonlinearities and Network Performance

A single-layered neural network with binary output can be alternatively trained with respect to its analog or binary output. The response of a neural network with binary output is determined by the sign, and not by the magnitude of the analog output components. Training the network with respect to its binary output increases the degrees of freedom in the distribution of information over the synaptic weights, since the trained network can produce the desired output from a much wider range of analog outputs. However, training the network with respect to its binary output entails certain analytical problems, due to the hard-limiting nonlinearity sgn(.). These analytical problems can be avoided by approximating the hard-limiters by soft-limiting nonlinearities. In practice, the neural network is trained by minimizing the objective function $E = \frac{1}{2}\sum_{k=1}^{m}\sum_{i=1}^{n_o}(y_{i,k} - \hat{y}_{i,k})^2$, where the estimates provided by the network consist of the elements

$$\hat{y}_{i,k} = \sigma(\bar{y}_{i,k}) = \sigma(\sum_{j=1}^{n_i} w_{ij}x_{j,k}) \; , i = 1, 2, \ldots, n_o \tag{6.4.1}$$

In (6.4.1), $\sigma(.)$ is a continuous, differentiable everywhere function which approximates the signum function. In this chapter, the two states of the network are $\{-1, +1\}$ and the hard-limiting nonlinearity sgn(.) is approximated in the training phase by the hyperbolic tangent, that is, $\sigma(x) = \tanh(x)$.

The Delta Rule can be the basis for the development of learning algorithms for training single-layered neural networks with respect to their binary output (Widrow and Lehr, 1990). However, the

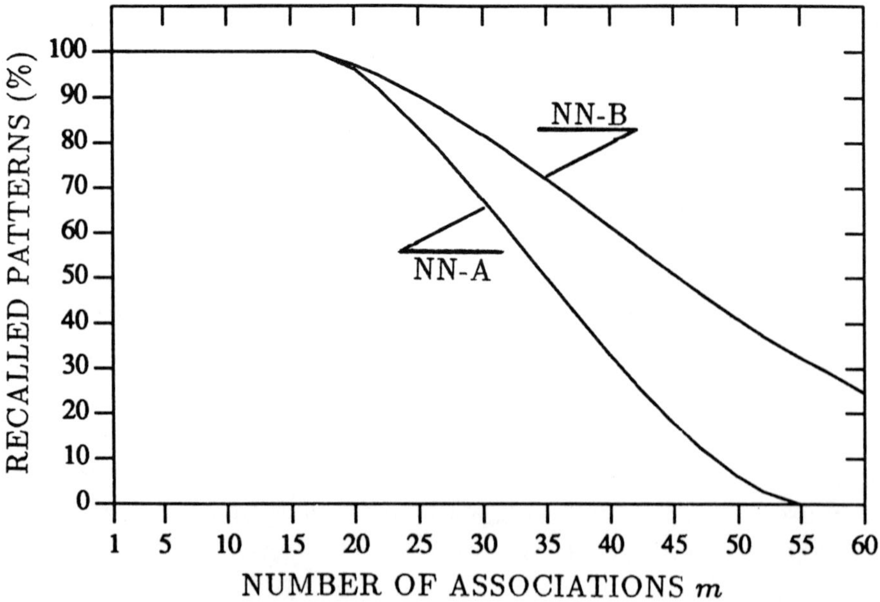

Figure 6.8: Percentage of perfectly recalled patterns by a single-layered neural network trained with respect to its analog output (NN-A) and the same network trained with respect to its binary output (NN-B) as a function of the number of stored associations.

learning algorithms based on the Delta Rule are characterized by slow convergence. An alternative optimization strategy presented in chapter 3 produced the ELEANNE 3, a second-order learning algorithm which performs the training of single-layered neural networks with respect to their binary output, and its simplified version, called ELEANNE 4. The matrix of synaptic weights resulting from training a single-layered neural network with respect to its binary output is not unique, regardless of the algorithm employed for its training. As a result, the performance of the trained neural network cannot be analytically predicted.

6.4.1 Experimental Results

This set of experiments studies the effect of the output nonlinearities on the performance of single-layered neural networks, by experimentally comparing the capacity of single-layered neural networks trained with respect to their binary and analog outputs. The capacity evaluation is based on the percentage of perfectly recalled patterns by a network fed with noisy key patterns when the number of stored associations increases. A single-layered neural network with $n_i = 50$ inputs and $n_o = 10$ output units was trained with respect to its analog or binary output using the ELEANNE 2 or ELEANNE 3, respectively. In this experiment, the number of associations stored in the networks varied from $m = 2$ to $m = 60$. In each case, the two networks were tested by using the same set of 500 input patterns, which were generated in such a way that the Hamming distance between each test pattern and one of the key patterns was 2. Figure 6.8 shows the percentage of perfectly recalled patterns obtained by testing the network trained with respect to its analog and binary output. According to Figure 6.8, the performance of a single-layered neural network is not affected by the particular training scheme chosen when the number of stored associations is small. However, when the number of stored associations exceeds a certain limit, the performance of the network trained with respect to its analog output drops rather quickly. In conclusion, the performance of a single-layered neural network trained with respect to its binary output is superior to that of the same network trained with respect to its analog output.

6.5 Discussion

This chapter focused on the performance of single-layered neural networks. The evaluation of learning schemes for single-layered neural networks began with a comparison of the performance of such networks trained with respect to their analog output using optimal least-squares techniques and the outer-product rule. It was analytically predicted and experimentally verified that the correlation of the key patterns significantly affects the generalization ability of neural networks trained using the outer-product rule, while optimally trained neural networks are practically unaffected by the correlation of the key patterns. It was also shown that the optimal training of any neural network improves the capacity of the network of the same architecture trained using the outer-product rule by a factor roughly proportional to $\log n$, where n is the length of the key patterns. The actual output of the neural networks considered in this chapter is formed by the binary outputs of hard-limiting nonlinearities. This chapter presented experimental verification of the intuitively reasonable argument that the training of such networks with respect to their binary output increases the amount of information that can be reliably stored in them during training.

Appendix A6.1

This Appendix determines the eigenvalues of the matrix $\mathbf{C}^*$, which is obtained in the case where all key patterns are identical. Consider the matrix $\mathbf{C}^*$ with elements

$$(\mathbf{C}^*)_{k\ell} = \begin{cases} 0 & \text{if } k = \ell \\ 1 & \text{if } k \neq \ell \end{cases} \tag{A6.1.1}$$

Define the following matrix,

$$\mathbf{D}(\lambda) = \lambda \mathbf{I} - \mathbf{C}^* = \begin{bmatrix} \lambda & -1 & & -1 \\ -1 & \lambda & & -1 \\ \vdots & \vdots & \vdots & \vdots \\ -1 & -1 & & \lambda \end{bmatrix} \tag{A6.1.2}$$

It can easily be seen that $\det \mathbf{D}(-1) = 0$. That is, $\lambda = -1$ is an eigenvalue of $\mathbf{C}^*$. In fact, it can be shown by induction that $\lambda = -1$ is an eigenvalue of multiplicity $m - 1$. To verify that $\lambda = m - 1$ is the mth eigenvalue of $\mathbf{C}^*$, consider the matrix

$$\mathbf{D}(m-1) = \begin{bmatrix} m-1 & -1 & & -1 \\ -1 & m-1 & & -1 \\ \vdots & \vdots & \vdots & \vdots \\ -1 & -1 & & m-1 \end{bmatrix} \tag{A6.1.3}$$

Since $\mathbf{D}(m-1)$ is an $m \times m$ matrix, its columns sum to the zero vector. In other words, the columns of $\mathbf{D}(m-1)$ are linearly dependent. Therefore, $\det \mathbf{D}(m-1) = 0$. In conclusion, $\lambda = m - 1$ is the

mth eigenvalue of $\mathbf{C}^*$. Considering the nontrivial case that $m \geq 2$, it is clear that $m-1 \geq 1$. Therefore, $\lambda = m-1$ is the dominant eigenvalue of $\mathbf{C}^*$.

Appendix A6.2

This Appendix presents the derivation of closed-form formulas for any finite approximation of $\mathbf{W}_m$. According to the definition of $\mathbf{C}$ in (6.2.10),

$$\mathbf{C} = \frac{1}{n_i}\mathbf{X}_m^*\,\mathbf{X}_m - \mathbf{I} \tag{A6.2.1}$$

Taking into account (A6.2.1),

$$\begin{aligned}\frac{1}{n_i}\mathbf{Y}_m\,\mathbf{C}\,\mathbf{X}_m^* &= \frac{1}{n_i^2}\mathbf{Y}_m\,\mathbf{X}_m^*\,\mathbf{X}_m\,\mathbf{X}_m^* - \frac{1}{n_i}\mathbf{Y}_m\,\mathbf{X}_m^* \\ &= \mathbf{W}_m[0]\,\mathbf{U}_m[0] - \mathbf{W}_m[0] = \mathbf{W}_m[0](\mathbf{U}_m[0]-\mathbf{I})\end{aligned} \tag{A6.2.2}$$

According to (A6.2.1),

$$\begin{aligned}\mathbf{C}^2 &= \frac{1}{n_i^2}\mathbf{X}_m^*\,\mathbf{X}_m\,\mathbf{X}_m^*\,\mathbf{X}_m - \frac{2}{n_i}\mathbf{X}_m^*\,\mathbf{X}_m + \mathbf{I} \\ &= \frac{1}{n_i}\mathbf{X}_m^*\,\mathbf{U}_m[0]\,\mathbf{X}_m - \frac{2}{n_i}\mathbf{X}_m^*\,\mathbf{X}_m + \mathbf{I}\end{aligned} \tag{A6.2.3}$$

Taking into account (A6.2.3),

$$\begin{aligned}\frac{1}{n_i}(\mathbf{Y}_m\,\mathbf{C}^2\,\mathbf{X}_m^*) &= \mathbf{W}_m[0]\,\mathbf{U}_m[0]^2 - 2\,\mathbf{W}_m[0]\,\mathbf{U}_m[0] + \mathbf{W}_m[0] \\ &= \mathbf{W}_m[0]\,(\mathbf{U}_m[0]-\mathbf{I})^2\end{aligned} \tag{A6.2.4}$$

The direct generalization of (A6.2.2) and (A6.2.4) leads to

$$\frac{1}{n_i}(\mathbf{Y}_m\,\mathbf{C}^L\,\mathbf{X}_m^*) = \mathbf{W}_m[0]\,(\mathbf{U}_m[0]-\mathbf{I})^L \tag{A6.2.5}$$

The combination of (6.2.17) and (A6.2.5) gives

$$\mathbf{W}_m[L] = \mathbf{W}_m[L-1] + (-1)^L\,\mathbf{W}_m[0]\,(\mathbf{U}_m[0]-\mathbf{I})^L \tag{A6.2.6}$$

The following analysis derives explicit formulas for the elements of any finite approximation of $\mathbf{W}_m$. It is well-known that

$$(\mathbf{W}_m[0])_{ij} = \frac{1}{n_i} \sum_{k=1}^{m} y_{i,k}\, x_{j,k} \qquad \text{(A6.2.7)}$$

In addition,

$$\begin{aligned}(\mathbf{W}_m[0]\,\mathbf{U}_m[0])_{ij} &= \sum_{q_1=1}^{n_i} (\mathbf{W}_m[0])_{iq_1}\,(\mathbf{U}_m[0])_{q_1 j} \\ &= \frac{1}{n_i^2} \sum_{q_1=1}^{n_i} \sum_{k=1}^{m} \sum_{k_1=1}^{m} y_{i,k_1}\, x_{q_1,k_1},\, x_{q_1,k}\, x_{j,k} \\ &= \frac{1}{n_i} \sum_{k=1}^{m} [y_{i,k} + \frac{1}{n_i} \sum_{q_1=1}^{n_i} \sum_{k_1 \neq k}^{m} y_{i,k_1}\, x_{q_1,k_1}\, x_{q_1,k}]\, x_{j,k}\end{aligned} \qquad \text{(A6.2.8)}$$

The above results can be the basis for evaluating the elements of any finite approximation of $\mathbf{W}_m$. For $L = 1$, (A6.2.6) amounts to

$$\mathbf{W}_m[1] = 2\,\mathbf{W}_m[0] - \mathbf{W}_m[0]\,\mathbf{U}_m[0] \qquad \text{(A6.2.9)}$$

The combination of (A6.2.7) and (A6.2.8) with (A6.2.9) results in

$$(\mathbf{W}_m[1])_{ij} = \frac{1}{n_i} \sum_{k=1}^{m} [y_{i,k} - \frac{1}{n_i} \sum_{q_1=1}^{n_i} \sum_{k_1 \neq k}^{m} y_{i,k_1}\, x_{q_1,k_1}\, x_{q_1,k}]\, x_{j,k} \qquad \text{(A6.2.10)}$$

Obviously, (A6.2.10) can also be written as follows

$$(\mathbf{W}_m[1])_{ij} = \frac{1}{n_i} \sum_{k=1}^{m} y_{i,k}[1]\, x_{j,k} \qquad \text{(A6.2.11)}$$

where

$$y_{i,k}[1] = y_{i,k} - \frac{1}{n_i} \sum_{q_1=1}^{n_i} \sum_{k_1 \neq k}^{m} y_{i,k_1}\, x_{q_1,k_1}\, x_{q_1,k} \qquad \text{(A6.2.12)}$$

The elements of $\mathbf{W}_m[2]$ can similarly be obtained as (Karayiannis, 1991a)

$$(\mathbf{W}_m[2])_{ij} = \frac{1}{n_i} \sum_{k=1}^{m} y_{i,k}[2]\, x_{j,k} \qquad \text{(A6.2.13)}$$

where

$$y_{i,k}[2] = y_{i,k}[1] + \frac{1}{n_i^2} \sum_{q_1=1}^{n_i} \sum_{q_2=1}^{n_i} \sum_{k_1 \neq k_2}^{m} \sum_{k_2 \neq k}^{m} y_{i,k_1}\, x_{q_1,k_1}\, x_{q_1,k_2}\, x_{q_2,k_2}\, x_{q_2,k} \tag{A6.2.14}$$

The above results can be generalized to any finite approximation of $\mathbf{W}_m$. The elements of $\mathbf{W}_m[L]$ are given by

$$(\mathbf{W}_m[L])_{ij} = \frac{1}{n_i} \sum_{k=1}^{m} y_{i,k}[L]\, x_{j,k} \tag{A6.2.15}$$

where $y_{i,k}[L]$ is given in terms of $y_{i,k}[L-1]$ as follows

$$\begin{aligned} y_{i,k}[L] \;&=\; y_{i,k}[L-1] + \frac{(-1)^L}{n_i^L} \\ &\sum_{q_1,\ldots,q_L=1}^{n_i} \sum_{k_1 \neq k_2}^{m} \cdots \sum_{k_L \neq k}^{m} y_{i,k_1} x_{q_1,k_1} x_{q_1,k_2} x_{q_2,k_2} \cdots x_{q_L,k_L} x_{q_L,k} \end{aligned} \tag{A6.2.16}$$

and $y_{i,k}[0] = y_{i,k}$.

The remainder of this Appendix presents the derivation of an alternative expression for the elements of $\mathbf{W}_m[L]$. The elements of $\mathbf{W}_m[0]$ are given by (A6.2.7). In addition,

$$\begin{aligned} (\mathbf{W}_m[0]\, \mathbf{U}_m[0])_{ij} \;&=\; \sum_{q_1=1}^{n_i} (\mathbf{W}_m[0])_{iq_1}\, (\mathbf{U}_m[0])_{q_1 j} \\ &=\; \frac{1}{n_i^2} \sum_{q_1=1}^{n_i} \sum_{k=1}^{m} \sum_{k_1=1}^{m} y_{i,k}\, x_{q_1,k}\, x_{q_1,k_1}\, x_{j,k_1} \\ &=\; \frac{1}{n_i} \sum_{k=1}^{m} y_{i,k}\, [x_{j,k} + \frac{1}{n_i} \sum_{q_1=1}^{n_i} \sum_{k_1 \neq k}^{m} x_{q_1,k}\, x_{q_1,k_1}\, x_{j,k_1}] \end{aligned} \tag{A6.2.17}$$

The combination of (A6.2.7) and (A6.2.17) with (A6.2.9) amounts to

$$(\mathbf{W}_m[1])_{ij} = \frac{1}{n_i} \sum_{k=1}^{m} y_{i,k}\, [x_{j,k} - \frac{1}{n_i} \sum_{q_1=1}^{n_i} \sum_{k_1 \neq k}^{m} x_{q_1,k}\, x_{q_1,k_1}\, x_{j,k_1}] \tag{A6.2.18}$$

Obviously, (A6.2.18) can also be written as follows

$$(\mathbf{W}_m[1])_{ij} = \frac{1}{n_i} \sum_{k=1}^{m} y_{i,k}\, x_{j,k}[1] \tag{A6.2.19}$$

where

$$x_{j,k}[1] = x_{j,k} - \frac{1}{n_i} \sum_{q_1=1}^{n_i} \sum_{k_1 \neq k}^{m} x_{q_1,k}\, x_{q_1,k_1}\, x_{j,k_1} \tag{A6.2.20}$$

The elements of $\mathbf{W}_m[2]$ can similarly be obtained as (Karayiannis, 1991a)

$$(\mathbf{W}_m[2])_{ij} = \frac{1}{n_i} \sum_{k=1}^{m} y_{i,k}\, x_{j,k}[2] \tag{A6.2.21}$$

where

$$x_{j,k}[2] = x_{j,k}[1] + \frac{1}{n_i^2} \sum_{q_1=1}^{n_i} \sum_{q_2=1}^{n_i} \sum_{k_2 \neq k_1}^{m} \sum_{k_1 \neq k}^{m} x_{q_1,k}\, x_{q_1,k_1}\, x_{q_2,k_1}\, x_{q_2,k_2}\, x_{j,k_2} \tag{A6.2.22}$$

The above results can be generalized to any finite approximation of $\mathbf{W}_m$. The elements of $\mathbf{W}_m[L]$ are given by

$$(\mathbf{W}_m[L])_{ij} = \frac{1}{n_i} \sum_{k=1}^{m} y_{i,k}\, x_{j,k}[L] \tag{A6.2.23}$$

where $x_{j,k}[L]$ is given in terms of $x_{j,k}[L-1]$ as follows

$$\begin{aligned} x_{j,k}[L] &= x_{j,k}[L-1] + \frac{(-1)^L}{n_i^L} \\ & \sum_{q_1,\ldots,q_L=1}^{n_i} \sum_{k_L \neq k_{L-1}}^{m} \cdots \sum_{k_1 \neq k}^{m} x_{q_1,k} x_{q_1,k_1} x_{q_2,k_1} x_{q_2,k_2} \cdots x_{q_L,k_L} x_{j,k_L} \end{aligned} \tag{A6.2.24}$$

and $x_{j,k}[0] = x_{j,k}$.

Appendix A6.3

This Appendix determines the relationship between the direction cosine $c(\mathbf{x}_k[L], \mathbf{x}_\ell)$ and the matrix $\mathbf{C}^{L+1}$. Consider any two stored patterns $\mathbf{x}_k[0] = \mathbf{x}_k$ and $\mathbf{x}_\ell$, such that $k \neq \ell$. The direction cosine provides a measure of the similarity between these patterns and is defined as follows

$$c(\mathbf{x}_k[0], \mathbf{x}_\ell) = \frac{1}{n_i} \sum_{j=1}^{n_i} x_{j,k}[0]\, x_{j,\ell} = \frac{1}{n_i} \sum_{j=1}^{n_i} x_{j,k}\, x_{j,\ell} \qquad \text{(A6.3.1)}$$

Clearly, for $k \neq \ell$, (A6.3.1) provides the nondiagonal elements of $\mathbf{C}$, that is,

$$c(\mathbf{x}_k[0], \mathbf{x}_\ell) = (\mathbf{C})_{k\ell}\, \forall\, k \neq \ell \qquad \text{(A6.3.2)}$$

Consider now an analogous measure of similarity between any pair $\mathbf{x}_k[1]$ and $\mathbf{x}_\ell$, i.e.,

$$c(\mathbf{x}_k[1], \mathbf{x}_\ell) = \frac{1}{n_i} \sum_{j=1}^{n_i} x_{j,k}[1]\, x_{j,\ell} \qquad \text{(A6.3.3)}$$

For $L = 1$, (6.2.22) gives

$$x_{j,k}[1] = x_{j,k} - \frac{1}{n_i} \sum_{q_1=1}^{n_i} \sum_{k_1 \neq k}^{m} x_{q_1,k}\, x_{q_1,k_1}\, x_{j,k_1} \qquad \text{(A6.3.4)}$$

The substitution of (A6.3.4) in (A6.3.3) amounts to

$$c(\mathbf{x}_k[1], \mathbf{x}_\ell) = \frac{1}{n_i} \sum_{j=1}^{n_i} x_{j,k} x_{j,\ell} - \frac{1}{n_i^2} \sum_{j=1}^{n_i} \sum_{q_1=1}^{n_i} \sum_{k_1 \neq k}^{m} x_{q_1,k}\, x_{q_1,k_1}\, x_{j,k_1}\, x_{j,\ell} \qquad \text{(A6.3.5)}$$

For $k \neq l$, the elements of $\mathbf{C}^2$ are obtained from (A6.2.3) as

$$\begin{aligned}
(\mathbf{C}^2)_{k\ell} &= \frac{1}{n_i^2} \sum_{k_1=1}^{m} (\mathbf{X}_m^* \mathbf{X}_m)_{kk_1}\, (\mathbf{X}_m^* \mathbf{X}_m)_{k_1\ell} - \frac{2}{n_i} (\mathbf{X}_m^* \mathbf{X}_m)_{k\ell} \\
&= \frac{1}{n_i^2} \sum_{j=1}^{n_i} \sum_{q_1=1}^{n_i} \sum_{k_1=1}^{m} x_{q_1,k}\, x_{q_1,k_1}\, x_{j,k_1}\, x_{j,\ell} - \frac{2}{n_i} \sum_{j=1}^{n_i} x_{j,k}\, x_{j,\ell} \\
&= \frac{1}{n_i^2} \sum_{j=1}^{n_i} \sum_{q_1=1}^{n_i} \sum_{k_1 \neq k}^{m} x_{q_1,k}\, x_{q_1,k_1}\, x_{j,k_1}\, x_{j,\ell} - \frac{1}{n_i} \sum_{j=1}^{n_i} x_{j,k}\, x_{j,\ell}
\end{aligned} \qquad \text{(A6.3.6)}$$

The comparison of (A6.3.5) and (A6.3.6) indicates that

$$c(\mathbf{x}_k[1], \mathbf{x}_\ell) = -(\mathbf{C}^2)_{k\ell}\, \forall\, k \neq \ell \qquad \text{(A6.3.7)}$$

The direct generalization of (A6.3.2) and (A6.3.7) results in

$$c(\mathbf{x}_k[L], \mathbf{x}_\ell) = (-1)^L (\mathbf{C}^{L+1})_{k\ell}\, \forall\, k \neq \ell \qquad \text{(A6.3.8)}$$

Appendix A6.4

This Appendix derives an explicit relationship between $x_{i,\alpha}[L]$ and $x_{i,\alpha}$. Such a relationship is particularly useful in deriving the response of a neural network whose matrix of synaptic weights is any finite approximation of $\mathbf{W}_m$. This derivation is based on the representation of the elements of $\mathbf{W}_m[L]$ by the combination of (6.2.19) and (6.2.20). Since the key and associated patterns are assumed to be identical, $y_{i,k} = x_{i,k}$ and, therefore, $y_{i,k}[L] = x_{i,k}[L]$. For $L = 1$, (6.2.20) provides

$$x_{i,\alpha}[1] = x_{i,\alpha} - \frac{1}{n} \sum_{q_1=1}^{n} \sum_{k_1 \neq \alpha}^{m} x_{i,k_1}\, x_{q_1,k_1}\, x_{q_1,\alpha} \tag{A6.4.1}$$

Also, for $L = 2$, (6.2.20) gives

$$\begin{aligned} x_{i,\alpha}[2] &= x_{i,\alpha}[1] + \frac{1}{n^2} \sum_{q_1=1}^{n} \sum_{q_2=1}^{n} \sum_{k_1 \neq k_2}^{m} \sum_{k_2 \neq k}^{m} x_{i,k_1}\, x_{q_1,k_1}\, x_{q_1,k_2}\, x_{q_2,k_2}\, x_{q_2,k} \\ &= x_{i,\alpha} - \frac{1}{n} \sum_{j=1}^{n} \sum_{k \neq \alpha}^{m} [x_{i,k} - \sum_{q_1=1}^{n} \sum_{k_1 \neq k}^{m} x_{i,k_1}\, x_{q_1,k_1}\, x_{q_1,k}]\, x_{j,k}\, x_{j,\alpha} \end{aligned} \tag{A6.4.2}$$

Taking into account (A6.4.1),

$$x_{i,\alpha}[2] = x_{i,\alpha} - \frac{1}{n} \sum_{j=1}^{n} \sum_{k \neq \alpha}^{m} x_{i,k}[1]\, x_{j,k}\, x_{j,\alpha} \tag{A6.4.3}$$

The direct generalization of (A6.4.1) and (A6.4.3) leads to

$$x_{i,\alpha}[L] = x_{i,\alpha} - \frac{1}{n} \sum_{j=1}^{n} \sum_{k \neq \alpha}^{m} x_{i,k}[L-1]\, x_{j,k}\, x_{j,\alpha} \tag{A6.4.4}$$

The analysis which follows determines the response of a neural network whose matrix of synaptic weights is any finite approximation of $\mathbf{U}_m$. Assuming that $\mathbf{x}_\alpha$ is the input of the network and $\mathbf{U}_m[L]$ the matrix of synaptic weights, the response of the network is determined by

$$x_i = \mathrm{sgn}(\sum_{j=1}^{n} (\mathbf{U}_m[L])_{ij}\, x_{j,\alpha}) \tag{A6.4.5}$$

where the elements of $\mathbf{U}_m[L]$ are given by (6.2.19) and (6.2.20) for $y_{i,k} = x_{i,k}$. Thus,

$$\begin{aligned}\sum_{j=1}^{n}(\mathbf{U}_m[L])_{ij}\, x_{j,\alpha} &= \frac{1}{n}\sum_{j=1}^{n}\sum_{k=1}^{m} x_{i,k}[L]\, x_{j,k}\, x_{j,\alpha} \\ &= \frac{1}{n}\sum_{j=1}^{n}(x_{i,\alpha}[L]\, x_{j,\alpha}\, x_{j,\alpha} + \sum_{k\neq\alpha}^{m} x_{i,k}[L]\, x_{j,k}\, x_{j,\alpha}) \\ &= x_{i,\alpha}[L] + \frac{1}{n}\sum_{j=1}^{n}\sum_{k\neq\alpha}^{m} x_{i,k}[L]\, x_{j,k}\, x_{j,\alpha} \qquad \text{(A6.4.6)}\end{aligned}$$

Substituting in (A6.4.6) $x_{i,\alpha}[L]$ from (A6.4.4),

$$\sum_{j=1}^{n}(\mathbf{U}_m[L])_{ij}\, x_{j,\alpha} = x_{i,\alpha} + n_{i,\alpha}[L] \qquad \text{(A6.4.7)}$$

where

$$n_{i,\alpha}[L] = \frac{1}{n}\sum_{j=1}^{n}\sum_{k\neq\alpha}^{m}(x_{i,k}[L] - x_{i,k}[L-1])\, x_{j,k}\, x_{j,\alpha} \qquad \text{(A6.4.8)}$$

Taking into account (6.2.20), (A6.4.8) amounts to

$$\begin{aligned}n_{i,\alpha}[L] &= \frac{(-1)^L}{n^{L+1}} \\ &\sum_{j,q_1,\ldots,q_L=1}^{n}\ \sum_{k_1\neq k_2}^{m}\cdots\sum_{k_L\neq k}^{m}\ \sum_{k\neq\alpha}^{m} x_{i,k_1}x_{q_1,k_1}x_{q_1,k_2}\ldots x_{q_L,k}x_{j,k}x_{j,\alpha}\end{aligned}$$
(A6.4.9)

Finally, the combination of (A6.4.5) and (A6.4.7) amounts to $x_i = \mathrm{sgn}(x_{i,\alpha} + n_{i,\alpha}[L])$, where the noise term $n_{i,\alpha}[L]$ is given by (A6.4.9).

Chapter 7

High-order Neural Networks and Networks with Composite Key Patterns

7.1 Introduction

The development of neural networks of order higher than one was approached independently by a number of research groups. Rumelhart et al examined high-order neural networks by introducing new types of units, different than the conventional ones, known as sigma-pi units (Rumelhart and McClelland, 1986). Motivated by the relationship between first-order neural networks and linear discriminant functions, other groups of researchers emphasized the connection between neural networks of order higher than one and nonlinear discriminant functions (Chen et al., 1986; Psaltis and Park, 1986; Psaltis et al., 1988). Giles et al. used high-order recurrent neural networks to infer regular grammars from positive and negative strings of training samples (Giles et al., 1990; Giles et al., 1991). This chapter focuses on the architecture, training, and properties of neural networks of order higher than one (Karayiannis, 1991a; Karayiannis and Venetsanopoulos, 1992e). It is shown that high-order neural networks of any order can be trained by any of the learning algorithms developed for first-order neural networks. This chapter also evaluates the efficiency of the outer-product rule when applied to the training of neural networks of order higher than one. The investigation of the architecture of high-order neural networks leads to the development of neural networks with composite key patterns that are the essential generalization of high-order neural networks. The training, performance, and potential applications of neural networks with composite key patterns are also investigated (Karayiannis, 1991a; Karayiannis and Venetsanopoulos, 1992e).

This chapter is organized as follows: Section 7.2 focuses on the architecture and training of high-order neural networks. Section 7.3 introduces neural networks with composite key patterns, with emphasis on their justification, training, and properties. The capacity of neural networks with composite key patterns is studied in section 7.4. Section 7.5 contains concluding remarks.

7.2 High-order Neural Networks

In this section neural networks of order higher than one are interpreted as first-order neural networks with key patterns of a special form. This interpretation suggests a more detailed study of optimal least-squares training of neural networks of order higher than one. Furthermore, the simplicity of the outer-product rule motivates an investigation of the relationship between the outer-product rule and optimal least-squares training, with particular emphasis on the efficiency of the approximation of optimal least-squares training by the outer-product rule in the case of neural networks of order higher than one (Karayiannis, 1991a).

7.2.1 On the Architecture of High-order Neural Networks

This study begins by considering a second-order neural network trained with respect to the associations $(\mathbf{y}_k, \mathbf{x}_k); k = 1, 2, \ldots, m$, where $\mathbf{x}_k^* = [x_{1,k}\, x_{2,k} \ldots x_{n_i,k}], k = 1, 2, \ldots, m$ are the *key patterns* and $\mathbf{y}_k^* = [y_{1,k}\, y_{2,k} \ldots y_{n_o,k}], k = 1, 2, \ldots, m$ the *associated patterns.*

The response of a second-order neural network to the input pattern $\mathbf{x}^* = [x_1\, x_2 \ldots x_{n_i}]$ is given by

$$y_i = \text{sgn}(\sum_{j_1=1}^{n_i} \sum_{j_2=1}^{n_i} w_{i;j_1,j_2}\, x_{j_1}\, x_{j_2}) \tag{7.2.1}$$

where $w_{i;j_1,j_2}$ are the synaptic weights of the network. Clearly, the network response depends on the weighted summation of the quadratic terms $x_{j_1}\, x_{j_2}, 1 \leq j_1, j_2 \leq n_i$. A neural network of second-order can be realized by a neural network with one layer of hidden units, shown in Figure 2.8. Obviously, this network has a special structure. The number of hidden units n_h is given in terms of the length of the input patterns n_i as $n_h = n_i^2$. The synaptic weights that connect the input and hidden layers are either zero or equal to selected elements of the input patterns. The units that form the hidden layer accept only one input. If the elements of the input $\mathbf{x}$ can only take the values ± 1, $x_i\, x_j = \text{sgn}(x_i\, x_j)$. Therefore, the hidden layer consists of conventional units used as the basic building elements for neural networks. In other words, the hidden units form the sum of their inputs and pass it through a hard-limiting nonlinearity. Obviously, there is only one set of synaptic weights to be determined in the training phase, i.e., the synaptic weights that connect the hidden and the output layers of units.

The training and performance evaluation of neural networks of order higher than one can be facilitated by introducing a special type of matrix product, known as the Kronecker product. In general, the Kronecker product between any rectangular matrices $\mathbf{A} = [a_{ij}]$ and $\mathbf{B}$ of size $p \times q$ and $r \times s$, respectively, is the $pr \times qs$ matrix $\mathbf{A} \otimes \mathbf{B} = [a_{ij}\mathbf{B}]$. Figure 7.1 shows the formation of the Kronecker product between the vectors $\mathbf{a}$ and $\mathbf{b}$ of size $n_1 \times 1$ and $n_2 \times 1$, respectively. It can easily be seen from Figure 2.8 that the outputs of the hidden units form a new vector $\bar{\mathbf{x}}^2$, which is the Kronecker product of the input pattern $\mathbf{x}$ with itself, i.e., $\bar{\mathbf{x}}^2 = \mathbf{x} \otimes \mathbf{x}$. The new vector $\bar{\mathbf{x}}^2$ is the input of the single-layered neural network of first order. Accordingly, the general structure of a neural network of second order is shown in Figure 7.2. Clearly, a second-order neural network consists of a block that forms the Kronecker product between vectors and a single-layered neural network of first order. The

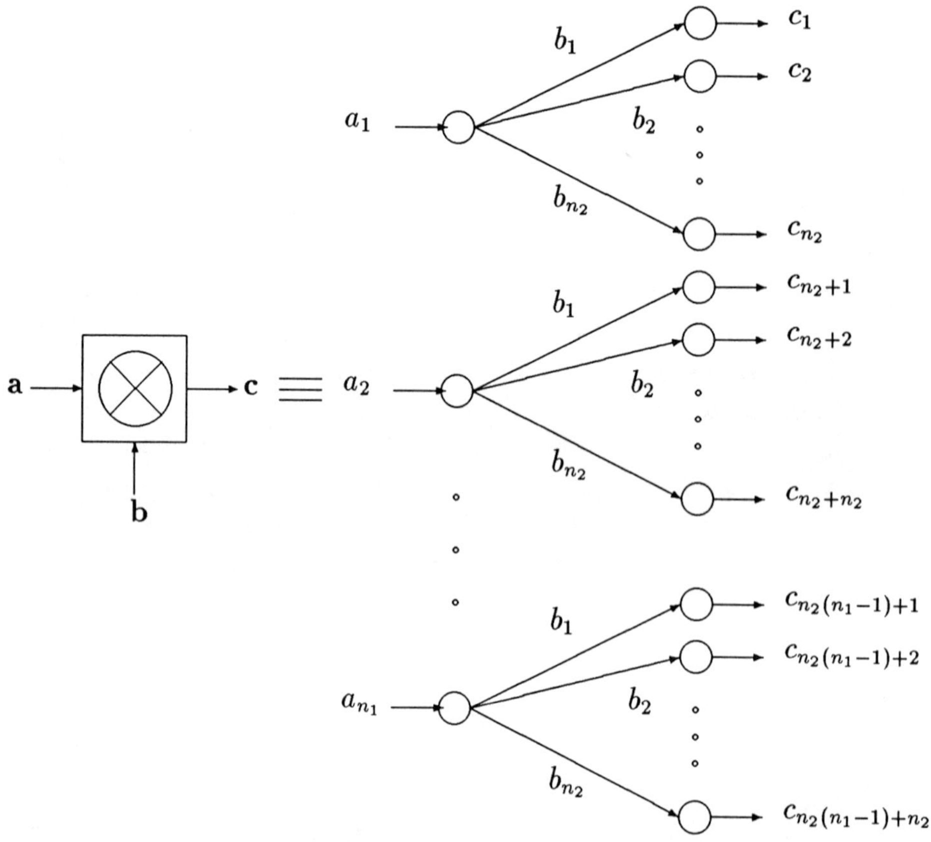

Figure 7.1: The Kronecker product $\mathbf{c} = \mathbf{a} \otimes \mathbf{b}$ between the vectors $\mathbf{a}$ and $\mathbf{b}$.

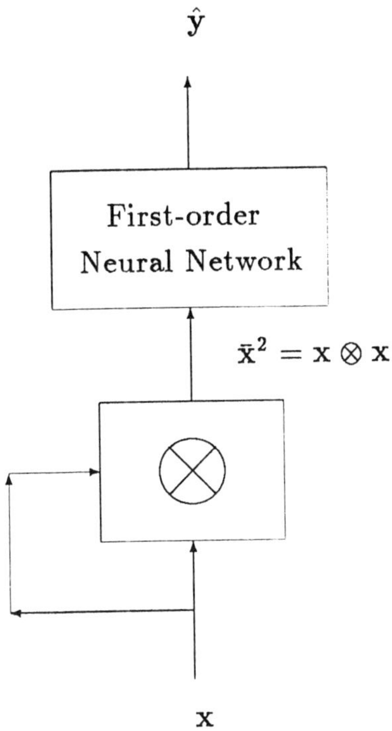

Figure 7.2: A neural network of second order.

actual input of this network is given in terms of the input pattern $\mathbf{x}$ by $\bar{\mathbf{x}}^2 = \mathbf{x} \otimes \mathbf{x}$.

The interpretation presented above indicates that the training of a neural network of second order with respect to the associations $(\mathbf{y}_k, \mathbf{x}_k)$, $k = 1, 2, \ldots, m$ is equivalent to training a neural network of first order with respect to the associations $(\mathbf{y}_k, \bar{\mathbf{x}}_k^2)$, $k = 1, 2, \ldots, m$, where the new key patterns $\bar{\mathbf{x}}_k^2$ are given in terms of the original key patterns $\mathbf{x}_k$ by $\bar{\mathbf{x}}_k^2 = \mathbf{x}_k \otimes \mathbf{x}_k$. The same reasoning can be extended to the training of neural networks of any order. According to the above discussion, a neural network of any order can be trained by applying any of the learning algorithms proposed for training single-layered neural networks of first order. For example, a high-order neural network can be trained by one of the algorithms presented in chapters 3 and 4. However, the size of the first-order neural network to be trained increases substantially with the order of the original network, as a result of the special formation of the input patterns. For example, training a second-order neural network involving key patterns of length n_i is equivalent to training a first-order neural network with key patterns of length $\bar{n}_i = n_i^2$. If n_o is the length of the associated patterns, the number of synaptic weights to be determined by the learning algorithm is $n_o \times n_i^2$ as opposed to the $n_o \times n_i$ synaptic weights of the corresponding neural network of first order. In conclusion, the application of any adaptive algorithm to the training of neural networks of order higher than one implies a tremendous increase in time and computational requirements.

The outer-product rule is the simplest and most widely used learning scheme for training neural networks of order higher than one. However, the outer-product rule is not an optimal learning scheme in any sense. The relationship between optimal least-squares training and the outer-product rule in the case of first-order neural networks was investigated in chapter 6. This investigation revealed that the outer-product rule is a rather crude approximation of optimal least-squares training. Therefore, it would be interesting to investigate the same relationship in the case of neural networks of order higher than one. The first step towards this end is an investigation of optimal least-squares training of high-order neural networks (Karayiannis, 1991a).

7.2.2 Optimal Least-squares Training of High-order Neural Networks

Consider two sets of vectors, or patterns, $\mathbf{x}_k$ and $\mathbf{y}_k$, $k = 1, 2, \ldots, m$ of size $n_i \times 1$ and $n_o \times 1$, respectively. The neural network can be seen as a system which associates the two sets of patterns in such a way that the response of the system to the key pattern $\mathbf{x}_k$ is the associated pattern $\mathbf{y}_k$. Optimal associative recall is the problem of determining the mapping which optimally associates these two sets of patterns. The application of optimal associative recall in the training of neural networks was initially proposed in the early seventies (Kohonen and Ruohonen, 1973). According to this formulation, the associated data $\mathbf{y}_k, k = 1, 2, \ldots, m$ can be related to the key data $\mathbf{x}_k, k = 1, 2, \ldots, m$ through the following linear mapping

$$\mathbf{y}_k = \mathbf{W}_m^1 \mathbf{x}_k \, \forall\, k = 1, 2, \ldots, m \tag{7.2.2}$$

where $\mathbf{W}_m^1$ is the $n_o \times n_i$ matrix of synaptic weights to be optimally determined. If $\mathbf{X}_m(\mathbf{Y}_m)$ is the matrix whose column vectors are $\mathbf{x}_k(\mathbf{y}_k), k = 1, 2, \ldots, m$, (7.2.2) amounts to

$$\mathbf{Y}_m = \mathbf{W}_m^1 \mathbf{X}_m \tag{7.2.3}$$

It was shown that the optimal solution of (7.2.3) in the least-squares sense is given by (Kohonen and Ruohonen, 1973)

$$\mathbf{W}_m^1 = \mathbf{Y}_m \mathbf{X}_m^+ = \mathbf{Y}_m (\mathbf{X}_m^* \mathbf{X}_m)^+ \mathbf{X}_m^* \tag{7.2.4}$$

where $\mathbf{X}_m^+$ is the generalized inverse of $\mathbf{X}_m$ (Rao and Mitra, 1971; Albert, 1972).

The assumption that the associated data are related to the key data through a linear mapping provided the basis for optimal least-squares training of neural networks of first order. In general, the associated data can be expressed as a nonlinear function of the key data. Poggio studied the problem of optimal nonlinear associative recall by determining the nonlinear mapping which optimally associates the two sets of data (Poggio, 1975). According to his formulation, the nonlinear mapping is of polynomial form, consisting of a linear term, a quadratic term, etc., up to degree r. In mathematical terms,

$$\begin{aligned} y_{i,k} &= \sum_{j_1=1}^{n_i} (\mathbf{W}_m^1)_{i,j_1} x_{j_1,k} + \sum_{j_1=1}^{n_i} \sum_{j_2=1}^{n_i} (\mathbf{W}_m^2)_{i;j_1,j_2} x_{j_1,k} x_{j_2,k} + \ldots \\ &+ \sum_{j_1=1}^{n_i} \sum_{j_2=1}^{n_i} \cdots \sum_{j_r=1}^{n_i} (\mathbf{W}_m^r)_{i;j_1,j_2,\ldots,j_r} x_{j_1,k} x_{j_2,k} \ldots x_{j_r,k} \end{aligned} \tag{7.2.5}$$

This nonlinear mapping can be written in a more compact form as follows

$$\mathbf{y}_k = \mathbf{W}_m^1 \bar{\mathbf{x}}_k^1 + \mathbf{W}_m^2 \bar{\mathbf{x}}_k^2 + \ldots + \mathbf{W}_m^r \bar{\mathbf{x}}_k^r \tag{7.2.6}$$

where $\mathbf{W}_m^p, p = 1, 2, \ldots, r$ are matrices of size $n_o \times n_i^p$ to be determined, while $\bar{\mathbf{x}}_k^p, p = 1, 2, \ldots, r$ are defined in terms of $\mathbf{x}_k$ by

$\bar{\mathbf{x}}_k^p = \bar{\mathbf{x}}_k^{p-1} \otimes \mathbf{x}_k$, with $\bar{\mathbf{x}}_k^1 = \mathbf{x}_k$. The general idea behind this approach is the following: The first-order estimate of the associated data is initially obtained, by optimally determining the parameters of the linear term. The error between the associated data and the optimal estimate of first order is then "corrected" by optimally determining the parameters of the quadratic term. In general, the optimal solution of order r "corrects" the error between the associated data and the estimate up to order $r-1$.

The approach followed here is also based on the optimal nonlinear association of two sets of data. The only difference is the form of the nonlinear mapping that relates the associated data to the argument, or the key data. The mapping introduced here is of polynomial form as well. However, when optimal least-squares training of a neural network of order r is considered, the parameters of all terms up to order $r-1$ are set to zero. In other words, the only parameters that are to be optimally estimated correspond to the term of degree r. Therefore, the association of order r directly provides the estimate of order r of the associated data, instead of "correcting" the error between the associated data and the estimate up to order $r-1$. The analysis which follows is facilitated by the use of the Kronecker product notation.

Consider the training of a neural network of second order with respect to the associations $(\mathbf{y}_k, \mathbf{x}_k), k = 1, 2, \ldots, m$. According to the above analysis, this network can be trained by optimally determining the $n_o \times n_i^2$ matrix $\mathbf{W}_m^2$ for which

$$\mathbf{y}_k = \mathbf{W}_m^2 \, \bar{\mathbf{x}}_k^2 \, \forall \, k = 1, 2, \ldots, m \tag{7.2.7}$$

where $\bar{\mathbf{x}}_k^2 = \mathbf{x}_k \otimes \mathbf{x}_k$. According to the definition of the Kronecker product, each element of $\bar{\mathbf{x}}_k^2$ is given by $(\bar{\mathbf{x}}_k^2)_j = (\mathbf{x}_k \otimes \mathbf{x}_k)_j = x_{j_1,k} \, x_{j_2,k}, j = j_2 + n_i(j_1 - 1)$. The nonlinearity of the mapping (7.2.7) becomes obvious by writing the explicit formula for each element of $\mathbf{y}_k$ as follows

$$y_{i,k} = \sum_{j_1=1}^{n_i} \sum_{j_2=1}^{n_i} (\mathbf{W}_m^2)_{ij} \, x_{j_1,k} \, x_{j_2,k} \, , \; j = j_2 + n_i(j_1 - 1) \tag{7.2.8}$$

The combination of the m equations described by (7.2.7) amounts to

$$\mathbf{Y}_m = \mathbf{W}_m^2 \, \bar{\mathbf{X}}_m^2 \tag{7.2.9}$$

where $\mathbf{Y}_m$ is the matrix whose column vectors are $\mathbf{y}_k, k = 1, 2, \ldots, m$ and $\bar{\mathbf{X}}_m^2$ is the $n_i^2 \times m$ matrix whose column vectors are $\bar{\mathbf{x}}_k^2 = \mathbf{x}_k \otimes \mathbf{x}_k, k = 1, 2, \ldots, m$, that is,

$$\begin{aligned}\bar{\mathbf{X}}_m^2 &= [\bar{\mathbf{x}}_1^2 \, \mathbf{x}_2^2 \, \ldots \, \bar{\mathbf{x}}_m^2] \\ &= [\mathbf{x}_1 \otimes \mathbf{x}_1 \; \mathbf{x}_2 \otimes \mathbf{x}_2 \; \ldots \; \mathbf{x}_m \otimes \mathbf{x}_m] = \mathbf{X}_m \bigcirc \mathbf{X}_m\end{aligned} \tag{7.2.10}$$

where $\bigcirc$ denotes a special type of product, known also as Khatri-Rao product (Rao and Mitra, 1971). The optimal solution of (7.2.9) in the least-squares sense is given by

$$\mathbf{W}_m^2 = \mathbf{Y}_m \, \bar{\mathbf{X}}_m^{2+} = \mathbf{Y}_m \, (\bar{\mathbf{X}}_m^{2*} \, \bar{\mathbf{X}}_m^2)^+ \, \bar{\mathbf{X}}_m^{2*} \tag{7.2.11}$$

If the associated and key patterns are identical, i.e., $\mathbf{Y}_m = \mathbf{X}_m$, the optimal matrix of synaptic weights is of size $n_i \times n_i^2$ and is given by $\mathbf{W}_m^2 = \mathbf{X}_m \, (\bar{\mathbf{X}}_m^{2*} \, \bar{\mathbf{X}}_m^2)^+ \, \bar{\mathbf{X}}_m^{2*}$. It is shown here that each element of $\bar{\mathbf{X}}_m^{2*} \, \bar{\mathbf{X}}_m^2$ can be obtained from the corresponding element of $\mathbf{X}_m^* \, \mathbf{X}_m$ according to a simple rule (Poggio, 1975). From the definition of $\bar{\mathbf{X}}_m^2$, each element of $\bar{\mathbf{X}}_m^{2*} \, \bar{\mathbf{X}}_m^2$ is given as follows

$$\begin{aligned}(\bar{\mathbf{X}}_m^{2*} \, \bar{\mathbf{X}}_m^2)_{k\ell} &= \bar{\mathbf{x}}_k^{2*} \, \bar{\mathbf{x}}_\ell^2 \\ &= (\mathbf{x}_k \otimes \mathbf{x}_k)^*(\mathbf{x}_\ell \otimes \mathbf{x}_\ell) = \mathbf{x}_k^* \, \mathbf{x}_\ell \otimes \mathbf{x}_k^* \, \mathbf{x}_l\end{aligned} \tag{7.2.12}$$

Since $\mathbf{x}_k^* \, \mathbf{x}_\ell$ is clearly a scalar quantity, equation (7.2.12) amounts to $(\bar{\mathbf{X}}_m^{2*} \, \bar{\mathbf{X}}_m^2)_{k\ell} = (\mathbf{x}_k^* \, \mathbf{x}_\ell)^2 = ((\mathbf{X}_m^* \, \mathbf{X}_m)_{k\ell})^2$. Therefore,

$$\bar{\mathbf{X}}_m^{2*} \, \bar{\mathbf{X}}_m^2 = (\mathbf{X}_m^* \, \mathbf{X}_m)^{(2)} \tag{7.2.13}$$

where the superscript "(2)" denotes that each one of the elements of $\mathbf{X}_m^* \, \mathbf{X}_m$ is raised to the second power.

The same analysis can be extended to neural networks of any order. It can be similarly verified that a neural network of order r can be interpreted as a neural network of a special structure with $r - 1$ layers of hidden units. The use of the Kronecker product results in the general structure of a neural network of order r shown in Figure 7.3. Once again, a neural network of order r consists of

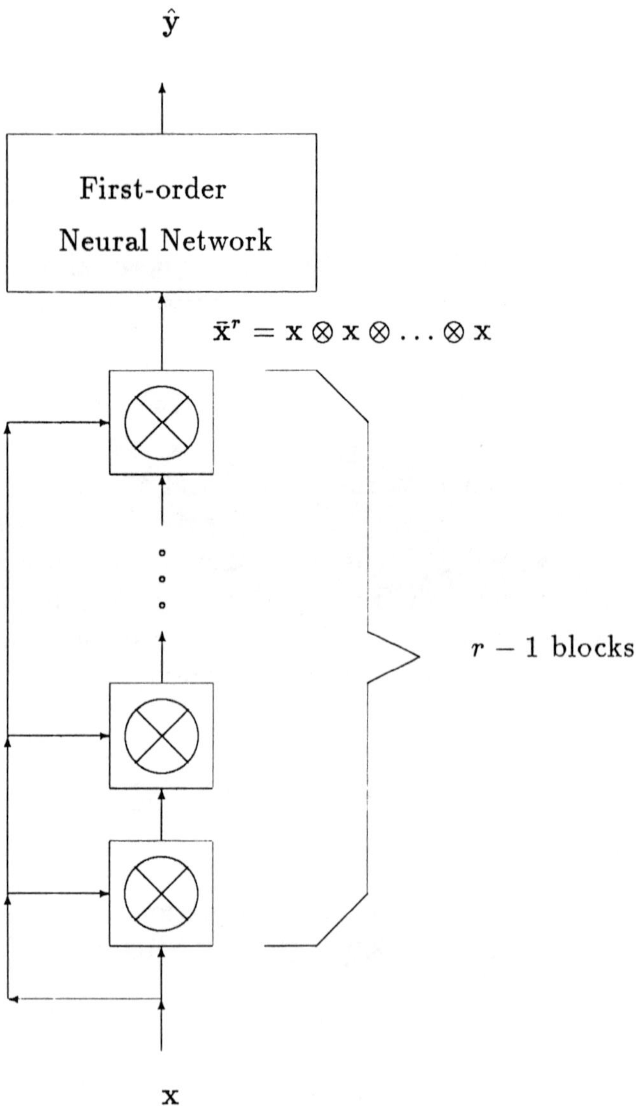

Figure 7.3: A neural network of order r.

$r-1$ blocks that form the Kronecker product between vectors and a single-layered neural network of first order. The actual input of this network is given in terms of the input pattern $\mathbf{x}$ by $\bar{\mathbf{x}}^r = \prod_{p=1}^r \otimes \mathbf{x} = \bar{\mathbf{x}}^{r-1} \otimes \mathbf{x}$. Therefore, training a neural network of order r with respect to the associations $(\mathbf{y}_k, \mathbf{x}_k), k = 1, 2, \ldots, m$ is equivalent to the training of a neural network of first order with respect to the associations $(\mathbf{y}_k, \bar{\mathbf{x}}_k^r), k = 1, 2, \ldots, m$. In this case, each pattern $\bar{\mathbf{x}}_k^r$ is formed in terms of the original key pattern $\mathbf{x}_k$ by $\bar{\mathbf{x}}_k^r = \prod_{p=1}^r \otimes \mathbf{x}_k = \bar{\mathbf{x}}_k^{r-1} \otimes \mathbf{x}_k$.

Consider the training of a neural network of order r with respect to the associations $(\mathbf{y}_k, \mathbf{x}_k), k = 1, 2, \ldots, m$. The direct generalization of the above analysis indicates that this network can be trained by determining the optimal $n_o \times n_i^r$ matrix $\mathbf{W}_m^r$ for which

$$\mathbf{y}_k = \mathbf{W}_m^r \, \bar{\mathbf{x}}_k^r \; \forall \, k = 1, 2, \ldots, m \tag{7.2.14}$$

The vector $\bar{\mathbf{x}}_k^r$ appearing in the right hand side of (7.2.14) is the Kronecker product repeated $r-1$ times between the r identical vectors $\mathbf{x}_k$, that is, $\bar{\mathbf{x}}_k^r = \prod_{p=1}^r \otimes \mathbf{x}_k = \bar{\mathbf{x}}_k^{r-1} \otimes \mathbf{x}_k \, \forall \, r \geq 2$, where $\bar{\mathbf{x}}_k^1 = \mathbf{x}_k$. By the definition of the Kronecker product, each element of $\bar{\mathbf{x}}_k^r$ is given by $(\bar{\mathbf{x}}_k^r)_j = \prod_{p=1}^r x_{j_p,k}, j = 1 + \sum_{p=1}^r n_i^{r-p}(j_p - 1)$. According to the mapping described by (7.2.14), each element of $\mathbf{y}_k$ is given by

$$y_{i,k} = \sum_{j_1=1}^{n_i} \sum_{j_2=1}^{n_i} \cdots \sum_{j_r=1}^{n_i} (\mathbf{W}_m^r)_{ij} \prod_{p=1}^r x_{j_p,k} \,, \; j = 1 + \sum_{p=1}^r n_i^{r-p}(j_p - 1) \tag{7.2.15}$$

The combination of the equations given by (7.2.14) amounts to

$$\mathbf{Y}_m = \mathbf{W}_m^r \, \bar{\mathbf{X}}_m^r \tag{7.2.16}$$

where $\bar{\mathbf{X}}_m^r$ is the $n_i^r \times m$ matrix whose column vectors are $\bar{\mathbf{x}}_k^r, k = 1, 2, \ldots, m$, that is, $\bar{\mathbf{X}}_m^r = [\bar{\mathbf{x}}_1^r \, \bar{\mathbf{x}}_2^r \, \ldots \, \bar{\mathbf{x}}_m^r]$. Taking into account the definition of $\bar{\mathbf{x}}_k^r, k = 1, 2, \ldots, m$,

$$\bar{\mathbf{X}}_m^r = [\bar{\mathbf{x}}_1^{r-1} \otimes \mathbf{x}_1 \, \bar{\mathbf{x}}_2^{r-1} \otimes \mathbf{x}_2 \, \ldots \, \bar{\mathbf{x}}_m^{r-1} \otimes \mathbf{x}_m] = \bar{\mathbf{X}}_m^{r-1} \bigcirc \mathbf{X}_m \, \forall r \geq 2 \tag{7.2.17}$$

where $\bar{\mathbf{X}}_m^1 = \mathbf{X}_m$. The optimal solution of (7.2.16) in the least-squares sense is given by

$$\mathbf{W}_m^r = \mathbf{Y}_m \, \bar{\mathbf{X}}_m^{r\,+} = \mathbf{Y}_m \, (\bar{\mathbf{X}}_m^{r\,*} \, \bar{\mathbf{X}}_m^r)^+ \, \bar{\mathbf{X}}_m^{r\,*} \tag{7.2.18}$$

When the associated patterns coincide with the key patterns, the optimal matrix of synaptic weights for the neural network of order r is of size $n_i \times n_i^r$ and is given by $\mathbf{W}_m^r = \mathbf{X}_m\,(\bar{\mathbf{X}}_m^{r\,*}\,\bar{\mathbf{X}}_m^r)^+\,\bar{\mathbf{X}}_m^{r\,*}$. Once again, each element of $\bar{\mathbf{X}}_m^{r\,*}\,\bar{\mathbf{X}}_m^r$ can be expressed in terms of $\mathbf{X}_m^*\,\mathbf{X}_m$, by generalizing (7.2.13) as follows (Poggio, 1975)

$$\bar{\mathbf{X}}_m^{r\,*}\,\bar{\mathbf{X}}_m^r = (\mathbf{X}_m^*\,\mathbf{X}_m)^{(r)} \qquad (7.2.19)$$

where the superscript "(r)" denotes that each one of the elements of $\mathbf{X}_m^*\,\mathbf{X}_m$ is raised to the rth power.

7.2.3 Optimal Least-squares Training and the Outer-product Rule

The outer-product rule is the simplest learning scheme proposed for training neural networks of order higher than one. Another advantage of the outer-product rule is the potential optical implementation of high-order neural networks trained using this learning scheme (Psaltis and Park, 1986; Psaltis et al., 1988). This section studies the efficiency of the approximation of optimal least-squares training by the outer-product rule (Karayiannis, 1991a). The effect of the increase of the network order on the efficiency of the approximation of the optimal matrix of synaptic weights by the outer-product rule is particularly emphasized.

Consider a neural network of order r and assume that the key patterns are uncorrelated, that is, $\mathbf{X}_m^*\,\mathbf{X}_m = n_i\,\mathbf{I}$, where $\mathbf{I}$ denotes the $m \times m$ identity matrix. Under this assumption, $\bar{\mathbf{X}}_m^{r\,*}\,\bar{\mathbf{X}}_m^r = (\mathbf{X}_m^*\,\mathbf{X}_m)^{(r)} = n_i^r\,\mathbf{I}$, and the optimal solution (7.2.18) becomes

$$\mathbf{W}_m^r = \mathbf{Y}_m\,(n_i^r\,\mathbf{I})^{-1}\,\bar{\mathbf{X}}_m^{r\,*} = \frac{1}{n_i^r}\,\mathbf{Y}_m\,\bar{\mathbf{X}}_m^{r\,*} = \frac{1}{n_i^r}\sum_{k=1}^{m}\mathbf{y}_k\,\bar{\mathbf{x}}_k^{r\,*} \qquad (7.2.20)$$

Each element of $\mathbf{W}_m^r$ is given by

$$(\mathbf{W}_m^r)_{ij} = \frac{1}{n_i^r}\sum_{k=1}^{m} y_{i,k}\prod_{p=1}^{r} x_{j_p,k}\,,\ j = 1 + \sum_{p=1}^{r} n_i^{r-p}(j_p - 1) \qquad (7.2.21)$$

The response of a neural network of order r to the input $\mathbf{x}$ is determined, in general, by the following equation

$$y_i = \text{sgn}(\sum_{j_1=1}^{n_i} \sum_{j_2=1}^{n_i} \cdots \sum_{j_r=1}^{n_i} (\mathbf{W}_m^r)_{ij} \prod_{p=1}^{r} x_{j_p}), j = 1 + \sum_{p=1}^{r} n_i^{r-p}(j_p - 1) \tag{7.2.22}$$

If the key patterns are uncorrelated, the elements of the matrix of synaptic weights $\mathbf{W}_m^r$ are given by (7.2.21). The combination of (7.2.21) and (7.2.22) provides the response of the neural network trained using the outer-product rule, given in this case as follows

$$\begin{aligned} y_i &= \text{sgn}(\frac{1}{n_i^r} \sum_{j_1=1}^{n_i} \sum_{j_2=1}^{n_i} \cdots \sum_{j_r=1}^{n_i} \sum_{k=1}^{m} y_{i,k} \prod_{p=1}^{r} x_{j_p,k}\, x_{j_p}) \\ &= \text{sgn}(\sum_{k=1}^{m} y_{i,k} \prod_{p=1}^{r} (\frac{1}{n_i} \sum_{j_p=1}^{n_i} x_{j_p,k}\, x_{j_p})) = \text{sgn}(\sum_{k=1}^{m} y_{i,k}\, c(\mathbf{x}_k, \mathbf{x})^r) \end{aligned} \tag{7.2.23}$$

where $c(\mathbf{x}_k, \mathbf{x})$ is the direction cosine between the key pattern $\mathbf{x}_k$ and the input $\mathbf{x}$, defined by

$$c(\mathbf{x}_k, \mathbf{x}) = \frac{1}{n_i} \sum_{j=1}^{n_i} x_{j,k}\, x_j \tag{7.2.24}$$

Optimal least-squares training of neural networks of any order coincides with the outer-product rule, provided that the key patterns are uncorrelated. This is the direct extension of a similar result proven for first-order neural networks in chapter 6. On the other hand, the combination of (7.2.23) and (7.2.24) suggests an alternative realization of neural networks of order higher than one trained using the outer-product rule, shown in Figure 7.4 (Psaltis et al., 1988).

The effect of the correlation of the key patterns on the efficiency of the outer-product rule in the case of neural networks of first order was also studied in chapter 6. This analysis was based on the observation that the correlation matrix $\mathbf{X}_m^* \mathbf{X}_m$ has a special form, that is,

$$\mathbf{X}_m^* \mathbf{X}_m = n_i (\mathbf{I} + \mathbf{C}) \tag{7.2.25}$$

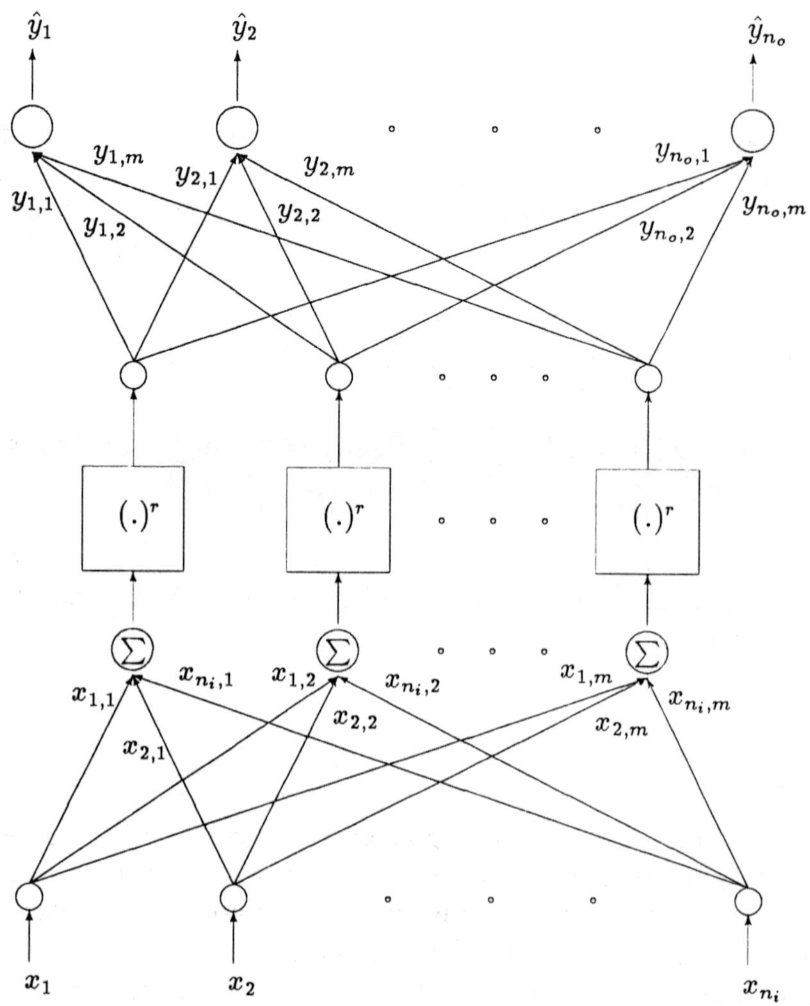

Figure 7.4: An equivalent realization of a neural network of order r, trained using the outer-product rule.

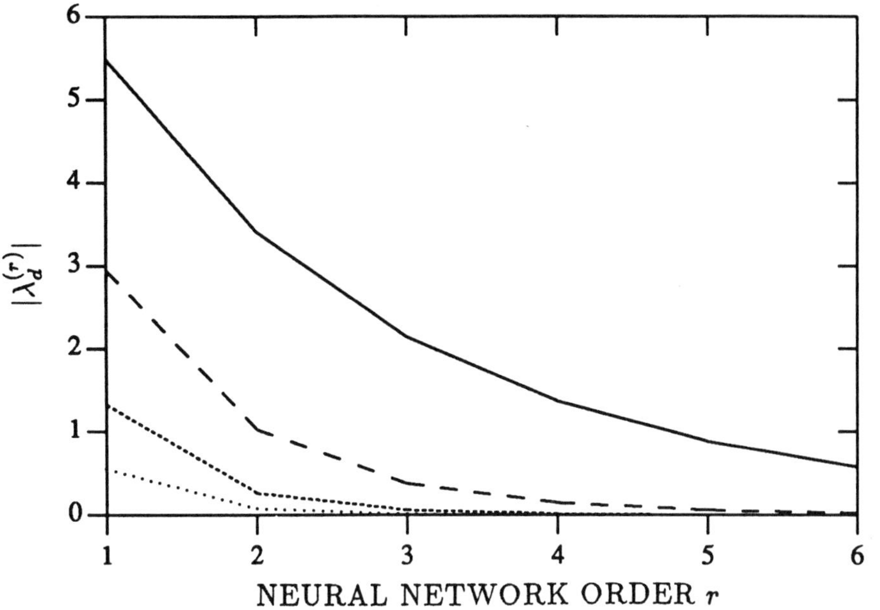

Figure 7.5: The magnitude of the dominant eigenvalue $\lambda_d^{(r)}$ of $\mathbf{C}^{(r)}, r = 1, 2, \ldots, 6$, for four different sets of key patterns.

where $\mathbf{I}$ is the $m \times m$ identity matrix and $\mathbf{C}$ is the $m \times m$ matrix with elements

$$(\mathbf{C})_{k\ell} = \begin{cases} 0 & \text{if } k = \ell \\ \frac{1}{n_i} \sum_{i=1}^{n_i} x_{i,k}\, x_{i,\ell} & \text{if } k \neq \ell \end{cases} \tag{7.2.26}$$

Based on the argument that the information describing the correlation of the key patterns is carried exclusively by $\mathbf{C}$, the dominant eigenvalue λ_d of $\mathbf{C}$ was proposed in chapter 6 as a quantitative measure for the correlation of the key patterns. In fact, the key patterns were classified as *weakly correlated* if $|\lambda_d| < 1$ and *strongly correlated* if $|\lambda_d| \geq 1$.

The assumption that the key patterns are uncorrelated is a strong one, very rarely satisfied in practical situations. Therefore, it is particularly interesting to study the effect of the correlation of the key patterns on the approximation of the optimal matrix of synaptic weights by the outer-product rule in the case of neural networks of order higher than one. Consider a neural network of order

r and assume that the key patterns are correlated. Clearly, the information regarding the correlation of the key patterns is carried in (7.2.18) by

$$\bar{\mathbf{X}}_m^{r\,*}\,\bar{\mathbf{X}}_m^r = (\mathbf{X}_m^*\,\mathbf{X}_m)^{(r)} = n_i^r\,(\mathbf{I} + \mathbf{C}^{(r)}) \tag{7.2.27}$$

where $\mathbf{C}$ is defined in (7.2.26). The form of (7.2.27) indicates that $\mathbf{C}^{(r)}$ carries the information describing the correlation of the key patterns. Assume that $\lambda_d^{(r)}$ is the dominant eigenvalue of $\mathbf{C}^{(r)}$. Although there exists no rigorous mathematical proof, it can be experimentally verified that $|\lambda_d^{(r)}| < |\lambda_d^{(1)}| = |\lambda_d|\,\forall\, r \geq 2$. Figure 7.5 shows the magnitude of the dominant eigenvalues $\lambda_d^{(r)}, r = 1, 2, \ldots, 6$ that correspond to four different sets of key patterns. Clearly, $|\lambda_d^{(r)}|$ decreases as the order r of the neural network increases. Three of the sets of key patterns considered in Figure 7.5 correspond to $|\lambda_d| > 1$, i.e., they consist of strongly correlated key patterns. However, regardless of the dominant eigenvalue $|\lambda_d|$ corresponding to the original set of key patterns, there exists an $r > 1$ such that $|\lambda_d^{(r)}| < 1$. In other words, a set of key patterns that are strongly correlated with regard to a neural network of first order may be weakly correlated with regard to a neural network of order higher than one. For the sake of analytical tractability, assume now that $|\lambda_d^{(r)}| < 1$. Then, the optimal matrix of synaptic weights (7.2.18) can be expanded as follows (Karayiannis, 1991a)

$$\begin{aligned}\mathbf{W}_m^r &= \frac{1}{n_i^r}\sum_{k=0}^{\infty}(-1)^k\,\mathbf{Y}_m\,(\mathbf{C}^{(r)})^k\,\bar{\mathbf{X}}_m^{r\,*} \\ &= \frac{1}{n_i^r}\,\mathbf{Y}_m\,\bar{\mathbf{X}}_m^{r\,*} + \frac{1}{n_i^r}\sum_{k=1}^{\infty}(-1)^k\,\mathbf{Y}_m\,(\mathbf{C}^{(r)})^k\,\bar{\mathbf{X}}_m^{r\,*}\end{aligned} \tag{7.2.28}$$

In the general case where the key patterns are correlated, the outer-product rule is an approximation of the infinite series given by (7.2.28) by the first term only. This is also an extension of an analogous result presented in chapter 6 for first-order neural networks. The effect of the increase of the network order on the efficiency of the approximation of the optimal matrix of synaptic weights by the outer-product rule can be evaluated by comparing the infinite norms $||\mathbf{C}^{(r)}||_\infty$ and $||\mathbf{C}^{(1)}||_\infty = ||\mathbf{C}||_\infty$. By definition, $\mathbf{C}$ is an $m \times m$ matrix with zero diagonal elements. With the exception of the trivial case where the key patterns are identical, the elements of $\mathbf{C}$ satisfy the inequality $|(\mathbf{C})_{k\ell}| < 1$. For $r > 1$, $|(\mathbf{C}^{(r)})_{k\ell}| < |(\mathbf{C})_{k\ell}| \, \forall \, k, \ell = 1, 2, \ldots, m$. Therefore,

$$S_k^{(r)} = \sum_{\ell \neq k}^{m} |(\mathbf{C}^{(r)})_{k\ell}| < \sum_{\ell \neq k}^{m} |(\mathbf{C})_{k\ell}| = S_k \, \forall \, k = 1, 2, \ldots, m \quad (7.2.29)$$

Assuming that $\max_k S_k = S_{k_0}$, it is obvious that $\max_k S_k^{(r)} = S_{k_0}^{(r)}$. Therefore,

$$||\mathbf{C}^{(r)}||_\infty = \max_k \sum_{\ell \neq k}^{m} |(\mathbf{C}^{(r)})_{k\ell}| < \max_k \sum_{\ell \neq k}^{m} |(\mathbf{C})_{k\ell}| = ||\mathbf{C}||_\infty \quad (7.2.30)$$

In conclusion, the outer-product rule provides a better approximation of the optimal matrix of synaptic weights in the case of neural networks of higher order, compared to the same approximation in the case of neural networks of first order.

The effect of the order of the neural network on this approximation is studied here. By the definition of the infinite norm,

$$\lim_{r \to \infty} ||\mathbf{C}^{(r)}||_\infty = \max_k \sum_{\ell \neq k}^{m} \lim_{r \to \infty} |(\mathbf{C}^{(r)})_{k\ell}| \quad (7.2.31)$$

Since $|(\mathbf{C})_{k\ell}| < 1 \, \forall \, k \neq \ell$, $\lim_{r \to \infty} |(\mathbf{C}^{(r)})_{k\ell}| = 0$. Therefore, $\lim_{r \to \infty} ||\mathbf{C}^{(r)}||_\infty = 0$. In conclusion, for large r, $\mathbf{C}^{(r)}$ asymptotically approaches the zero matrix. Subsequently, for large r,

$$\bar{\mathbf{X}}_m^{r\,*} \bar{\mathbf{X}}_m^r \to n_i^r \, \mathbf{I} \quad (7.2.32)$$

The above analysis indicates the *orthogonalization effect* caused by increasing the order of the network. That is, the use of neural networks of higher order reduces the effect of the correlation of the key patterns on the approximation of the optimal matrix of synaptic weights by the outer-product rule. The larger the order of the neural network, the better the approximation of the optimal matrix of synaptic weights by the outer-product rule. In the limit, the outer-product rule coincides with optimal least-squares training.

7.2.4 Experimental Results

The first set of experiments illustrates the effect of the correlation of the key patterns on the generalization ability of neural networks of various orders trained using the outer-product rule. The number of associations stored in the network was chosen to be $m = 10$, while the length of the key patterns was $n = 150$. For the particular length of key patterns chosen in this experiment, $m = 10$ is approximately the maximum number of associations that guarantees reliable operation of the first-order neural network. Several sets of key patterns were generated in such a way that the magnitude of the dominant eigenvalue λ_d of $\mathbf{C}$ was in the range $0.1 - 7.0$. For each set of key patterns, 1000 input patterns were created with the normalized Hamming distance between each input pattern and one of the key patterns uniformly distributed between 0 and 1/3. This set of patterns was used to test the response of a neural network of first, second, third, and fourth order. For each of these networks, Figure 7.6 shows the percentage of perfectly recalled patterns as a function of $|\lambda_d|$, or equivalently the correlation of the key patterns. According to Figure 7.6, the generalization ability of the first-order neural network drops rather quickly with an increase in the correlation of the key patterns. This is an experimental verification of the analysis presented in chapter 6, which indicated that the outer-product rule is optimal when the key patterns are uncorrelated and provides a satisfactory approximation of the optimal matrix of synaptic weights when $|\lambda_d| \ll 1$. However, the generalization ability of neural networks of order higher than one is satisfactory even when $|\lambda_d| \gg 1$, that is, the key patterns are strongly correlated. According to Figure 7.6, the larger the order of the network, the weaker the effect of the correlation of the key patterns on the generalization ability of the network. This is an excellent experimental verification of the orthogonalization effect caused by the increase of the network order, which was analytically predicted in this section.

The second set of experiments evaluates the generalization ability of neural networks of various orders. The outer-product rule was used for training a neural network of first, second, and third order with respect to $m = 10$ associations. In all of these experiments,

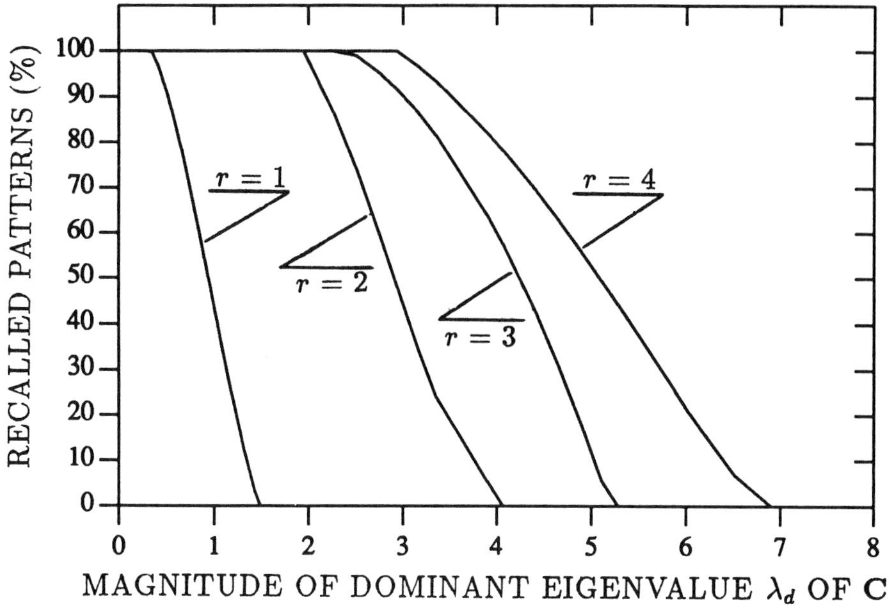

Figure 7.6: Percentage of perfectly recalled patterns by neural networks of different order trained using the outer-product rule as a function of the dominant eigenvalue $|\lambda_d|$ of $\mathbf{C}$.

the pattern length was $n = 150$. The response of the networks was tested using the same set of 75×300 test patterns as inputs. The normalized Hamming distance between each of the test patterns and one of the key patterns was uniformly distributed between 0 and 0.5. Figure 7.7 shows the percentage of perfectly recalled patterns by the networks as a function of the normalized Hamming distance between the input pattern and one of the key patterns. Clearly, any neural network guarantees perfect recall of one of the key patterns when the normalized Hamming distance does not exceed a certain threshold. However, the generalization ability of neural networks of any order drops rather quickly when the normalized Hamming distance between the input pattern and one of the key patterns exceeds this threshold. On the other hand, Figure 7.7 indicates that the neural networks of second and third order guarantee perfect recall of all the associated patterns when the first-order network fails to recall a significant portion of them. This is an experimental verification

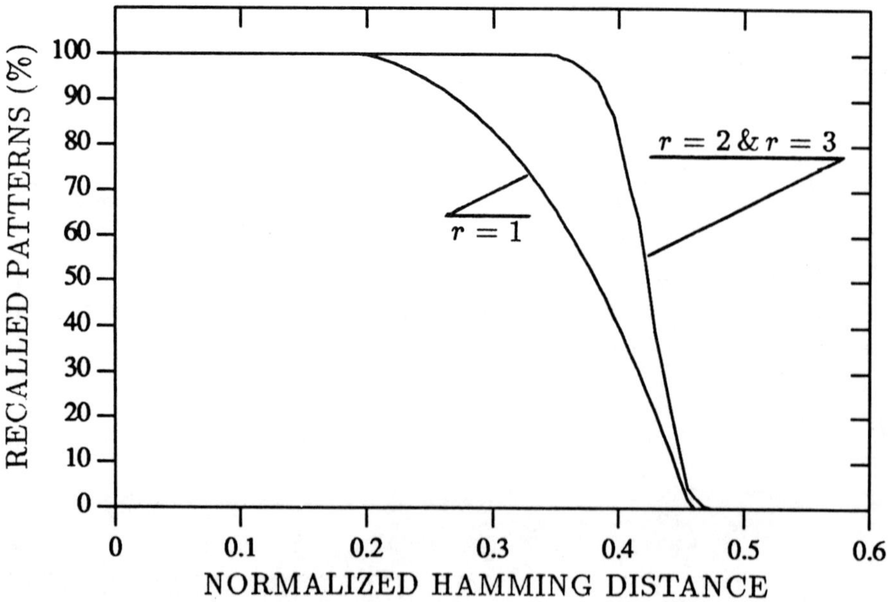

Figure 7.7: Percentage of perfectly recalled patterns by a neural network of first, second, and third order as a function of the normalized Hamming distance between the input and one of the key patterns.

that increasing the order of the network improves its generalization ability. This latter conclusion is in full agreement with the analysis presented in this section which indicated that the higher the order of the neural network, the more efficient the approximation of the optimal matrix of synaptic weights by the outer-product rule. Finally, Figure 7.7 reveals that, in this case, the performance of the neural networks of second and third order is very similar. This is attributed to the relatively small number of associations m stored in the networks. In fact, this choice of m was made in order that the first-order neural network operated under favorable conditions. Nevertheless, it was experimentally verified that the superiority of the third-order neural network becomes clear when the number of the associations stored in the networks increases.

7.3 Neural Networks with Composite Key Patterns

The analysis presented in section 7.2 indicates that any neural network of order higher than one can be viewed as a first-order neural network with *composite key patterns.* Consider, for instance, a second-order neural network that stores the associations $(\mathbf{y}_k, \mathbf{x}_k), k = 1, 2, \ldots, m$. The same network can be viewed as a first-order neural network that stores the associations $(\mathbf{y}_k, \bar{\mathbf{x}}_k), k = 1, 2, \ldots, m$, where $\bar{\mathbf{x}}_k$ is the composite key pattern formed as the Kronecker product between the single key pattern $\mathbf{x}_k$ with itself, that is, $\bar{\mathbf{x}}_k = \bar{\mathbf{x}}_k^2 = \mathbf{x}_k \otimes \mathbf{x}_k$. The basic idea behind the development of neural networks of order higher than one is that the key data relate to the associated data through a nonlinear mapping. Although the mapping given by (7.2.7) is the simplest one, it is not unique. An alternative nonlinear mapping is described here. Let S_k be the set consisting of the elements of $\mathbf{x}_k$, that is, $S_k = \{x_{i,k} : i = 1, 2, \ldots, n_i\}$. Consider two new vectors $\mathbf{x}_k^1$ and $\mathbf{x}_k^2$ of size $n_1 \times 1$ and $n_2 \times 1$, respectively, where $n_p \leq n_i, p = 1, 2$. Let S_k^1 and S_k^2 be the sets consisting of the elements of $\mathbf{x}_k^1$ and $\mathbf{x}_k^2$, respectively. In mathematical terms, $S_k^p = \{x_{i,k}^p : i = 1, 2, \ldots, n_p\}, p = 1, 2$. The new vectors $\mathbf{x}_k^1$ and $\mathbf{x}_k^2$ are formed in terms of $\mathbf{x}_k$ in such a way that $S_k^p \subseteq S_k, p = 1, 2$ and also $S_k^1 \cup S_k^2 = S_k$. Consider now the composite key pattern $\bar{\mathbf{x}}_k$, defined by $\bar{\mathbf{x}}_k = \mathbf{x}_k^1 \otimes \mathbf{x}_k^2$. Clearly, the neural network of second order results as the special case of this formulation for which $n_p = n_i \, \forall \, p = 1, 2$ and $S_k^p = S_k \, \forall \, p = 1, 2$.

A simple application of such a formation of the composite key pattern is described here. Consider the single key pattern $\mathbf{x}_k$ and the composite key pattern $\bar{\mathbf{x}}_k = \mathbf{x}_k^1 \otimes \mathbf{x}_k^2$, where $\mathbf{x}_k^1 = \mathbf{x}_k$ and the elements of $\mathbf{x}_k^2$ are given by $x_{i,k}^2 = x_{i,k} \, \forall \, i = 1, 2, \ldots, n', 1 \leq n' \leq n_i$. Clearly, the composite key pattern $\bar{\mathbf{x}}_k$ is a $n_i \, n' \times 1$ vector and, therefore, the network that stores the associations $(\mathbf{y}_k, \bar{\mathbf{x}}_k), k = 1, 2, \ldots, m$ consists of $\bar{n}_i = n_i \, n'$ inputs. Consider now the trivial case that $n' = 1$. Then, $\mathbf{x}_k^2$ consists of only one element and the corresponding composite key pattern $\bar{\mathbf{x}}_k$ is the $n_i \times 1$ vector $\bar{\mathbf{x}}_k = x_{1,k} \mathbf{x}_k$. The corresponding neural network consists of $\bar{n}_i = n_i$ inputs. Consider also the other extreme case where $n' = n_i$. In this case, $\mathbf{x}_k^1 =$

$\mathbf{x}_k^2 = \mathbf{x}_k$ and the composite key pattern is the $n_i^2 \times 1$ vector $\bar{\mathbf{x}}_k = \mathbf{x}_k \otimes \mathbf{x}_k$. Clearly, this particular formation of the composite key pattern corresponds to a neural network of second order. In general, this formation of the composite key patterns allows the development of neural networks with $\bar{n}_i = n_i\, n'$ inputs, where $1 \leq n' \leq n_i$. Since $n_i \leq \bar{n}_i \leq n_i^2$, this formulation bridges the gap between neural networks of first and second order, by allowing the development of a large variety of networks in between. In the same context, another possibility is the formation of the composite key patterns by $\mathbf{x}_k = \mathbf{x}_k^1 \otimes \mathbf{x}_k^2$, where $x_{i,k}^1 = x_{i,k}\,,\, i = 1, 2, \ldots, n'$ and $x_{i,k}^2 = x_{(i+n'),k}\,,\, i = 1, 2, \ldots, n_i - n'$, where $1 \leq n' < n_i$. The number of inputs of the corresponding neural network is $\bar{n}_i = n'(n_i - n')$. According to this formation of the composite key patterns, the maximum number of inputs is $\bar{n}_i = n_i^2/4$ and corresponds to $n' = n_i/2$. In the nontrivial case that $n_i \geq 4$, $n_i \leq \bar{n}_i \leq n_i^2$.

The direct generalization is the development of neural networks with composite key patterns formed as the Kronecker product between two or more possibly distinct patterns (Karayiannis, 1991a). The associative learning/recall in humans provides an interpretation of this new formulation. When a human receives a *stimulus,* a set of *attributes* are activated (Nakano, et al., 1988). For example, when the stimulus is a certain person, the attributes are height, weight, age, hair color, etc. Conversely, the attributes height, weight, age, hair color, etc. are crucial when a human tries to recall a certain person-stimulus. The importance of the combination of various attributes in associative recall is illustrated by the following example. If the stimulus is "lion", some of the associated attributes are "animal", "dangerous", and "brown". Similarly, if the stimulus is "rabbit", the corresponding attributes are "animal", "not dangerous", and "white". Clearly, the attribute "animal" is not enough for the discrimination between "lion" and "rabbit", unless at least one of the other two attributes are available. In the context of associative learning, the stimulus could be the associated pattern $\mathbf{y}_k$, while the attributes could be the single key patterns $\mathbf{x}_k^p, p = 1, 2, \ldots, r$. The composite key pattern $\bar{\mathbf{x}}_k$ can be formed as the Kronecker product between the single key patterns $\mathbf{x}_k^p, p = 1, 2, \ldots, r$ of length $n_p, p = 1, 2, \ldots, r$, respectively. In mathematical terms, each com-

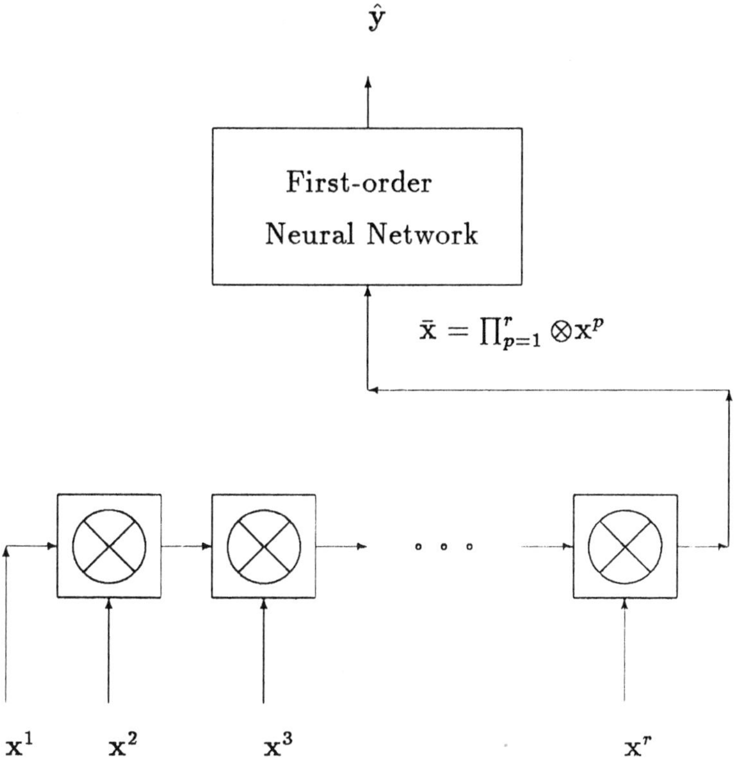

Figure 7.8: A neural network with composite key patterns.

posite key pattern can be formed as

$$\bar{\mathbf{x}}_k = \mathbf{x}_k^1 \otimes \mathbf{x}_k^2 \otimes \ldots \otimes \mathbf{x}_k^r = \prod_{p=1}^{r} \otimes \mathbf{x}_k^p \, \forall \, k = 1, 2, \ldots, m \qquad (7.3.1)$$

The structure of the corresponding neural network with composite key patterns is shown in Figure 7.8. The overall system consists of $r-1$ blocks that form the Kronecker product between the input patterns and a neural network of first order. It can easily be seen that the training of this neural network is reduced to the training of a single-layered neural network of first order with respect to the associations $(\mathbf{y}_k, \bar{\mathbf{x}}_k), k = 1, 2, \ldots, m$, where each $\bar{\mathbf{x}}_k$ is formed in terms of $\mathbf{x}_k^p, p = 1, 2, \ldots, r$ by (7.3.1).

7.3.1 Training Neural Networks with Composite Key Patterns

Consider that each pattern $\mathbf{y}_k, k = 1, 2, \ldots, m$ is associated with the corresponding composite key pattern $\bar{\mathbf{x}}_k, k = 1, 2, \ldots, m$, defined in terms of $\mathbf{x}_k^p, p = 1, 2, \ldots, r$ by (7.3.1). The optimal matrix of synaptic weights $\bar{\mathbf{W}}_m$ for this network with composite key patterns can be determined by requiring that

$$\mathbf{y}_k = \bar{\mathbf{W}}_m \, \bar{\mathbf{x}}_k \; \forall \, k = 1, 2, \ldots, m \tag{7.3.2}$$

The set of equations defined by (7.3.2) is equivalent to the following matrix equation

$$\mathbf{Y}_m = \bar{\mathbf{W}}_m \, \bar{\mathbf{X}}_m \tag{7.3.3}$$

where $\mathbf{Y}_m = [\mathbf{y}_1 \, \mathbf{y}_2 \, \ldots \, \mathbf{y}_m]$ and $\bar{\mathbf{X}}_m = [\bar{\mathbf{x}}_1 \, \bar{\mathbf{x}}_2 \, \ldots \, \bar{\mathbf{x}}_m]$. The optimal solution of (7.3.3) in the least-squares sense is given by

$$\bar{\mathbf{W}}_m = \mathbf{Y}_m \, \bar{\mathbf{X}}_m^+ = \mathbf{Y}_m \, (\bar{\mathbf{X}}_m^* \, \bar{\mathbf{X}}_m)^+ \, \bar{\mathbf{X}}_m^* \tag{7.3.4}$$

The properties of the matrix $\bar{\mathbf{X}}_m^* \bar{\mathbf{X}}_m$ are particularly important for evaluating the performance of neural networks with composite key patterns. These properties are studied here. By the definition of $\bar{\mathbf{X}}_m$, the elements of $\bar{\mathbf{X}}_m^* \bar{\mathbf{X}}_m$ are given as follows

$$\begin{aligned} (\bar{\mathbf{X}}_m^* \bar{\mathbf{X}}_m)_{k\ell} &= \bar{\mathbf{x}}_k^* \, \bar{\mathbf{x}}_\ell \\ &= (\prod_{p=1}^{r} \otimes \mathbf{x}_k^p)^* \, (\prod_{p=1}^{r} \otimes \mathbf{x}_\ell^p) = \prod_{p=1}^{r} \otimes (\mathbf{x}_k^{p*} \, \mathbf{x}_\ell^p) \end{aligned} \tag{7.3.5}$$

From the properties of the Kronecker product,

$$(\bar{\mathbf{X}}_m^* \, \bar{\mathbf{X}}_m)_{k\ell} = (\mathbf{x}_k^{1*} \, \mathbf{x}_\ell^1) \, (\mathbf{x}_k^{2*} \, \mathbf{x}_\ell^2) \, \ldots \, (\mathbf{x}_k^{r*} \, \mathbf{x}_\ell^r) \tag{7.3.6}$$

Let $\mathbf{X}_m^p$ be the matrix whose column vectors are $\mathbf{x}_k^p, k = 1, 2, \ldots, m$, that is, $\mathbf{X}_m^p = [\mathbf{x}_1^p \, \mathbf{x}_2^p \, \ldots \, \mathbf{x}_m^p] \, \forall \, p = 1, 2, \ldots, r$. Obviously,

$$(\mathbf{X}_m^{p*} \, \mathbf{X}_m^p)_{k\ell} = \mathbf{x}_k^{p*} \, \mathbf{x}_\ell^p \, \forall \, p = 1, 2, \ldots, r \tag{7.3.7}$$

The comparison between (7.3.6) and (7.3.7) indicates that

$$(\bar{\mathbf{X}}_m^* \bar{\mathbf{X}}_m)_{k\ell} = (\mathbf{X}_m^{1*} \mathbf{X}_m^1)_{k\ell} (\mathbf{X}_m^{2*} \mathbf{X}_m^2)_{k\ell} \ldots (\mathbf{X}_m^{r*} \mathbf{X}_m^r)_{k\ell} \qquad (7.3.8)$$

An elegant expression for the matrix $\bar{\mathbf{X}}_m^* \bar{\mathbf{X}}_m$ in terms of $\mathbf{X}_m^{p*} \mathbf{X}_m^p$, $p = 1, 2, \ldots, r$ can be deduced from (7.3.8) by using a special type of matrix product, known as Hadamard product (Rao and Mitra, 1971). The Hadamard product between two matrices $\mathbf{A}$ and $\mathbf{B}$ of size $m \times n$ is defined as the $m \times n$ matrix $\mathbf{C} = \mathbf{A} * \mathbf{B}$ whose elements are obtained by multiplying pointwise the corresponding elements of $\mathbf{A}$ and $\mathbf{B}$, that is, $(\mathbf{C})_{k\ell} = (\mathbf{A})_{k\ell} (\mathbf{B})_{k\ell} \forall 1 \leq k \leq m, 1 \leq l \leq n$. Using the Hadamard product notation, (7.3.8) amounts to

$$\bar{\mathbf{X}}_m^* \bar{\mathbf{X}}_m = (\mathbf{X}_m^{1*} \mathbf{X}_m^1) * (\mathbf{X}_m^{2*} \mathbf{X}_m^2) * \ldots * (\mathbf{X}_m^{r*} \mathbf{X}_m^r) \qquad (7.3.9)$$

If the elements of $\mathbf{X}_m^p$ can only take the values ± 1, the matrix $\mathbf{X}_m^{p*} \mathbf{X}_m^p$ has a special form. In fact, the elements of $\mathbf{X}_m^{p*} \mathbf{X}_m^p$ are given as follows

$$(\mathbf{X}_m^{p*} \mathbf{X}_m^p)_{k\ell} = \begin{cases} \mathbf{x}_k^{p*} \mathbf{x}_k^p = n_p & \text{if } k = \ell \\ \mathbf{x}_k^{p*} \mathbf{x}_l^p < n_p & \text{if } k \neq \ell \end{cases} \qquad (7.3.10)$$

Therefore,

$$\mathbf{X}_m^{p*} \mathbf{X}_m^p = n_p (\mathbf{I} + \mathbf{C}_p) \, \forall \, p = 1, 2, \ldots, r \qquad (7.3.11)$$

where $\mathbf{I}$ is the $m \times m$ identity matrix and $\mathbf{C}_p$ is the $m \times m$ matrix with elements

$$(\mathbf{C}_p)_{k\ell} = \begin{cases} 0 & \text{if } k = \ell \\ \frac{1}{n_p} \mathbf{x}_k^{p*} \mathbf{x}_\ell^p & \text{if } k \neq \ell \end{cases} \qquad (7.3.12)$$

Clearly, $\mathbf{C}_p$ is an $m \times m$ matrix with zero diagonal elements and nondiagonal elements which satisfy $|(\mathbf{C}_p)_{k\ell}| < 1 \, \forall \, k \neq \ell$. The combination of (7.3.9) and (7.3.11) amounts to

$$\bar{\mathbf{X}}_m^* \bar{\mathbf{X}}_m = \bar{n} (\mathbf{I} + \bar{\mathbf{C}}) \qquad (7.3.13)$$

where $\bar{n} = n_1 n_2 \ldots n_r$ and

$$\bar{\mathbf{C}} = \mathbf{C}_1 * \mathbf{C}_2 * \ldots * \mathbf{C}_r \qquad (7.3.14)$$

The above analysis facilitates the investigation of the correlation matrix $\bar{\mathbf{C}}$ of the network with composite key patterns with respect to the correlation of any set of patterns $\mathbf{x}_k^p, p = 1, 2, \ldots, r$ used to form the composite key patterns. First of all, consider the strong assumption that the patterns $\mathbf{x}_k^p, k = 1, 2, \ldots, m$ are uncorrelated. In this case, $\mathbf{C}_p = 0$, or equivalently, $\mathbf{X}_m^{p*} \mathbf{X}_m^p = n_p \mathbf{I}$. Taking into account the definition of $\bar{\mathbf{C}}$ by (7.3.14), this assumption implies that $\bar{\mathbf{C}} = 0$ and, therefore, $\bar{\mathbf{X}}_m^* \bar{\mathbf{X}}_m = \bar{n}\, \mathbf{I}$.

Consider the same case but with no assumptions regarding the correlation of the patterns $\mathbf{x}_k^p, p = 1, 2, \ldots, r$. By the definition of the Hadamard product, the elements of $\bar{\mathbf{C}}$ are given by $(\bar{\mathbf{C}})_{k\ell} = \prod_{p=1}^r (\mathbf{C}_p)_{k\ell} \,\forall\, 1 \leq k, \ell \leq m$. Since $|(\mathbf{C}_p)_{k\ell}| < 1 \,\forall\, k \neq \ell$,

$$|(\bar{\mathbf{C}})_{k\ell}| < |(\mathbf{C}_p)_{k\ell}| \,\forall\, p = 1, 2, \ldots, r \tag{7.3.15}$$

According to (7.3.15), any element of $\bar{\mathbf{C}}$ is smaller than the corresponding element of any correlation matrix $\mathbf{C}_p$ which describes the correlation of the patterns $\mathbf{x}_k^p, p = 1, 2, \ldots, r$. In other words, the composite key patterns are less correlated than any set of single key patterns. In addition, (7.3.15) indicates that any element of $\bar{\mathbf{C}}$ approaches zero for sufficiently large r. In other words, $\bar{\mathbf{C}}$ approaches the zero matrix if r is sufficiently large. Therefore, $\bar{\mathbf{X}}_m^* \bar{\mathbf{X}}_m \rightarrow \bar{n}\, \mathbf{I}$.

Assume that $\bar{\mathbf{X}}_m^* \bar{\mathbf{X}}_m = \bar{n}\, \mathbf{I}$. Under this assumption, the optimal matrix of synaptic weights (7.3.4) becomes

$$\bar{\mathbf{W}}_m = \mathbf{Y}_m \, (\bar{n}\, \mathbf{I})^{-1} \, \bar{\mathbf{X}}_m^* = \frac{1}{\bar{n}} \mathbf{Y}_m \, \bar{\mathbf{X}}_m^* = \frac{1}{\bar{n}} \sum_{k=1}^m \mathbf{y}_k \, \bar{\mathbf{x}}_k^* \tag{7.3.16}$$

The elements of the matrix of synaptic weights $\bar{\mathbf{W}}_m$ are given by

$$(\bar{\mathbf{W}}_m)_{ij} = \frac{1}{\bar{n}} \sum_{k=1}^m y_{i,k} \, (\bar{\mathbf{x}}_k)_j \tag{7.3.17}$$

Assume now that the composite pattern $\bar{\mathbf{x}} = \mathbf{x}^1 \otimes \mathbf{x}^2 \otimes \ldots \otimes \mathbf{x}^r$ is the input of the neural network with matrix of synaptic weights defined by (7.3.17). The response of the network is described by the following equation

$$y_i = \mathrm{sgn}(\sum_{j=1}^{\bar{n}} (\bar{\mathbf{W}}_m)_{ij} \, (\bar{\mathbf{x}})_j) = \mathrm{sgn}(\frac{1}{\bar{n}} \sum_{k=1}^m \sum_{j=1}^{\bar{n}} y_{i,k} \, (\bar{\mathbf{x}}_k)_j \, (\mathbf{x})_j) \tag{7.3.18}$$

Taking into account the definition of the Kronecker product, $(\bar{\mathbf{x}}_k)_j = \prod_{p=1}^{r} x_{j_p,k}^p, j = 1 + \sum_{p=1}^{r} n_p^r (j_p - 1)$, where the following notation is used

$$n_p^{p+s} = \begin{cases} n_{p+1}\, n_{p+2} \,\ldots\, n_{p+s} & \text{if } s > 0 \\ 1 & \text{if } s = 0 \end{cases} \tag{7.3.19}$$

In addition, $(\bar{\mathbf{x}})_j = \prod_{p=1}^{r} x_{j_p}^p, j = 1 + \sum_{p=1}^{r} n_p^r (j_p - 1)$. With these definitions, (7.3.18) becomes

$$\begin{aligned} y_i &= \operatorname{sgn}(\frac{1}{\bar{n}} \sum_{k=1}^{m} y_{i,k} \sum_{j_1=1}^{n_1} \cdots \sum_{j_r=1}^{n_r} \prod_{p=1}^{r} x_{j_p,k}^p \, x_{j_p}^p) \\ &= \operatorname{sgn}(\sum_{k=1}^{m} y_{i,k} \prod_{p=1}^{r} (\frac{1}{n_p} \sum_{j_p=1}^{n_p} x_{j_p,k}^p \, x_{j_p}^p)) \end{aligned} \tag{7.3.20}$$

The direction cosine between each input pattern $\mathbf{x}^p$ and each one of the corresponding key patterns $\mathbf{x}_k^p$ is defined as follows

$$c(\mathbf{x}_k^p, \mathbf{x}^p) = \frac{1}{n_p} \sum_{j_p=1}^{n_p} x_{j_p,k}^p \, x_{j_p}^p \tag{7.3.21}$$

With this definition, (7.3.20) becomes

$$y_i = \operatorname{sgn}(\sum_{k=1}^{m} y_{i,k} \prod_{p=1}^{r} c(\mathbf{x}_k^p, \mathbf{x}^p)) \tag{7.3.22}$$

The remainder of this section identifies the relationship between neural networks with multiple key patterns and high-order neural networks. Consider that all key patterns $\mathbf{x}_k^p, p = 1, 2, \ldots, r$ are identical, that is, $\mathbf{x}_k^p = \mathbf{x}_k \, \forall \, p = 1, 2, \ldots, r$. In this case, $\bar{\mathbf{x}}_k = \prod_{p=1}^{r} \otimes \mathbf{x}_k = \bar{\mathbf{x}}_k^r$. In addition, $\bar{\mathbf{X}}_m = [\bar{\mathbf{x}}_1 \, \bar{\mathbf{x}}_2 \, \ldots \, \mathbf{x}_m] = [\bar{\mathbf{x}}_1^r \, \bar{\mathbf{x}}_2^r \, \ldots \, \bar{\mathbf{x}}_m^r] = \mathbf{X}_m^r$ and also $\mathbf{X}_m^p = \mathbf{X}_m \, \forall \, p = 1, 2, \ldots, r$. Therefore, (7.3.9) amounts to

$$\begin{aligned} \bar{\mathbf{X}}_m^* \, \bar{\mathbf{X}}_m &= (\mathbf{X}_m^{1*} \mathbf{X}_m^1) * (\mathbf{X}_m^{2*} \mathbf{X}_m^2) * \ldots * (\mathbf{X}_m^{r*} \mathbf{X}_m^r) \\ &= (\mathbf{X}_m^* \, \mathbf{X}_m)^{(r)} \end{aligned} \tag{7.3.23}$$

Not surprisingly, the optimal matrix of synaptic weights of this network is identical to the optimal matrix of synaptic weights of the neural network of order r. In conclusion, neural networks of order higher than one can be interpreted as the special case of neural networks with composite key patterns.

7.3.2 Experimental Results

This set of experiments evaluates the effect of the correlation of the key patterns on the generalization ability of the neural network with composite key patterns formed as $\bar{\mathbf{x}}_k = \mathbf{x}_k^1 \otimes \mathbf{x}_k^2$, where $\mathbf{x}_k^1 = \mathbf{x}_k$ and $x_{i,k}^1 = x_{i,k} \, \forall \, i = 1, 2, \ldots, n_1$. The number of associations stored in the network was $m = 10$, while the length of the original key patterns $\mathbf{x}_k$ was $n = 150$. Several sets of key patterns were generated in such a way that the magnitude of the dominant eigenvalue λ_d of $\mathbf{C}$ was in the range 0.1 - 5.0. For each set of key patterns, 1000 patterns were used as input patterns. The normalized Hamming distance between each of the input patterns and one of the key patterns was uniformly distributed between 0 and 1/3. Figure 7.9 shows the percentage of perfectly recalled patterns by the neural network considered in these experiments as a function of the dominant eigenvalue λ_d of $\mathbf{C}$ for $n_1 = 1$, $n_1 = n/2$, and $n_1 = n$. A comparison of Figures 7.6 and 7.9 indicates that the curve which corresponds to $n_1 = 1$ is identical to that corresponding to the neural network of first order. This experimental outcome can be analytically justified by investigating the relationship between the correlation matrices which correspond to these neural networks. It can easily be verified that each element of the correlation matrix $\bar{\mathbf{C}}$ of the key patterns $\bar{\mathbf{x}}_k, k = 1, 2, \ldots, m$ is related to the corresponding element of the correlation matrix $\mathbf{C}$ of the original key patterns $\mathbf{x}_k, k = 1, 2, \ldots, m$ by $(\bar{\mathbf{C}})_{k\ell} = x_{1,k} \, x_{1,\ell} \, (\mathbf{C})_{k\ell}$. Since $x_{1,k} \, x_{1,\ell}$ can be either $+1$ or -1, $|(\bar{\mathbf{C}})_{k\ell}| = |(\mathbf{C})_{k\ell}| \, \forall \, k, \ell$. Therefore, the correlation of the key patterns has the same effect on the performance of both neural networks. For obvious reasons, the curve which corresponds to $n_1 = n$ is identical to that of the neural network of second order. When $n_1 = n/2$, the performance of the corresponding neural network is very close to that of the second-order neural network. In fact, the maximum $|\lambda_d|$ which guarantees perfect recall is the same for both networks. The only difference is that the performance of the neural network which corresponds to $n_1 = n/2$ drops faster than the performance of the neural network of second order, when $|\lambda_d|$ increases. According to Figure 7.9, the increase of n_1 from 1 to $n/2$ has a strong impact on the robustness of the corresponding neural network to correlation of the original key patterns. However, the

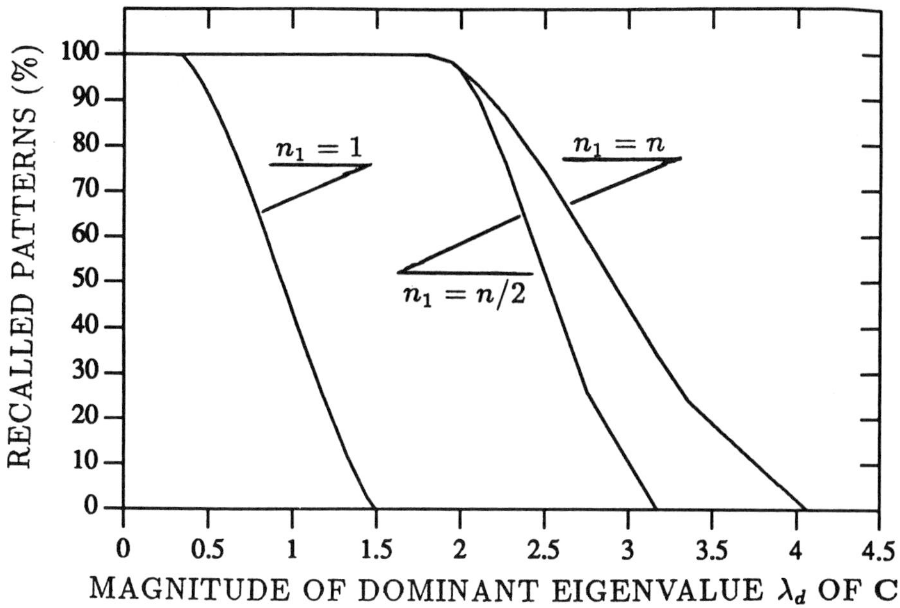

Figure 7.9: Percentage of perfectly recalled patterns by a neural network with composite key patterns formed as $\bar{\mathbf{x}}_k = \mathbf{x}_k^1 \otimes \mathbf{x}_k^2$, where $\mathbf{x}_k^1 = \mathbf{x}_k$ and $x_{i,k}^2 = x_{i,k} \, \forall \, i = 1, 2, \ldots, n_1$, as a function of the dominant eigenvalue $|\lambda_d|$ of $\mathbf{C}$ for different values of n_1.

improvement in the performance of the network with respect to its sensitivity to the correlation of the key patterns is rather marginal when n_1 increases from $n/2$ to n.

7.4 Capacity Considerations

This section studies the capacity of neural networks with composite key patterns trained using the outer-product rule. Consider a network which stores the associations $(\mathbf{x}_k, \bar{\mathbf{x}}_k)$, $k = 1, 2, \ldots, m$. Assume that any composite key pattern $\bar{\mathbf{x}}_k$ is formed as the Kronecker product of the original key pattern $\mathbf{x}_k$, of length $n_i = n$, and a portion of the same pattern. In mathematical terms,

$$\bar{\mathbf{x}}_k = \mathbf{x}_k^1 \otimes \mathbf{x}_k^2 \, \forall \, k = 1, 2, \ldots, m \qquad (7.4.1)$$

where $\mathbf{x}_k^1 = \mathbf{x}_k$ and $x_{i,k}^2 = x_{i,k} \, \forall \, i = 1, 2, \ldots, n_1$, where $n_1 \leq n$. Obviously, if $n_1 = n$, this formation of the input patterns corresponds to the neural network of second order. Assume that the composite pattern $\bar{\mathbf{x}}_\alpha = \mathbf{x}_\alpha^1 \otimes \mathbf{x}_\alpha^2$ is the input pattern of this network. The corresponding output is given by

$$x_i = \text{sgn}(\sum_{j=1}^{\bar{n}} (\bar{\mathbf{W}})_{ij} \, (\bar{\mathbf{x}}_\alpha)_j) \tag{7.4.2}$$

where $\bar{n} = n \, n_1$ and $\bar{\mathbf{W}}$ is the matrix of synaptic weights provided by the outer-product rule, with elements

$$(\bar{\mathbf{W}})_{ij} = \frac{1}{\bar{n}} \sum_{k=1}^{m} x_{i,k} \, (\bar{\mathbf{x}}_k)_j \tag{7.4.3}$$

The combination of (7.4.2) and (7.4.3) gives

$$\begin{aligned} x_i &= \text{sgn}(\frac{1}{\bar{n}} \sum_{k=1}^{m} \sum_{j=1}^{\bar{n}} x_{i,k} \, (\bar{\mathbf{x}}_k)_j \, (\bar{\mathbf{x}}_\alpha)_j) \\ &= \text{sgn}(\frac{1}{\bar{n}} \sum_{j=1}^{\bar{n}} x_{i,\alpha} \, (\bar{\mathbf{x}}_\alpha)_j \, (\bar{\mathbf{x}}_\alpha)_j + \frac{1}{\bar{n}} \sum_{k \neq \alpha}^{m} \sum_{j=1}^{\bar{n}} x_{i,k} \, (\bar{\mathbf{x}}_k)_j \, (\bar{\mathbf{x}}_\alpha)_j) \end{aligned} \tag{7.4.4}$$

Therefore, $x_i = \text{sgn}(x_{i,\alpha} + n_{i,\alpha})$, where $n_{i,\alpha}$ is the noise term, given by

$$n_{i,\alpha} = \frac{1}{\bar{n}} \sum_{k \neq \alpha}^{m} \sum_{j=1}^{\bar{n}} x_{i,k} \, (\bar{\mathbf{x}}_k)_j \, (\bar{\mathbf{x}}_\alpha)_j \tag{7.4.5}$$

Assume that the elements $x_{i,k}$ of $\mathbf{x}_k$ are independent random variables, taking the values $+1$ and -1 with equal probability. It can easily be verified that $n_{i,\alpha}$ is a zero-mean random variable. Therefore, its variance is given by

$$\begin{aligned} \bar{\sigma}^2 &= E\{n_{i,\alpha}^2\} \\ &= E\{\sum_{k \neq \alpha}^{m} \frac{1}{\bar{n}^2} \sum_{k' \neq \alpha}^{m} \sum_{j=1}^{\bar{n}} \sum_{j'=1}^{\bar{n}} x_{i,k} x_{i,k'} (\bar{\mathbf{x}}_k)_j (\bar{\mathbf{x}}_{k'})_{j'} (\bar{\mathbf{x}}_\alpha)_j (\bar{\mathbf{x}}_\alpha)_{j'}\} \end{aligned} \tag{7.4.6}$$

Since $x_{i,k}$ are independent random variables and $x_{i,k}\, x_{i,k} = 1$, (7.4.6) can be simplified as follows

$$\bar{\sigma}^2 = E\{n_{i,\alpha}^2\} = \frac{1}{\bar{n}^2} \sum_{k \neq \alpha}^{m} \sum_{j=1}^{\bar{n}} \sum_{j'=1}^{\bar{n}} E\{(\bar{\mathbf{x}}_k)_j\, (\bar{\mathbf{x}}_k)_{j'}\, (\bar{\mathbf{x}}_\alpha)_j\, (\bar{\mathbf{x}}_\alpha)_{j'}\} \quad (7.4.7)$$

Using the definition of $\bar{\mathbf{x}}_k$ in (7.4.1), (7.4.7) becomes

$$\bar{\sigma}^2 = \frac{1}{\bar{n}^2} \sum_{k \neq \alpha}^{m} \sum_{j_1, j_1'=1}^{n} \sum_{j_2, j_2'=1}^{n_1} E\{\prod_{p=1}^{2} x_{j_p,k}\, x_{j_p',k}\, x_{j_p,\alpha}\, x_{j_p',\alpha}\} \quad (7.4.8)$$

Clearly, the variance of the noise term $n_{i,\alpha}$ is given by

$$\bar{\sigma}^2 = \frac{m-1}{\bar{n}^2} N(n, n_1) = \frac{m-1}{(n\, n_1)^2} N(n, n_1) \quad (7.4.9)$$

where $N(.)$ is the total number of terms for which

$$E\{\prod_{p=1}^{2} x_{j_p,k}\, x_{j_p',k}\, x_{j_p,\alpha}\, x_{j_p',\alpha}\} = 1 \quad (7.4.10)$$

Since it was assumed that $x_{i,k}$ are independent random variables, (7.4.10) holds if (i) $j_1 = j_1'$ and $j_2 = j_2'$, (ii) $j_1 = j_2$ and $j_1' = j_2'$, and (iii) $j_1 = j_2'$ and $j_2 = j_1'$. Since $n_1 \leq n$, there are (i) $n\, n_1$ terms such that $j_1 = j_1'$ and $j_2 = j_2'$, (ii) n_1^2 terms such that $j_1 = j_2$ and $j_1' = j_2'$, and (iii) n_1^2 terms such that $j_1 = j_2'$ and $j_2 = j_1'$. However, the case $j_1 = j_2 = j_1' = j_2'$ is included in each of the cases considered above. Since there are n_1 terms such that $j_1 = j_2 = j_1' = j_2'$,

$$N(n, n_1) = n\, n_1 + n_1^2 + n_1^2 - 2\, n_1 = n_1(n + 2\, n_1 - 2) \quad (7.4.11)$$

The variance $\bar{\sigma}^2$ is given by the combination of (7.4.9) and (7.4.11) as follows

$$\bar{\sigma}^2 = \frac{m-1}{(n\, n_1)^2}\, n_1\, (n + 2\, n_1 - 2) \approx \frac{m}{n}\, \frac{n + 2\, n_1 - 2}{n_1\, n} \quad (7.4.12)$$

If m is the maximum allowable number of associations and $\bar{c} = m/n$ the corresponding capacity ratio, then

$$\bar{\sigma}^2 = \bar{c}\, \frac{n + 2\, n_1 - 2}{n_1\, n} \quad (7.4.13)$$

Let $c_1 = m/n$ be the capacity ratio of the first-order neural network trained using the outer-product rule. It is well-known that the variance σ_1^2 of the noise term which corresponds to this network is given by (Amari and Maginu, 1988)

$$\sigma_1^2 = \frac{m-1}{n} \approx \frac{m}{n} = c_1 \tag{7.4.14}$$

The two networks are equivalent in terms of their capacity if $\bar{\sigma}^2 = \sigma_1^2$. Under this assumption, the combination of (7.4.13) and (7.4.14) gives

$$\frac{\bar{c}}{c_1} = \frac{n_1 \, n}{n + 2\, n_1 - 2} \tag{7.4.15}$$

This is the relationship between the capacity $\bar{c}$ of the neural network with composite key patterns considered in this case and the capacity c_1 of the first-order neural network. This relationship is now investigated for some special cases. If $n_1 = 1$, (7.4.15) indicates that $\bar{c} = c_1$. In this case there is no improvement in the capacity of the network. This is not surprising, since in this case the composite key patterns are formed as $\bar{\mathbf{x}}_k = x_{1,k} \mathbf{x}_k$. In other words, each composite key pattern is obtained by multiplying the original key pattern $\mathbf{x}_k$ by -1 or $+1$, depending on the value of the element $x_{1,k}$. If $n_1 = n$, the neural network with composite key patterns considered here coincides with the neural network of second order. In this case,

$$\frac{\bar{c}}{c_1} = \frac{n^2}{3n - 2} \tag{7.4.16}$$

This is precisely the relationship between the capacity of the neural networks of first and second order, proposed in (Psaltis, et al., 1988). For large n, (7.4.16) can be approximated by $\bar{c}/c_1 \approx n^2/3n = n/3$. Consider now that $n \gg 1$, while $n_1 \ll n$. In this case,

$$\frac{\bar{c}}{c_1} = \frac{n_1}{1 + \frac{2}{n}(n_1 - 1)} \approx n_1 \tag{7.4.17}$$

Obviously, the capacity ratio $\bar{c}$ is obtained by multiplying the capacity ratio c_1 of the first-order neural network with the number of the elements of the second pattern. Finally, consider that $n_1 = n/2$.

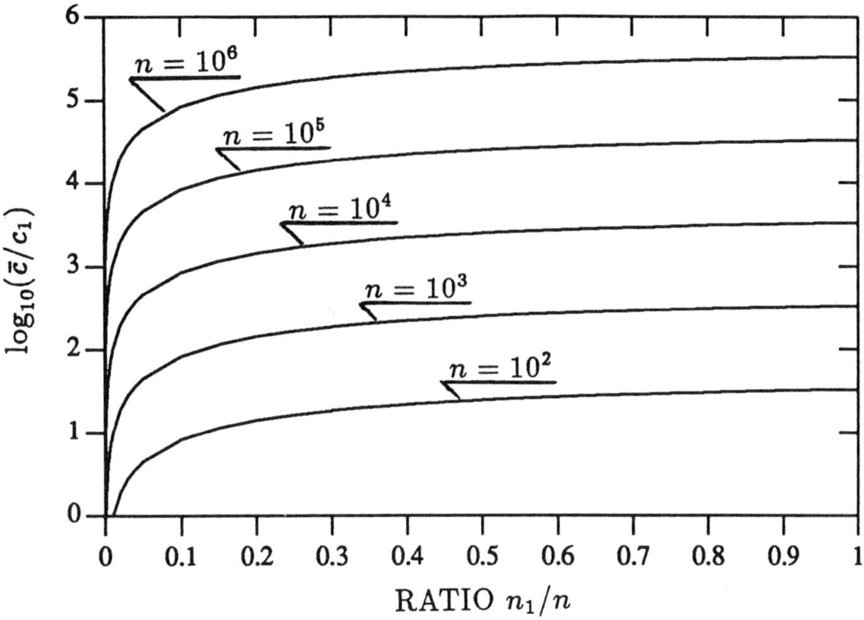

Figure 7.10: Relationship between the capacity $\bar{c}$ of a neural network with composite key patterns formed as $\bar{\mathbf{x}}_k = \mathbf{x}_k^1 \otimes \mathbf{x}_k^2$, where $\mathbf{x}_k^1 = \mathbf{x}_k$ and $x_{i,k}^2 = x_{i,k} \, \forall \, i = 1, 2, \ldots, n_1$ and the capacity ratio c_1 of a neural network of first order. $\log_{10}(\bar{c}/c_1)$ as a function of n_1/n for different values of n.

Then,

$$\frac{\bar{c}}{c_1} = \frac{n^2/2}{2(n-1)} \tag{7.4.18}$$

If $n \gg 1$, (7.4.18) can be approximated by $\bar{c}/c_1 \approx n^2/4n = n/4$. The relationship between the capacity $\bar{c}$ of the neural network with composite key patterns considered here and the capacity c_1 of the neural network of first order is exhibited in Figure 7.10, which shows $\log_{10}(\bar{c}/c_1)$ as a function of the ratio n_1/n for different n. According to this Figure, the increase in the capacity $\bar{c}$ of this network is substantial when the ratio n_1/n increases from $1/n$ to 0.1. However, the increase in the capacity $\bar{c}$ becomes slower as the ratio n_1/n increases from 0.1 to 1.0.

The remainder of this section studies the capacity of neural networks with composite key patterns formed as the Kronecker product between two or more possibly distinct patterns. Assume that this network is trained with respect to the associations $(\mathbf{x}_k, \bar{\mathbf{x}}_k)$, $k = 1, 2, \ldots, m$, where each composite key pattern $\bar{\mathbf{x}}_k$ is defined by (7.3.1). The capacity of this network is studied by considering its response to one of the key patterns, say $\bar{\mathbf{x}}_\alpha = \mathbf{x}_\alpha^1 \otimes \mathbf{x}_\alpha^2 \otimes \ldots \otimes \mathbf{x}_\alpha^r$. The response of the network to the input pattern $\bar{\mathbf{x}} = \bar{\mathbf{x}}_\alpha$ is determined by

$$x_i = \text{sgn}(\sum_{j=1}^{\bar{n}} (\bar{\mathbf{W}})_{ij} \, (\bar{\mathbf{x}}_\alpha)_j) \tag{7.4.19}$$

Assuming that the neural network is trained using the outer-product rule,

$$\begin{aligned} x_i &= \text{sgn}(\frac{1}{\bar{n}} \sum_{k=1}^{m} \sum_{j=1}^{\bar{n}} x_{i,k} \, (\bar{\mathbf{x}}_k)_j \, (\bar{\mathbf{x}}_\alpha)_j) \\ &= \text{sgn}(\frac{1}{\bar{n}} \sum_{j=1}^{\bar{n}} x_{i,\alpha} \, (\bar{\mathbf{x}}_\alpha)_j \, (\bar{\mathbf{x}}_\alpha)_j + \frac{1}{\bar{n}} \sum_{k \neq \alpha}^{m} \sum_{j=1}^{\bar{n}} x_{i,k} \, (\bar{\mathbf{x}}_k)_j \, (\bar{\mathbf{x}}_\alpha)_j) \end{aligned} \tag{7.4.20}$$

Therefore, $x_i = \text{sgn}(x_{i,\alpha} + n_{i,\alpha})$, where $n_{i,\alpha}$ is the noise term, given by

$$n_{i,\alpha} = \frac{1}{\bar{n}} \sum_{k \neq \alpha}^{m} \sum_{j=1}^{\bar{n}} x_{i,k} \, (\bar{\mathbf{x}}_k)_j \, (\bar{\mathbf{x}}_\alpha)_j \tag{7.4.21}$$

Assume that the elements $x_{i,k}$ of $\mathbf{x}_k$ as well as the elements $x_{i,k}^p$ of $\mathbf{x}_k^p, p = 1, 2, \ldots, r$ are independent random variables, taking the values $+1$ and -1 with equal probability. Under this assumption, the variance of $n_{i,\alpha}$ is given by

$$\begin{aligned} \bar{\sigma}_r^2 &= E\{n_{i,\alpha}^2\} \\ &= E\{\frac{1}{\bar{n}^2} \sum_{k \neq \alpha}^{m} \sum_{k' \neq \alpha}^{m} \sum_{j=1}^{\bar{n}} \sum_{j'=1}^{\bar{n}} x_{i,k} x_{i,k'} \, (\bar{\mathbf{x}}_k)_j \, (\bar{\mathbf{x}}_{k'})_{j'} \, (\bar{\mathbf{x}}_\alpha)_j \, (\bar{\mathbf{x}}_\alpha)_{j'}\} \end{aligned} \tag{7.4.22}$$

Since the elements of all the patterns appearing in (7.4.22) are assumed to be independent random variables, (7.4.22) can be simplified as follows

$$\bar{\sigma}_r^2 = \frac{1}{\bar{n}^2} \sum_{k \neq \alpha}^{m} \sum_{j=1}^{\bar{n}} \sum_{j'=1}^{\bar{n}} E\{(\bar{\mathbf{x}}_k)_j \, (\bar{\mathbf{x}}_k)_{j'} \, (\bar{\mathbf{x}}_\alpha)_j \, (\bar{\mathbf{x}}_\alpha)_{j'}\} \tag{7.4.23}$$

By the definition of the Kronecker product,

$$(\bar{\mathbf{x}}_k)_j = \prod_{p=1}^{r} x^p_{j_p,k}\,,\; j = 1 + \sum_{p=1}^{r} n^r_p (j_p - 1) \tag{7.4.24}$$

The combination of (7.4.23) and (7.4.24) results in

$$\bar{\sigma}^2_r = \frac{1}{\bar{n}^2} \sum_{k \neq \alpha}^{m} \sum_{j_1, j'_1 = 1}^{n_1} \cdots \sum_{j_r, j'_r = 1}^{n_r} E\{\prod_{p=1}^{r} x^p_{j_p,k}\, x^p_{j'_p,k}\, x^p_{j_p,\alpha}\, x^p_{j'_p,\alpha}\} \tag{7.4.25}$$

In general, the evaluation of the variance $\bar{\sigma}^2_r$ is a particularly complicated problem (Psaltis et al., 1988). The capacity of high-order neural networks can be estimated by evaluating $\bar{\sigma}^2_r = \sigma^2_r$ in the case where the patterns $\mathbf{x}^p_k, p = 1, 2, \ldots, r$ are identical, that is, $\mathbf{x}^p_k = \mathbf{x}_k \,\forall\, p = 1, 2, \ldots, r$. In the case where $r = 1, 2,$ and 3 the variance of the corresponding noise term is $\sigma^2_1 = (m-1)n/n^2$, $\sigma^2_2 = (m-1)(3n^2 - 2)/n^4$, and $\sigma^2_3 = (m-1)(15n^3 - 30n^2 + 16n)/n^6$, respectively (Psaltis et al., 1988). However, the exact evaluation of the variance σ^2_r is a particularly difficult task for $r > 3$. Nevertheless, the variance σ^2_r is upper bounded as follows (Psaltis et al., 1988)

$$\sigma^2_r \leq \frac{1}{n^{2r}} (m-1)\, n^r \,\frac{(2r)!}{2^r r!} = \frac{m-1}{n^r} \frac{(2r)!}{2^r r!} \tag{7.4.26}$$

For r small, compared to n, this upper bound can be used to estimate the capacity of neural networks of order higher than three (Psaltis et al., 1988).

The exact evaluation of the variance $\bar{\sigma}^2_r$ can also be performed if the patterns $\mathbf{x}^p_k, p = 1, 2, \ldots, r$ are distinct, that is, $\mathbf{x}^p_k \neq \mathbf{x}^q_k, k = 1, 2, \ldots, m \,\forall\, p \neq q$. Under this assumption,

$$E\{\prod_{p=1}^{r} x^p_{j_p,k}\, x^p_{j_{p'},k}\, x^p_{j_p,\alpha}\, x^p_{j_{p'},\alpha}\} = \prod_{p=1}^{r} \delta_{j_p j'_p} \tag{7.4.27}$$

where δ_{pq} denotes the Kronecker delta, defined as $\delta_{pq} = 1$ if $p = q$ and $\delta_{pq} = 0$ if $p \neq q$. In this case, the variance $\bar{\sigma}^2_r$ is given by

$$\bar{\sigma}^2_r = \frac{1}{\bar{n}^2} \sum_{k \neq \alpha}^{m} \sum_{j_1=1}^{n_1} \sum_{j_2=1}^{n_2} \cdots \sum_{j_r=1}^{n_r} \sum_{j'_1=1}^{n_1} \sum_{j'_2=1}^{n_2} \cdots \sum_{j'_r=1}^{n_r} \prod_{p=1}^{r} \delta_{j_p j'_p} \tag{7.4.28}$$

It can easily be seen that

$$\bar{\sigma}_r^2 = \frac{1}{\bar{n}^2}(m-1)\, n_1\, n_2 \ldots n_r = \frac{(m-1)\,\bar{n}}{\bar{n}^2} = \frac{m-1}{\bar{n}} \approx \frac{m}{\bar{n}} \quad (7.4.29)$$

The neural network with composite key patterns is now compared to the first-order neural network in terms of capacity. Consider the first-order neural network and assume that the key patterns $\mathbf{x}_k, k = 1, 2, \ldots, m$ are of length n. Let c_1 be the capacity ratio of the first-order neural network trained using the outer-product rule. The capacity ratio of this network is related to the variance σ_1^2 of the corresponding noise term by (7.4.14). For the sake of this comparison, assume that each composite key pattern $\bar{\mathbf{x}}_k$ consists of single patterns $\mathbf{x}_k^p, p = 1, 2, \ldots, r$ of length n, that is, $n_1 = n_2 = \ldots = n_r = n$. Under this assumption, $\bar{n} = n^r$. In this case, (7.4.29) amounts to

$$\bar{\sigma}_r^2 = \frac{m-1}{n^r} \approx \frac{m}{n^r} = \frac{\bar{c}_r}{n^{r-1}} \quad (7.4.30)$$

where $\bar{c}^r = m/n$ is the capacity ratio for the network with composite key patterns. This network is equivalent to the first-order neural network with respect to their capacity if $\bar{\sigma}_r^2 = \sigma_1^2$. Under this assumption, the combination of (7.4.14) and (7.4.30) indicates that

$$\frac{\bar{c}^r}{c_1} = n^{r-1} \quad (7.4.31)$$

This is the relationship between the capacity of the neural network with composite key patterns and the first-order neural network. Among the existing capacity bounds for neural networks of first order, a reasonable one is $c_1 = 1/(2\log n)$ (McEliece et al., 1987; Amari and Maginu, 1988). According to (7.4.31), the capacity of the neural network with composite key patterns is given by $\bar{c}_r = n^{r-1}/(2\log n)$.

7.4.1 Experimental Results

This last set of experiments evaluates the capacity of a neural network with composite key patterns formed as $\bar{\mathbf{x}}_k = \mathbf{x}_k^1 \otimes \mathbf{x}_k^2$, where $\mathbf{x}_k^1 = \mathbf{x}_k$ and $x_{i,k}^2 = x_{i,k}\, \forall\, i = 1, 2, \ldots, n_1$. The number of associations stored in the network varied from $m = 2$ to $m = 1500$. The

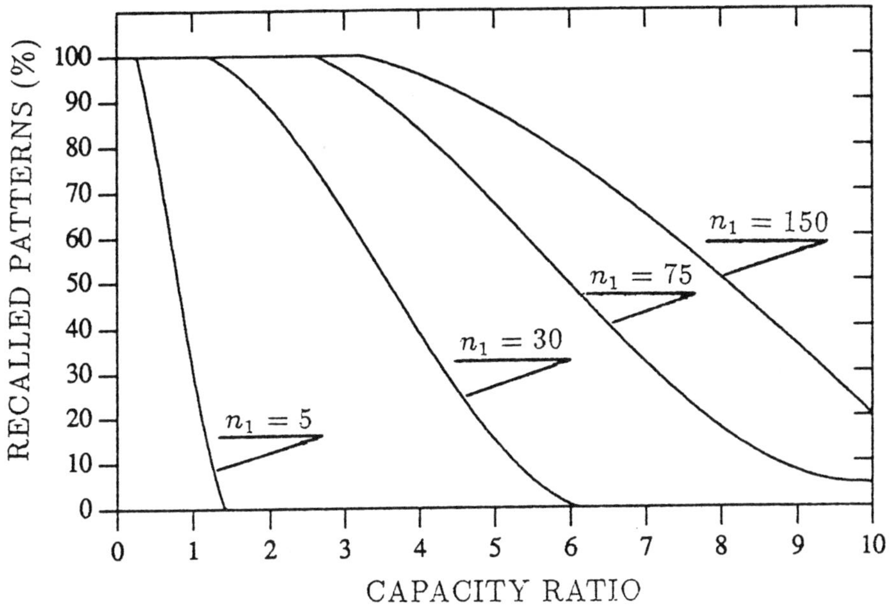

Figure 7.11: Percentage of perfectly recalled patterns by a neural network with composite key patterns formed as $\bar{\mathbf{x}}_k = \mathbf{x}_k^1 \otimes \mathbf{x}_k^2$, where $\mathbf{x}_k^1 = \mathbf{x}_k$ and $x_{i,k}^2 = x_{i,k} \, \forall i = 1, 2, \ldots, n_1$, as a function of the capacity ratio $c = m/n$ for different values of n_1.

length of the stored patterns was $n = 150$. For each experiment, 1000 noisy patterns were generated to be used as input patterns. The normalized Hamming distance between the input patterns and one of the stored patterns was uniformly distributed between 0 and 1/4. Figure 7.11 shows the percentage of perfectly recalled patterns as a function of the capacity ratio $\bar{c} = m/n$ for different values of n_1. Figure 7.11 indicates that, regardless of n_1, the network can guarantee perfect recall of the stored patterns only if the capacity ratio does not exceed a certain threshold. For each network, this threshold provides the maximum allowable number of associations which guarantees perfect recall. According to Figure 7.11, the maximum number of associations that can be reliably stored in the network increases as n_1 increases. This is intuitively reasonable and also consistent with the analysis presented in section 7.4. However, Figure 7.11 contains valuable information which is not revealed by the analysis. This information relates to the slope of the curves that represent the recall efficiency of the network as a function of the capacity ratio for different values of n_1. When the number of associations exceeds the maximum allowable one, the percentage of perfectly recalled patterns drops faster for small n_1. When n_1 increases, the recall efficiency of the network decreases slowly. Not surprisingly, the slowest decrease corresponds to $n_1 = n$, that is, the neural network of second order. Figure 7.11 can also be used to verify some of the quantitative results derived in section 7.4. This set of experiments evaluates the capacity of the neural networks with composite key patterns when the inputs of the network are noisy versions of the key patterns. In addition, the key patterns generated for this experiment were correlated. Therefore, the comparison of the analytical and experimental results provides a basis for the justification of the assumptions made in the analysis, i.e., the input of the network is one of the key patterns and the key patterns are uncorrelated. Section 7.4 investigated the relationship between the capacity bound $\bar{c}$ of the neural network considered in these experiments and the capacity ratio c_1 of the first-order neural network trained using the outer-product rule. According to this analysis, $\bar{c}$ is given in terms of c_1 by (7.4.15). Clearly, the evaluation of a specific capacity bound for the neural network with composite key patterns assumes the derivation of a specific capacity bound c_1. There are various bounds proposed in the literature for the capacity ratio c_1 of first-order neural networks trained using the outer-product rule. An "optimistic" bound is $c_1 = 1/(2 \log n) = 0.1$, while a "pessimistic" one is $c_1 = 1/(4 \log n) = 0.05$ (McEliece et al., 1987). It was shown in chapter 6 that a reasonable compromise between these two extremes is $c_1 = 1/(2 \log n + \log \log n) = 0.086$. When $n_1 = n/5$, Figure 7.11 indicates that the capacity ratio that guarantees perfect recall is $\bar{c} \approx 1.2$. In this case, $\bar{c} \approx (n/7)c_1 = 21c_1$. It can easily be verified that $0.05 \times 21 < \bar{c} < 0.086 \times 21$, that is, the capacity ratio of this network is upper bounded by $(n/5) \times 1/(2 \log n + \log \log n)$ and lower bounded by $(n/5) \times 1/(4 \log n)$. An analogous result can be verified in the case where $n_1 = n/2$. The capacity bound provided by Figure 7.11 for this network is $\bar{c} \approx 2.7$. According to the analysis presented in section 7.4, $\bar{c} \approx (n/4)c_1 = 37.5c_1$. Once again, $0.05 \times 37.5 < \bar{c} < 0.086 \times 37.5$. Finally, this set of experiments provides a capacity bound for second-order neural networks. According to Figure 7.11, the capacity bound is $\bar{c} \approx 3.2$. This experimental outcome should be compared to the capacity bound which resulted as a by-product of the analysis presented in this section (Psaltis et al, 1988). In $n_1 = n$, $\bar{c} = (n/3)c_1 = 50c_1$. Once again, $0.05 \times 50 < \bar{c} < 0.086 \times 50$. Therefore, the capacity of second-order neural networks is upper bounded by $(n/3) \times 1/(2 \log n + \log \log n)$ and lower bounded by $(n/3) \times 1/(4 \log n)$.

7.5 Discussion

The performance of first-order neural networks can be improved up to certain bounds by employing sophisticated learning schemes for their training. The analysis presented in this chapter indicated that neural networks of order higher than one achieve superior performance even if their training is based on the outer-product rule, i.e., the simplest training scheme. In fact, the outer-product rule provides a reliable approximation of optimal least-squares training of neural networks of order higher than one. The price to be paid for this superior performance is the substantial increase in the size of the neural network. Clearly, there is a tradeoff between the size of a neural network, depending on its order, and the complexity of the learning scheme used for its training. It is well-known that the outer-product rule is the training scheme most frequently, if not exclusively, used by researchers attempting optical implementation of neural networks. As the science and technology related to the optical implementation of neural networks advances, the development of high-order neural networks seems to be the most promising path. This argument is supported not only by the poor performance of first-order neural networks trained using the outer-product rule but also by the potential of optical implementations of very large neural networks. The development of neural networks with composite key patterns addressed the crucial problem of forming the input, i.e., the "stimulus", of neural networks. Such neural networks can be used whenever the output of the neural network depends on more than one parameter, or feature, and the application demands a neural network with high performance. The outer-product rule is the simplest learning scheme, but not the only one that can be used for the training of neural networks with composite key patterns. This chapter described how such a neural network can be trained by employing any of the existing algorithms for single-layered neural networks.

Chapter 8

Applications of Neural Networks: A Case Study

8.1 Introduction

The interest of researchers from different scientific disciplines in the research area of neural networks grows as the list of applications of neural networks becomes longer. The recent interest in neural networks is reinforced by the complexity and diversity of the problems that can be successfully treated using neural networks. At this stage of research, the successful application of neural networks to new problems is crucial for the evolution of this research field in the future. In addition to its practical value, any new application gives rise to new efforts in theoretical aspects of this research field, such as the evaluation of the performance of existing neural network architectures and learning schemes, the search for alternative neural network architectures, and the development of fast and efficient learning algorithms.

The successful applications of neural networks motivate the identification of new problems which are potential candidates for neural network applications. Creative thinking and imagination can be useful guides in selecting suitable problems. However, engineering judgment and knowledge of the capabilities and shortcomings of neural networks are crucial for the efficient formulation of a candidate problem. Such a formulation is efficient when it leads to systems that achieve the best possible performance using neural networks of the smallest possible size.

This chapter presents a general methodology for the development of neural network systems. This methodology shows how to utilize the existing learning algorithms and the knowledge about the performance of various neural network architectures. The detailed presentation of a case study of a neural network application also provides some hints about the transformation of real-world problems into potential neural network applications. This chapter presents the development of a neural network system for environmental protection (Karayiannis, 1991a; Karayiannis and Venetsanopoulos, 1990d).

This chapter is organized as follows: Section 8.2 proposes a general methodology for the development of systems based on neural networks. Section 8.3 presents an application of neural networks in environmental protection. The problem considered here is the evaluation of a set of pollution sources in terms of their contribution to the pollution of a certain ecosystem. Section 8.4 contains concluding remarks.

8.2 General Methodology for the Development of Neural Network Systems

There exist many diverse applications of neural networks that have been reported in the literature. The problems treated using neural networks differ in the type and volume of the data involved as well as the final objective. However, there are some basic steps that should be followed in all neural network applications. This section presents a general methodology for the development of systems based on neural networks (Karayiannis, 1991a).

The development of such systems can be achieved in the four levels presented here:

- *Level 1: Formulation of the Problem.* This level examines whether the problem under consideration can be solved using neural networks. If the answer is positive, this level formulates the problem in a way amenable to treatment by neural networks.

- *Level 2: System Design.* This level includes the transformation of the available data into a form acceptable by neural networks and the subsequent selection of the rules governing the training of the neural network. The choices made in this level are strongly affected by the formulation of the problem selected in Level 1.

- *Level 3: System Realization and Implementation.* This level includes the selection of the neural network architecture and the training scheme. This choice is based on the requirements imposed by the specific application and requires a knowledge of the performance characteristics of certain neural network architectures and training schemes. The realization of the system is completed by training the neural network(s) involved. The system can be implemented in software or hardware. Depending on the specific application, a hybrid implementation is also possible.

- *Level 4: System Verification.* The efficiency of the system developed is tested and the results are evaluated and, if possible, compared with results provided by alternative systems performing the same task. If the performance of the system is not satisfactory, the formulation of the problem (Level 1) is reconsidered. If the formulation of the problem is satisfactory or there is no alternative formulation, an alternative system design (Level 2) is investigated and subsequent levels follow.

This procedure is summarized in the flowchart shown in Figure 8.1.

An important problem revealed by this methodology is the efficient transformation of the available data into admissible inputs/outputs of neural networks. In addition to its role in the performance of the resulting system, this transformation mainly determines the size of the neural network(s) involved. This transformation depends on the nature of the problem and the properties of

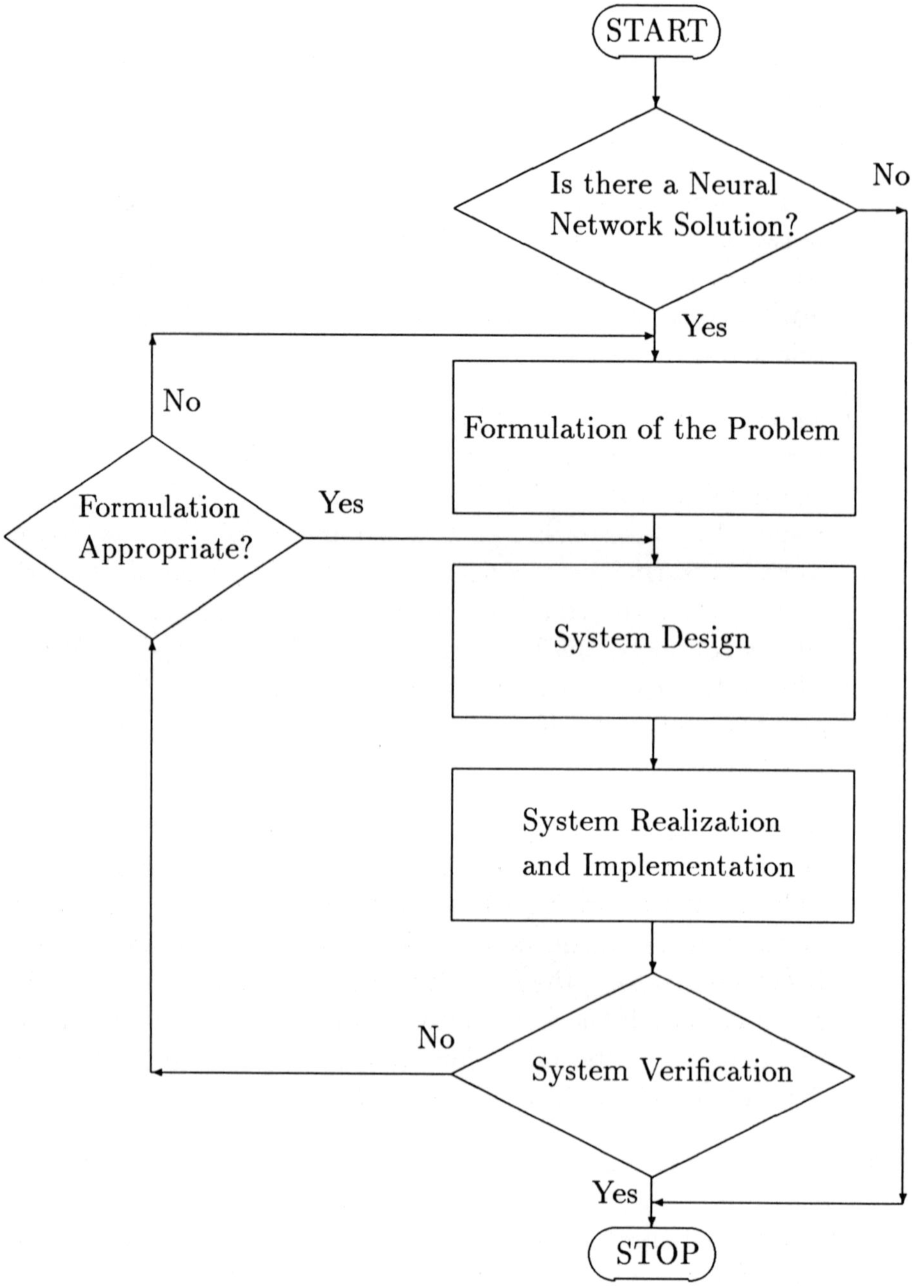

Figure 8.1: A general methodology for the development of neural network systems.

the neural network(s) used. Therefore, the use of existing coding schemes may not be appropriate for neural network applications. The requirement for efficient transformation schemes motivates new research efforts in the area. The proper choice of neural network architectures and training schemes for a given application is based on the performance of neural networks in conjunction with the requirements imposed by the problem under consideration. The impact of this choice on the performance of the resulting system emphasizes the importance of research efforts devoted to the performance evaluation of neural networks as well as the development of fast and efficient learning algorithms for neural networks. Finally, system verification is crucial in comparing the neural network with alternative systems developed for the same task on the basis of conventional techniques. In the following, these concepts are illustrated by using the methodology presented in this section for developing of a neural network system for decision making.

8.3 Application of Neural Networks in Environmental Protection

Decision making, a key issue in many facets of human activity, is not a simple task, given the volume of data and the tradeoffs usually found in practical situations. One of the main objectives of the scientific community during past decades was the development of *expert systems* that can reliably perform decision making in place of humans. Neural networks provide an alternative for the development of such systems. A class of problems involving decision making relates to the interpretation of data on the basis of a certain set of rules. In such a case, the neural network learns the rules governing the decision making by experiment. In another class of problems, the decision making can be facilitated by the simulation of systems with complicated nonlinear dynamics by neural networks. In fact, the existing applications of neural networks in decision making deal with only a small portion of the potential applications. In this section, neural networks are used for the evaluation of the relative effect of various sources on the pollution of a certain ecosystem. This is a rather controversial question that often arises in environmental protection. This choice was motivated not only by the importance of the issue but also by the decision making capability required.

One of the major problems in environmental protection is the development of a set of legal guidelines that would force the sources of pollution to process their harmful wastes before they reach the environment. Provided that such guidelines exist, the protection of the environment depends on the effective enforcement of the law. The necessary condition for effective law enforcement is the reliable detection of certain sources of environmental pollution that violate existing guidelines by not processing, or inadequately processing, their harmful wastes. In conclusion, the development of systems that can reliably detect the major source(s) of environmental pollution is crucial for the protection of our environment.

A simple application of the problem described above is the detection of the major source(s) of chemical pollutants (CP) in a river. Suppose that the river receives the wastes, processed or not, of a number of pollution sources, such as factories and domestic sewage. As a result, a set of chemical pollutants can be detected in the river. According to existing guidelines, the pollution sources should treat their wastes in such a way that each of the chemicals does not exceed a certain threshold. Suppose that the river is checked daily and the resulting record indicates which of the chemical pollutants exceeds the allowable threshold. The combination of chemical pollutants that can be detected in the wastes of each pollution source is assumed to be known. The problem is to determine the pollution source, or combination of sources, responsible for the pollution of the river.

Level 1: Formulation of the Problem

Among the advantages offered by expert systems are speed, reliability, and consistency. The use of human expertise in environmental problems is not always reliable, due not only to the inherent difficulties but also the political, economical, and emotional factors affecting human empirical decision-making. In such a case, the machine has an a priori advantage over any human expert: it is *impartial.*

The development of an expert system for reliable decision making in environmental problems is particularly difficult, mainly due to the complicated, and generally unknown, dynamics of the system

$$\textit{pollution sources} \rightarrow \textit{ecosystem} \rightarrow \textit{pollution} \tag{8.3.1}$$

The generality of the model (8.3.1) and the variety of factors affecting the pollution of the ecosystem make it difficult to study such environmental problems. The model (8.3.1) cannot be simplified by considering the effect of each pollution source or a combination of sources separately, since it is practically impossible to selectively inhibit sources of ecosystem pollution.

The basic idea behind the approach presented here is the simulation of the ecosystem by a neural network. The training of such a neural network can be achieved by exploiting the information available on the system (8.3.1). This information includes the daily appearance in the ecosystem of chemical pollutants exceeding a normal threshold and the chemical pollutants detected in the wastes of each pollution source. A key issue in this approach is the formation of the set of input/output patterns used for training the neural network from the available information.

Level 2: System Design

Assume that n_p chemical pollutants are detected in the ecosystem. The pollution of the ecosystem is determined each day by the specific chemical pollutants that exceed a certain threshold. Therefore, the pollution of the ecosystem during the kth day can be described by the following *pollution pattern*

$$\mathbf{p}_k = [p_{1k}\, p_{2k}\, \cdots\, p_{n_p k}] \tag{8.3.2}$$

where each element of $\mathbf{p}_k$ is defined by

$$p_{ik} = \begin{cases} +1 & \text{if the } i\text{th CP exceeds a threshold during the } k\text{th day} \\ -1 & \text{otherwise} \end{cases} \tag{8.3.3}$$

Assume that the ecosystem receives the wastes of n_s pollution sources. The role of each source in the pollution of the ecosystem is determined by the chemical pollutants that can be detected in its wastes. In mathematical terms, each pollution source is characterized by the *waste pattern*

$$\mathbf{s}_j = [s_{1j}\, s_{2j}\, \ldots\, s_{n_p j}]\,,\ j = 1, 2, \ldots, n_s \tag{8.3.4}$$

where each element of $\mathbf{s}_j$ is defined by

$$s_{ij} = \begin{cases} +1 & \text{if the } i\text{th CP is detected in the } j\text{th pollution source} \\ -1 & \text{otherwise} \end{cases} \tag{8.3.5}$$

In addition, define an auxiliary waste pattern $\mathbf{s}_0$ indicating that there are no chemicals detected in a certain input to the ecosystem. Clearly, the elements of this particular waste pattern are

$$s_{i0} = -1\,,\ i = 1, 2, \ldots, n_p \tag{8.3.6}$$

Assuming that the ecosystem is a river, such a waste pattern could, for instance, describe the situation where the river receives water from an absolutely clean tributary.

Consider the simulation of the daily appearance in the ecosystem of chemical pollutants exceeding a certain threshold by a neural network. Obviously, the outputs of this network are the pollution patterns recorded every day in the ecosystem. The crucial information regarding the formation of the pollution pattern of each day is the pollution pattern of the previous day as well as the possible input of chemical pollutants contained in the wastes of n_s pollution sources. This observation implies the following formation of the set of input-output patterns used for training the neural network. When the output of the network is the pollution pattern $\mathbf{p}_{k+1}$, the input of the network is formed by combining the pollution pattern of the previous day $\mathbf{p}_k$ and the waste patterns of all the pollution sources $\mathbf{s}_j, j = 1, 2, \ldots, n_s$. Assume that the existence of chemical pollutants in the ecosystem was recorded for a period of m days. According to the above discussion, the network is trained with respect to the set $(\mathbf{y}_k, \mathbf{x}_k), k = 1, 2, \ldots, m-1$, where the input patterns $\mathbf{x}_k$ are

$$\mathbf{x}_k^* = [\mathbf{x}_k^0\, \mathbf{x}_k^1\, \mathbf{x}_k^2\, \ldots\, \mathbf{x}_k^{n_s}]\,,\ k = 1, 2, \ldots, m-1 \tag{8.3.7}$$

where $\mathbf{x}_k^0 = \mathbf{p}_k$, $\mathbf{x}_k^r = \mathbf{s}_r \, \forall r = 1, 2, \ldots, n_s$ and the output patterns $\mathbf{y}_k$ are

$$\mathbf{y}_k^* = \mathbf{p}_{k+1} \,, \; k = 1, 2, \ldots, m-1 \tag{8.3.8}$$

After its training, the network should be able to simulate the appearance in the ecosystem of chemical pollutants exceeding a normal threshold every day. With such a network available, the relative effect of various pollution sources can be evaluated by simply checking the outputs of the neural network when various combinations of pollution sources are considered to be active. The selective activity of pollution sources, something impossible in real life, can be simulated in this case by appropriately forming the inputs of the neural network. The formation of the inputs of the neural network provides the rules for the decision making in this application. Since the waste patterns of the pollution sources are input to the neural network every day, the training of the neural network assigns a set of weights to each of the pollution sources. These sets of weights determine the relative effect of the pollution sources on the pollution of the ecosystem. Assume that the neural network can reproduce the daily appearance in the ecosystem of chemical pollutants exceeding a normal threshold with only one pollution source active. Then, this particular source must be the major contributor to the pollution of the ecosystem while the contribution of the rest of the pollution sources can be considered to be negligible. This reasoning can be summarized in the following *least disturbance principle:* The major effect on the pollution of the ecosystem comes from the pollution source that causes the least disturbance to the performance of the neural network when it is acting alone. If the consideration of only one pollution source provides no conclusive evaluation, the least disturbance principle can be extended to any combination of two or more pollution sources.

Consider now the contribution of each pollution source to the pollution of the ecosystem. In this case, the performance of the network is evaluated under the assumption that the contribution of all the rest of the pollution sources is eliminated. Accordingly, the input corresponding to a particular day is formed by combining the pollution pattern of that day with the waste pattern of the pollution source considered. In other words, the waste pattern $\mathbf{s}_0$ is assigned to all pollution sources except the one under consideration. In mathematical terms, when the jth pollution source is considered the inputs of the network are

$$\mathbf{x}_k(j)^* = [\mathbf{x}_k^0(j)\,\mathbf{x}_k^1(j)\,\mathbf{x}_k^2(j)\ \ldots\ \mathbf{x}_k^{n_s}(j)]\,,\ k = 1, 2, \ldots, m-1 \quad (8.3.9)$$

where $\mathbf{x}_k^0(j) = \mathbf{p}_k$ and

$$\mathbf{x}_k^r(j) = \begin{cases} \mathbf{s}_j & \text{if } r = j \\ \mathbf{s}_0 & \text{otherwise} \end{cases} \quad (8.3.10)$$

Assume that $\hat{\mathbf{y}}_k(j), k = 1, 2, \ldots, m-1$ are the outputs obtained when the patterns $\mathbf{x}_k(j), k = 1, 2, \ldots, m-1$ are fed into the network, i.e., the pseudo-pollution patterns, resulting under the assumption that only the jth pollution source is active. The contribution of any particular pollution source can be evaluated by observing the difference between its corresponding pseudo-pollution patterns and the real ones. Following this reasoning, a *characteristic number* is assigned to each pollution source. As an example, the characteristic number N_j assigned to the jth pollution source is obtained by counting the outputs $\hat{\mathbf{y}}_k(j)$ that differ from the expected outputs $\mathbf{y}_k$. Equivalently, if $d_k(j)$ is the Hamming distance between $\hat{\mathbf{y}}_k(j)$ and $\mathbf{y}_k$, N_j is the number of $d_k(j), k = 1, 2, \ldots, m-1$ that are nonzero. Clearly, $0 \leq N_j \leq m-1\, \forall\, j = 1, 2, \ldots, n_s$. The relative effect of each pollution source can be determined by comparing the characteristic numbers of the pollution sources $N_j, j = 1, 2, \ldots, n_s$ according to the following rule: The smaller the characteristic number, the stronger the effect of the corresponding pollution source.

In practice, there may be situations where the comparison of the characteristic numbers $N_j, j = 1, 2, \ldots, n_s$ cannot provide a reliable basis for a conclusive answer. In such a case, the evaluation of the pollution sources can be approached by considering the combined effect of two or more pollution sources. In this case, the waste pattern $\mathbf{s}_0$ is assigned to all pollution source except the two under consideration. In mathematical terms, when the combined effect of the ith and jth pollution sources is considered the inputs of the network are

$$\mathbf{x}_k(i,j)^* = [\mathbf{x}_k^0(i,j)\,\mathbf{x}_k^1(i,j)\,\mathbf{x}_k^2(i,j)\ \ldots\ \mathbf{x}_k^{n_s}(i,j)]\,,\ k = 1, 2, \ldots, m-1 \quad (8.3.11)$$

where $\mathbf{x}_k^0(i,j) = \mathbf{p}_k$ and

$$\mathbf{x}_k^r(i,j) = \begin{cases} \mathbf{s}_i & \text{if } r = i \\ \mathbf{s}_j & \text{if } r = j \\ \mathbf{s}_0 & \text{otherwise} \end{cases} \tag{8.3.12}$$

Assume that $\hat{\mathbf{y}}_k(i,j), k = 1,2,\ldots,m-1$ are the outputs obtained when the patterns $\mathbf{x}_k(i,j), k = 1,2,\ldots,m-1$ are fed into the network. The combined effect of two pollution sources can be evaluated by assigning a characteristic number to each of the $\frac{1}{2}n_s\,(n_s-1)$ combinations of two pollution sources. As an example, the characteristic number N_{ij} corresponding to the combined effect of the ith and jth pollution sources is obtained by counting the outputs $\hat{\mathbf{y}}_k(i,j)$ that differ from the expected outputs $\mathbf{y}_k$. Clearly, $N_{ij} = N_{ji}$ and also $0 \leq N_{ij} \leq m-1$. The definition of this set of characteristic numbers implies that $N_{ij} = 0$ if $i = j$. Therefore, N_{ij} form a symmetric matrix $\mathbf{N} = [N_{ij}]$ with zero diagonal elements. This matrix can be the basis for the evaluation of a pollution source by taking into account its combined effect with each one of the other pollution sources. However, the evaluation of each pollution source also requires the results obtained under the assumption that only this particular source is active. The overall evaluation can be revealed by the symmetric matrix $\bar{\mathbf{N}} = [\bar{N}_{ij}]$, defined by

$$\bar{N}_{ij} = \begin{cases} N_i & \text{if } i = j \\ N_{ij} & \text{if } i \neq j \end{cases} \tag{8.3.13}$$

The evaluation of each pollution source can be facilitated by considering a new set of *generalized characteristic numbers* $\bar{N}_i, i = 1,2,\ldots,n_s$, defined as

$$\bar{N}_i = \sum_{j=1}^{n_s} \bar{N}_{ij} = \sum_{j=1}^{n_s} \bar{N}_{ji} \tag{8.3.14}$$

The new set of characteristic numbers evaluates each pollution source assuming that it is acting alone or in combination with any other pollution source.

Level 3: System Realization and Implementation

The formation of the input and output patterns by (8.3.7) and (8.3.8), respectively, indicates that the neural network which simulates the daily appearance of chemical pollutants in the ecosystem consists of $n_i = (1 + n_s)n_p$ inputs and $n_o = n_p$ output units. For a given number of input and output units, there exist many feed-forward neural network architectures that can be used. The simplest choice is a single-layered, feed-forward neural network. However, a multi-layered feed-forward neural network may be used as well. The simplest multi-layered neural network, that is, a feed-forward network with one layer of hidden units, is the alternative considered here. The structure of such a neural network is shown in Figure 2.9. Since the number of input/output units is specified by the application, the number of hidden units is the only free choice allowed by this neural network architecture. The appearance of chemical pollutants in the ecosystem will be simulated by a feed-forward neural network with $n_i = n_p(1 + n_s)$ inputs, $n_h = n_i = n_p(1 + n_s)$ hidden units, and $n_o = n_p$ output units.

The neural network that simulates the appearance of chemical pollutants in the river contains one layer of hidden units. Therefore, it can be trained by using either the ELEANNE 5 or ELEANNE 6 algorithms, proposed in chapter 3 for training multi-layered neural networks. It was experimentally verified that the convergence of ELEANNE 6 is comparable to that of ELEANNE 5. But the fact that ELEANNE 6 requires fewer arithmetic operations per adaptation cycle than ELEANNE 5, provided that $n_o > 1$, suggests that it should be used for training this network.

Level 4: System Verification

This section presents the application of the procedure described above to a specific set of data (Block, et al., 1964). In this example the ecosystem considered is a river, with $n_s = 6$ pollution sources (PS), namely domestic sewage (PS1), a metal-plating plant (PS2), a milk product factory (PS3), a pulp and paper mill (PS4), a slaughter house (PS5), and a textile mill (PS6). For a period of $m = 15$

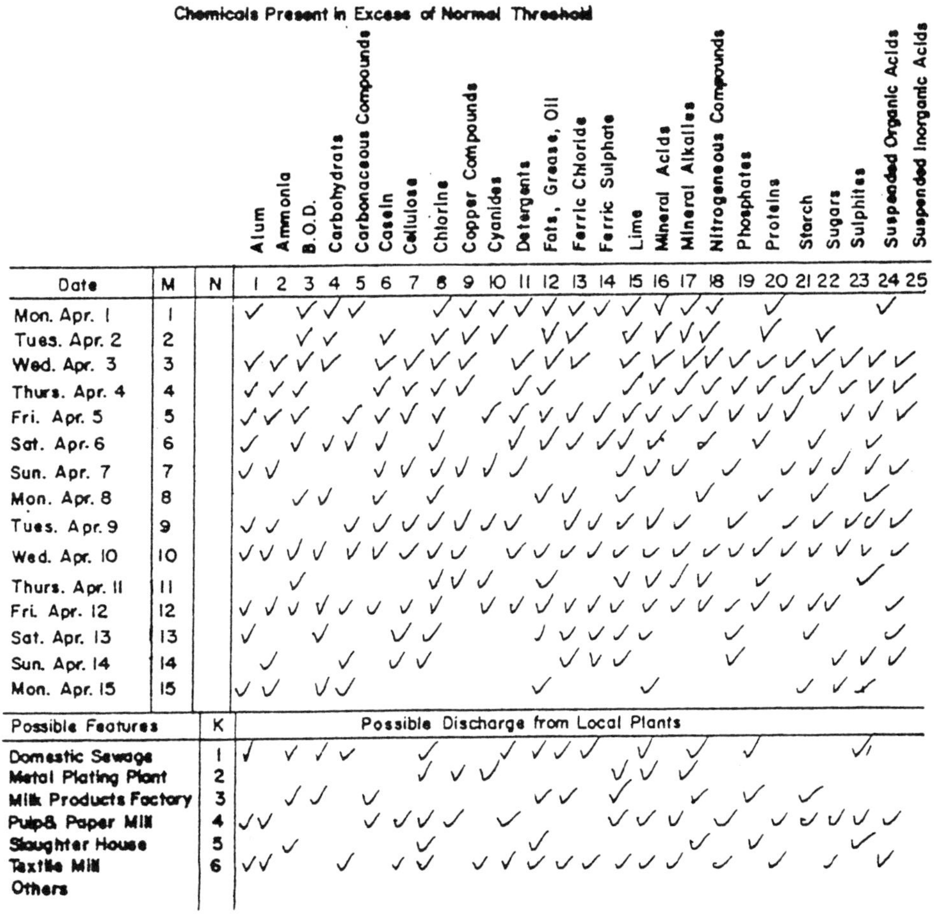

Figure 8.2: Recorded data regarding the pollution of a river by six pollution sources for a period of 15 days.

days, the river is tested every day for the presence of 25 chemical pollutants in excess of a normal threshold. The data recorded are shown in Figure 8.2, which is reproduced here from the original paper (Block, et al., 1964). The upper table in Figure 8.2 shows the chemical pollutants that exceeded a normal threshold on any of the 15 days. The lower table shows the chemicals detected in the wastes of the six pollution sources.

In this experiment, the effect of the pollution sources is evaluated in terms of two different sets of chemical pollutants, that is, the first ten and twenty five chemicals in the order that they appear in Figure 8.2. The formation of these sets of chemicals is somewhat arbitrary, since the main purpose of this experiment is to test the general procedure presented above. In real situations, the selection of the particular chemical pollutants used for the evaluation of the effect of the pollution sources would be made by chemists and/or environmentalists.

For this particular example, the information required for training the neural network can be obtained from Figure 8.2 by simple inspection. The waste patterns corresponding to the six pollution sources are formed on the basis of the lower table of Figure 8.2 by using the definitions of the waste patterns in (8.3.4) and (8.3.5). The pollution patterns of the $m = 15$ days are formed on the basis of the upper table of Figure 8.2 by using the definition of the pollution patterns in (8.3.2) and (8.3.3). Figure 8.2 indicates whether a certain chemical pollutant exceeds its threshold by a check symbol. Therefore, the element of $\mathbf{s}_j, j = 1, 2, \ldots, 6$ ($\mathbf{p}_k, k = 1, 2, \ldots, 15$) corresponding to a pollutant is +1 if there is a check symbol in the corresponding position of the lower (upper) table of Figure 8.2 and −1 otherwise. For example, provided that the $n_p = 10$ first chemical pollutants are considered in the evaluation, the waste pattern corresponding to the first pollution source is $\mathbf{s}_1 = [+1\ -1\ +1\ +1\ +1\ -1\ -1\ +1\ -1\ -1]$. The pollution pattern of the first day is obtained from Figure 8.2 as $\mathbf{p}_1 = [+1\ -1\ +1\ +1\ +1\ -1\ -1\ +1\ +1\ +1]$.

Consider here the evaluation of the six pollution sources of this example in terms of the $n_p = 10$ first chemical pollutants. In this case, the neural network that simulates the daily appearance of

PS	PS1	PS2	PS3	PS4	PS5	PS6	$\bar{N}_i$
PS1	1	1	0	0	2	0	4
PS2	1	2	1	1	3	1	9
PS3	0	1	2	1	0	0	4
PS4	0	1	1	2	1	0	5
PS5	2	3	0	1	3	0	9
PS6	0	1	0	0	0	1	2

Figure 8.3: Characteristic matrix for the river polluted by six pollution sources as described in Figure 8.2. The evaluation of the pollution sources is based on the $n_p = 10$ first chemical pollutants of Figure 8.2.

pollutants in the river consists of $n_i = 70$ inputs, $n_h = 70$ hidden units, and $n_o = 10$ output units. The resulting characteristic matrix is shown in Figure 8.3. The diagonal elements of this matrix provide the basis for the evaluation of the pollution sources when only one of the sources is considered to be active. According to Figure 8.3, $N_1 = N_6 < N_2 = N_3 = N_4 < N_5$. This set of characteristic numbers indicates that PS1 and PS6 are mainly responsible for the pollution of the river. On the other hand, PS5 has a relatively minor effect on the pollution of the river. However, this set of characteristic numbers provides no information regarding the relative effect of PS1 and PS6, or PS2, PS3, and PS4. This experimental outcome verifies the argument that the consideration of two active pollution sources is often necessary for the reliable evaluation of the relative effect of the sources on the pollution of the river. This additional information is included in the set of generalized characteristic numbers $\bar{N}_j, j = 1, 2, \ldots, 6$, shown in the right column of Figure 8.3. According to Figure 8.3, $\bar{N}_6 < \bar{N}_1 = \bar{N}_3 < \bar{N}_4 < \bar{N}_2 = \bar{N}_5$. The evaluation provided by this set of numbers agrees with the previous one in that the major contributor to the pollution of the river is PS6 while PS5 contributes the least. The additional information provided by this set of numbers is that the relative effect of PS1 is lower than the one of PS6, the effect of PS1 and PS3 is lower than the one of PS6 and higher than the one of PS4 and finally, PS2 and PS5 have the lowest effect on the pollution of the river. It was verified that

PS	PS1	PS2	PS3	PS4	PS5	PS6	N_i
PS1	2	0	0	0	0	0	2
PS2	0	9	3	1	3	1	17
PS3	0	3	7	1	3	1	15
PS4	0	1	1	1	1	0	4
PS5	0	3	3	1	8	1	16
PS6	0	1	1	0	1	2	5

Figure 8.4: Characteristic matrix for the river polluted by six pollution sources as described in Figure 8.2. The evaluation of the pollution sources is based on the $n_p = 25$ first chemical pollutants of Figure 8.2.

there is a remarkable agreement between the evaluation provided by the procedure presented above and the one based on the empirical interpretation of the data (Karayiannis, 1991a).

Finally, consider the evaluation of the six pollution sources in terms of all the $n_p = 25$ chemical pollutants appearing in Figure 8.2. In this case the neural network consists of $n_i = 175$ inputs, $n_h = 175$ hidden units, and $n_o = 25$ output units. The resulting characteristic matrix is shown in Figure 8.4. The comparison of the diagonal elements of the characteristic matrix indicates that $N_4 < N_1 = N_6 < N_3 < N_5 < N_2$. Clearly, the consideration that only one pollution source is active indicates that the major effect comes from PS4 followed by PS1 and PS6. In addition, the least effect comes from PS2 and PS5. However, there is no conclusive answer regarding the relative effect of PS1 and PS6. Consider here the evaluation of the pollution sources based on the set of generalized characteristic numbers $\bar{N}_j, j = 1, 2, \ldots, 6$. According to Figure 8.4, $\bar{N}_1 < \bar{N}_4 < \bar{N}_6 < \bar{N}_3 < \bar{N}_5 < \bar{N}_2$. This new evaluation reverses the relative effect of PS1 and PS4, indicating that PS1 is the major source of pollution. On the other hand, this set of numbers indicates that the least effect comes from PS2 and PS5. However, their relative effect has been reversed compared to that suggested by the previous evaluation. It should be emphasized that the consideration of all the $n_p = 25$ chemical pollutants assigns a less significant role to PS6, the main source of pollution according to all the previous evaluations. According to the characteristic matrix shown in Figure 8.4, $\bar{N}_{1j} = \bar{N}_{j1} = 0 \,\forall\, j \neq 1$. This observation serves as an additional verification of the strong impact of PS1 on the pollution of the river, already revealed by the set of generalized characteristic numbers $\bar{N}_j, j = 1, 2, \ldots, 6$. Finally, the observation that $\bar{N}_{34} = \bar{N}_{43} = 0$ reveals the strong impact of the combined action of PS3 and PS4 on the pollution of the river. This is consistent with the previous evaluation that characterized PS4 as the strongest contributor after PS1 and also indicated that PS3 has a higher impact on the pollution of the river compared to both PS5 and PS2.

8.4 Discussion

This chapter presented the application of neural networks to a problem involving decision making. This application was successful with regard to the criterion established in the introduction, i.e., the development of systems with the best possible performance using neural networks of the smallest possible size. The largest neural network used in this application consisted of 175 input and 25 output units. In a sense, the applications presented in this chapter were based on the parallelism that characterizes the structure of even the smallest neural networks; however, that parallelism was not massive. The success of such relatively small neural network systems provides an idea of what can be expected in the future, when the implementation of really massive neural networks will be facilitated by dedicated software/hardware.

Chapter 9

Applications of Neural Networks: A Review

9.1 Introduction

The popular neural network conferences provide a reliable indication of the trends in neural network applications. A review of the most recent conference proceedings reveals the declining, expanding, and emerging areas of neural network applications. A scan of the literature indicates that there is currently a slight decline in attempts to solve optimization problems using neural networks, while the area of character and digit recognition has successfully reached maturity. Since it is apparent that neural networks can be applied effectively in character recognition, most of the problems in this area relate to efficient design, manufacturing, and marketing of commercial neural network based character recognition systems. The check verification system developed and marketed by Mead and his team is an excellent example of the successful transition of neural networks from the laboratory to the real world. Most areas of neural network applications are currently in the transition from state-of-the-art paradigms to large-scale systems for real-world applications. This last category includes applications of neural networks in speech processing, image processing and computer vision, pattern classification and recognition, system control, and robotics. Finally, an emerging area of neural network applications is the development of systems combining fuzzy logic and neural networks.

There is a cross-fertilization between advances in theory and neural network applications. Certainly, applications of neural networks benefit from theoretical developments in neural network research. In return, problems arising in neural network applications provide useful feed-back to researchers in the area. For example, the development of large-scale systems for real-world applications emphasized the importance of certain issues, such as fast and efficient learning, generalization and scaling properties of various neural architectures, and implementation of neural systems.

This chapter presents a description of selected neural network applications. The selection of the applications reviewed in this chapter was based on their impact on neural network research and their contribution in establishing artificial neural networks as a reliable tool in many diverse real-world applications.

This chapter is organized as follows: Section 9.2 presents applications of neural networks in optimization, including solutions to NP-complete problems and ill-posed problems. Section 9.3 describes the principles of transform coding and also presents some related image compression approaches based on neural networks. Section 9.4 presents a variety of neural network applications in handwritten character and digit recognition. Section 9.5 describes NETtalk, a neural network trained to convert text to speech. Section 9.6 describes a neural network system developed for the classification of sonar signals. Section 9.7 presents neural network systems developed for the early diagnosis of heart attack and the diagnosis of low back pain. Section 9.8 compares various neural network systems trained to predict secondary structures of proteins. Section 9.9 presents some examples of neural network systems developed for weather forecasting. Section 9.10 evaluates the reliability of neural networks used in financial predictions, by presenting a test of the efficient market hypothesis and an experiment on short-term stock market prediction. Finally, section 9.11 presents a review of the extensive literature on neural network applications.

9.2 Optimization Problems

The use of neural networks in optimization problems became popular after Hopfield and Tank presented a "neural" system which could find satisfactory solutions to the traveling salesman problem (Hopfield and Tank, 1985). In addition to the neural solutions proposed for NP-complete problems such as the traveling salesman problem (Wilson and Pawley, 1988; Hedge et al., 1988; Brandt et al., 1988; Van den Bout and Miller, 1988; Van den Bout and Miller, 1989; Xu and Tsai, 1991), the graph partitioning problem (Ramanujam and Sadayappan, 1988), and the job-shop sequencing problem (Foo and Takefuji, 1988a; Foo and Takefuji, 1988b), neural networks have also been used to solve other problems which are directly or indirectly formulated in terms of function minimization. These problems include A/D converter design (Tank and Hopfield, 1986; Anastassiou, 1988), signal decomposition and linear programming (Tank and Hopfield, 1986), matrix inversion (Jang et al., 1988), analog decoding (Platt and Hopfield, 1986; Erlauson and Abu-Mostafa, 1989), concentrator assignment (Tagliarini and Page, 1987), job sequencing (Fang et al., 1990), and graph mapping (Iyer et al., 1992).

9.2.1 A/D Converter

A simple example of transforming an objective function into a Lyapunov function associated with a Hopfield network is the neural implementation of an A/D converter (Tank and Hopfield, 1986). Tank and Hopfield suggested that such an A/D converter can be realized by an electrical circuit implementing the Hopfield model. According to their formulation of the problem, the binary representation of a certain number x by a $(B+1)$-bit converter is formed by the amplifier output voltages $V_B\, V_{B-1}\, \ldots\, V_1\, V_0$ as follows

$$x \approx \sum_{i=0}^{B} 2^i\, V_i \qquad (9.2.1)$$

The coefficients $V_i, i = 0, 1, \ldots, B$ which satisfy (9.2.1) can be determined by minimizing

$$\begin{aligned} E_1 &= \frac{1}{2}(x - \sum_{i=0}^{B} 2^i V_i)^2 \\ &= \frac{1}{2}x^2 - \sum_{i=0}^{B} 2^i x V_i + \sum_{j=0}^{B} \sum_{i \neq j}^{B} 2^{i+j} V_i V_j + \sum_{i=0}^{B} 2^{2i} V_i^2 \end{aligned} \tag{9.2.2}$$

The representation of x by $V_B V_{B-1} \ldots V_1 V_0$ is binary if each V_i is either 0 or 1. However, the minimization of (9.2.2) does not guarantee that each V_i approaches 0 or 1. Tank and Hopfield suggested that this condition can be ensured by the minimization of the objective function

$$E_2 = -\frac{1}{2} \sum_{i=0}^{B} a_i^2 V_i (V_i - 1) \tag{9.2.3}$$

Clearly, E_2 attains zero, its minimum value, if each V_i is either zero or one. Tank and Hopfield selected a_i on the basis of the additional requirement that the diagonal elements of the matrix of synaptic weights $[T_{ij}]$ are zero (Tank and Hopfield, 1986). A comparison of (9.2.2) and (9.2.3) indicates that this condition is satisfied by $a_i = 2^i$. The overall objective function $E = E_1 + E_2$ was finally formed by ignoring the constant term $\frac{1}{2}x^2$ as follows

$$E = -\frac{1}{2} \sum_{j=0}^{B} \sum_{i \neq j}^{B} T_{ij} V_i V_j - \sum_{i=0}^{B} I_i V_i \tag{9.2.4}$$

where

$$T_{ij} = -2^{i+j} \; \forall \, i \neq j \tag{9.2.5}$$

and

$$I_i = -2^{(2i-1)} + 2^i x \tag{9.2.6}$$

The objective function (9.2.4) may be viewed as the Lyapunov function corresponding to a Hopfield network. Under this interpretation, T_{ij} are the synaptic weights which form the feed-back term and I_i are the external inputs to the network. Clearly, only the inputs

	1	2	3	4	5
A	0	1	0	0	0
B	0	0	0	1	0
C	1	0	0	0	0
D	0	0	0	0	1
E	0	0	1	0	0

Table 9.1: Permutation matrix corresponding to a path.

depend on the number x that is to be digitized. Tank and Hopfield suggested that this network, after evolving in time from an arbitrary initial state, will reach an equilibrium state minimizing (9.2.4). Because of the formulation of the objective function (9.2.4), the equilibrium state of the network will be a digital representation of x.

9.2.2 The Traveling Salesman Problem

The traveling salesman problem can be stated as follows: Given a set of N cities in a plane, a traveling salesman must traverse the shortest possible path which passes through each city exactly once. The problem is frequently described by a permutation matrix, i.e., a matrix whose rows and columns represent the cities and positions in the tour, respectively. Table 9.1 shows the permutation matrix corresponding to the tour CAEBD between the $N = 5$ cities A, B, C, D, and E. The path corresponding to this particular tour is $\ell = \ell_{CA} + \ell_{AE} + \ell_{EB} + \ell_{BD}$, where ℓ_{PQ} is the Euclidean distance between cities P and Q.

The constrained search for the shortest path between N cities is an NP-complete problem, which has been extensively studied over the years (Baum, 1986; Bruck and Goodman, 1988). Although a solution to this finite problem could be found by an extensive search, the required time would be of order $O(N!)$. Thus, such an approach quickly becomes impractical as N, the number of cities, increases. Since attempts to find an algorithm providing the shortest path in time bounded by some polynomial in N have failed, researchers have searched for heuristics providing good solutions to this problem

(Baum, 1986).

Hopfield and Tank proposed a solution to the traveling salesman problem based on the minimization of a certain objective function (Hopfield and Tank, 1985). They also suggested that a satisfactory solution to this problem can be found by transforming this objective function into a Lyapunov function and allowing the corresponding neural model to evolve in time. The output of the neurons in this model are denoted by V_{Xi}, where the first index represents a city and the second index represents a position in the tour. The network must converge to a stable state which corresponds to a valid permutation matrix, i.e., a matrix whose rows and columns each contain a single 1. This requirement guarantees that every city is visited only once. Hopfield and Tank suggested that this requirement is satisfied by the minimization of the following objective function

$$\begin{aligned} E_1 &= \frac{A}{2}\sum_X\sum_i\sum_{j\neq i} V_{Xi}V_{Xj} + \frac{B}{2}\sum_i\sum_X\sum_{Y\neq X} V_{Xi}V_{Yi} \\ &+ \frac{C}{2}(\sum_X\sum_i V_{Xi} - N)^2 \end{aligned} \tag{9.2.7}$$

where A, B, and C are positive constants which determine the effect of each term of (9.2.7) in the solution. The first term of (9.2.7) is zero if each row of the permutation matrix contains at most a single 1. Similarly, the second term of (9.2.7) is zero is each column of the permutation matrix contains at most a single 1. The third term ensures that the total number of 1's in the matrix is N. Taken together, these three terms guarantee that each row contains a single 1, i.e., each city is visited, and each column contains a single 1, i.e., each city is visited only once. The condition that the salesman covers the shortest path while visiting the cities is guaranteed by the minimization of the objective function

$$E_2 = \frac{D}{2}\sum_X\sum_{Y\neq X}\sum_i d_{XY}\, V_{Xi}\,(V_{Y,i+1} + V_{Y,i-1}) \tag{9.2.8}$$

where d_{XY} is the distance between the cities X and Y and D is another positive constant. The subscripts in (9.2.8) are taken modulo N, such that the Nth city in the tour is adjacent to the first. Finally, the traveling salesman problem can be solved by minimizing the objective function

$$\begin{aligned} E &= \frac{A}{2}\sum_X\sum_i\sum_{j\neq i} V_{Xi}V_{Xj} + \frac{B}{2}\sum_i\sum_X\sum_{Y\neq X} V_{Xi}V_{Yi} \\ &+ \frac{C}{2}(\sum_X\sum_i V_{Xi} - N)^2 \\ &+ \frac{D}{2}\sum_X\sum_{Y\neq X}\sum_i d_{XY}\, V_{Xi}\,(V_{Y,i+1} + V_{Y,i-1}) \end{aligned} \tag{9.2.9}$$

According to the formulation proposed by Hopfield and Tank, a solution to the minimization of (9.2.9) can be found by the time evolution of (Hopfield and Tank, 1985)

$$\begin{aligned} \frac{du_{Xi}}{dt} &= -\frac{u_{Xi}}{\tau} - A\sum_{j\neq i} V_{Xj} - B\sum_{Y\neq X} V_{Yi} \\ &- C(\sum_X\sum_j V_{Xj} - N) - D\sum_Y d_{XY}\,(V_{Y,i+1} + V_{Y,i-1}) \end{aligned} \tag{9.2.10}$$

where

$$V_{Xi} = g(u_{Xi}) = \frac{1}{2}\left(1 + \tanh(\frac{u_{Xi}}{u_0})\right) \tag{9.2.11}$$

Hopfield and Tank presented the results of their approach in the case of $N = 10$ cities, where there exist $N!/2N = 10!/20 = 181,440$ distinct paths. They reported that 16 out of 20 trials, starting from different initial states, converged to legitimate tours. Moreover, 50 % of the trials produced one of the 2 shortest paths. They also reported their results in the case of $N = 30$ cities, where there exist $N!/2N = 30!/60 \approx 4.4 \times 10^{30}$ distinct paths. For the set of cities used in this experiment, the shortest path is believed to be 4.26. The network commonly converged to paths of length less than 7, and occasionally less than 6. They also argued that the network provided a very good solution to this problem, by excluding poor paths by a factor of 10^{-22} to 10^{-23} (Hopfield and Tank, 1985).

The initial excitement following the Hopfield and Tank publication was moderated when other researchers identified several disadvantages of their approach. For example, Wilson and Pawley reported extensive experimental results which contradicted, to some extent, the results presented by Hopfield and Tank and pointed out several disadvantages of their approach (Wilson and Pawley, 1988). Since Hopfield and Tank had reported results for the traveling salesman problem for $N = 30$ cities, Wilson and Pawley attempted to more than double the number of cities to $N = 64$ in order to test the effectiveness of the approach when the number of cities increases. After their repeated failure to determine parameters producing results for $N = 64$ cities, they repeated the experiments by using 10 randomly selected sets of $N = 10$ cities. When they used the pure Hopfield and Tank approach they found that only 8 % of the trails produced legitimate paths. Even after the modification of the Hopfield and Tank approach involving Willshaw initialization plus noise, only 18 % of the trials produced legitimate paths (Wilson and Pawley, 1988). It should also be remarked here that a legitimate path does not necessarily satisfy the condition that the length of the path is the shortest possible.

Hedge et al. presented a cookbook method for setting the parameters in the Hopfield and Tank approach (Hedge et al., 1988). They indicated that the relation between two of these parameters appears to be linear. This publication established a relationship between these two parameters and the problem size, which explains why the Hopfield and Tank approach becomes less effective as the number of cities increases. Some researchers attempted to improve the formulation proposed by Hopfield and Tank on the basis of the argument that the determination of a legitimate path by the original approach depends rather strongly on the parameters A, B, C, and D that are provided by the user. An alternative formulation significantly reduced the parameters that must be externally provided and emphasized the search for the shortest path (Brandt, et al., 1988). According to this formulation, the solution to the traveling salesman problem can be found by minimizing the objective function

$$E_2 = \frac{D}{2} \sum_{X} \sum_{Y \neq X} \sum_{i} d_{XY} \, V_{Xi} \, (V_{Y,i+1} + V_{Y,i-1}) \qquad (9.2.12)$$

under the constraints that

$$\sum_i V_{Xi} = 1 \text{ for every city } X \tag{9.2.13}$$

and

$$\sum_X V_{Xi} = 1 \text{ for every stop } i \tag{9.2.14}$$

The minimization of (9.2.12) guarantees the search for the shortest path, while the constraints (9.2.13) and (9.2.14) guarantee the search for legitimate paths. In practice, this constrained minimization was transformed to the unconstrained minimization of the objective function

$$\begin{aligned} E &= \frac{1}{2}\sum_X \sum_{Y\neq X} \sum_i d_{XY}\, V_{Xi}\,(V_{Y,i+1} + V_{Y,i-1}) \\ &+ \frac{\gamma}{2}\left[\sum_X (1 - \sum_i V_{Xi})^2 + \sum_i (1 - \sum_X V_{Xi})^2\right] \end{aligned} \tag{9.2.15}$$

Van den Bout and Miller attempted to improve the Hopfield and Tank approach by employing simulated annealing in conjunction with an alternative objective function of the form (Van den Bout and Miller, 1988; Van den Bout and Miller, 1989)

$$E = \frac{d_p}{2}\sum_i \sum_X \sum_{Y\neq X} V_{Xi}\, V_{Yi} + \frac{1}{2}\sum_X \sum_{Y\neq X} \sum_i d_{XY}\, V_{Xi}\,(V_{Y,i+1} + V_{Y,i-1}) \tag{9.2.16}$$

The first term of (9.2.16) maintains feasibility by acting as a repulsive force that discourages two cities from occupying the same position on a tour, while the second term of (9.2.16) is a measure of the total tour distance. The two terms in (9.2.16) are related by the parameter d_p, which is set to a value slightly larger than the largest distance between any two cities multiplied by two. In the objective function (9.2.16) there are no terms to penalize a city which is not visited or is visited more than once. This constraint is taken into account in this approach by assuming that, at a given simulated temperature T, the probability that city X occupies tour position i obeys a Boltzmann-Gibbs distribution. In this approach, the neural outputs are normalized as follows

$$V_{Xi} = \frac{\exp\{-E_{Xi}/T\}}{\sum_j \exp\{-E_{Xj}/T\}} \tag{9.2.17}$$

where

$$E_{Xi} = d_p \sum_{Y \neq X} V_{Yi} + \sum_{Y \neq X} d_{XY}\, V_{Xi}\,(V_{Y,i+1} + V_{Y,i-1}) \tag{9.2.18}$$

Since $\sum_i V_{Xi} = 1$, this formulation guarantees that each city will be visited only once. This approach found the optimum tour between a set of $N = 30$ cities in 30 % of the trials.

9.2.3 Ill-posed Inverse Problems

An inverse problem is characterized as ill-posed when there is no guarantee of the *existence*, *uniqueness*, and *stability* of the solution based on direct inversion. The solution of an inverse problem is not guaranteed to be stable if a small perturbation of the data can produce a large change in the solution. Regularization theory provides a formal basis for the development of stable solutions of ill-posed problems (Tikhonov and Arsenin, 1977). The stabilizing functional approach is one of the basic methodologies for the development of regularized solutions (Karayiannis and Venetsanopoulos, 1990e). According to this approach, an ill-posed problem can be formulated as the constrained minimization of a certain functional, called a stabilizing functional. This constrained minimization is usually transformed into the unconstrained minimization of a quadratic function.

Several inverse problems in image processing and low level computer vision can be formulated as the minimization of a quadratic function. This class of problems includes visual surface reconstruction (Grimson, 1981; Terzopoulos, 1984), image restoration (Karayiannis and Venetsanopoulos, 1989; Karayiannis and Venetsanopoulos, 1990e) and image interpolation (Karayiannis and Venetsanopoulos, 1992e). Several researchers have indicated that such problems can be efficiently solved by employing the Hopfield network, probably inspired by the first waves of excitement following the publications suggesting that the time evolution of the Hopfield model can find the minima of quadratic functions (Hutchinson

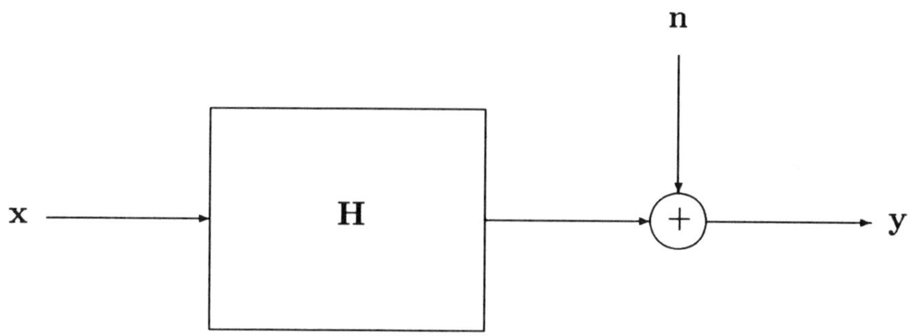

Figure 9.1: The image formation process.

and Koch, 1986). However, the Lyapunov function corresponding to the Hopfield model contains a term corresponding to the decay of neuron activity, in addition to the quadratic and linear terms corresponding to the feed-back and bias terms. According to the analysis presented in chapter 2, the additional term can be considered negligible only if the output of the neurons is determined by a nonlinearity which asymptotically approximates a hard-limiting nonlinearity.

The image formation process is commonly formulated by the following linear model

$$\mathbf{y} = \mathbf{H}\mathbf{x} + \mathbf{n} \tag{9.2.19}$$

If the model (9.2.19) is discrete, $\mathbf{x}$ and $\mathbf{y}$ are vectors representing the original object and the observed image, respectively, and $\mathbf{n}$ is additive noise due to the image recording system. $\mathbf{H}$ is a linear operator decscribing the degradation associated with the image formation process. In a discrete model, $\mathbf{H}$ is a matrix often referred to in the literature as a point spread matrix. Figure 9.1 shows schematically the image formation process. Image restoration is the process of inferring the best possible estimate for the original object $\mathbf{x}$, given the observed image $\mathbf{y}$.

A stable estimate of the original object can be obtained by minimizing

$$E = ||\mathbf{y} - \mathbf{H}\mathbf{x}||^2 + \alpha||\mathbf{D}\mathbf{x}||^2 \qquad (9.2.20)$$

where α is a positive real number, called the regularizing parameter and $||\mathbf{D}\mathbf{x}||^2$ is the discrete approximation of a quadratic functional. The minimization of (9.2.20) guarantees the stability of the inversion. The visual quality of the resulting image estimate depends strongly on the selection of the operator $\mathbf{D}$. A maximum a posteriori (MAP) estimation interpretation of a certain class of regularized solutions provided criteria for the selection of $\mathbf{D}$, by establishing an equivalence between the image representation by a stochastic partial differential equation (SPDE) image model and the constrained minimization of $||\mathbf{D}\mathbf{x}||^2$ (Karayiannis and Venetsanopoulos, 1990e).

Zhou et al. suggested that the Hopfield model can be used to perform the minimization of a quadratic function resulting in image restoration by artificially introducing a hard-limiting nonlinearity in the representation of the image pixels (Zhou et al., 1988). Zhou et al. considered the following two-dimensional version of the Hopfield model

$$u_{i,k}(t+1) = \sum_{j=1}^{N^2} \sum_{\ell=1}^{M} T_{i,k;j,\ell}\, v_{j,\ell}(t) + I_{i,k} \qquad (9.2.21)$$

where the synaptic weights satisfy the symmetry $T_{i,k;j,\ell} = T_{j,\ell;i,k}$ $\forall 1 \leq i,j \leq N^2, 1 \leq \ell, k \leq M$ and $v_{i,k}$ is the state of the (i,k) neuron, given by

$$v_{i,k} = g(u_{i,k}) \qquad (9.2.22)$$

In this approach, $g(x) = 1$ if $x \geq 0$ and $g(x) = 0$ if $x < 0$. As a result, the state of a neuron can be either "on", represented by 1, or "off", represented by 0.

The $N^2 \times 1$ vector $\mathbf{x}$ is obtained by lexicographically ordering the elements of the $N \times N$ image matrix $\mathbf{X}$. The elements of $\mathbf{x}$ are obtained from the image pixels as $x_m = (\mathbf{X})_{ij}, m = j + N(i - 1)$. A similar representation is used for transforming the $N \times N$ matrix $\mathbf{Y}$ into the vector $\mathbf{y}$. Zhou et al. condisidered an alternative image representation in an attempt to perform the minimization of the functional (9.2.20) by employing the Hopfield network (Zhou et al., 1988). The nonlinearity required by the Hopfield model was introduced in their formulation of the problem by a simple trick in the representation of the image pixels. The images considered in this approach are of size $N \times N$ with M gray levels. Each pixel is represented by

$$x_m = (\mathbf{X})_{ij} = \sum_{k=1}^{M} v_{m,k},\ m = j + N(i-1) \tag{9.2.23}$$

According to this representation, the gray level of each image pixel is obtained by adding the outputs of neurons which can be either 0 or 1. According to (9.2.23), the gray level of each pixel x_m is equal to the number of neurons $v_{m,k}$, $k = 1, 2, \ldots, M$ with nonzero output. This representation adds redundancy to the image representation, since the $N \times N$ pixels are represented by $N^2 \times M$ neurons. For example, a $N \times N = 256 \times 256$ image with $M = 256$ gray levels requires $N^2 \times M = 256^3 \approx 16.7 \times 10^6$ neurons. The quadratic functional (9.2.20) which is minimized in this approach can be written as

$$E = \sum_{p=1}^{N^2} (y_p - \sum_{i=1}^{N^2} h_{pi}\, x_i)^2 + \alpha \sum_{p=1}^{N^2} (\sum_{i=1}^{N^2} d_{pi}\, x_i)^2 \tag{9.2.24}$$

The substitution of $x_i = \sum_{k=1}^{M} v_{i,k}$ in (9.2.24) provides

$$E = -\frac{1}{2} \sum_{i=1}^{N^2} \sum_{j=1}^{N^2} \sum_{k=1}^{M} \sum_{\ell=1}^{M} T_{i,k;j\ell}\, v_{j,\ell} - \sum_{i=1}^{N^2} \sum_{k=1}^{M} I_{i,k}\, v_{i,k} \tag{9.2.25}$$

where

$$T_{i,k;j\ell} = -2 \sum_{p=1}^{N^2} h_{pi}\, h_{pj} - 2\, \alpha \sum_{p=1}^{N^2} d_{pi}\, d_{pj} \tag{9.2.26}$$

and

$$I_{i,k} = 2 \sum_{p=1}^{N^2} y_p\, h_{pi} \tag{9.2.27}$$

The resulting synaptic weights $T_{i,k;j\ell}$ are independent of k and ℓ, while the bias term $I_{i,k}$ is independent of k. This is clearly the result of the redundancy introduced by the representation of the image pixels by (9.2.23). Zhou et al. suggested that the objective function (9.2.25) can be viewed as the Lyapunov function of the two-dimensional version of the Hopfield network defined in (9.2.21). Accordingly, the restored image can be obtained as the equilibrium state of this network (Zhou et al., 1988).

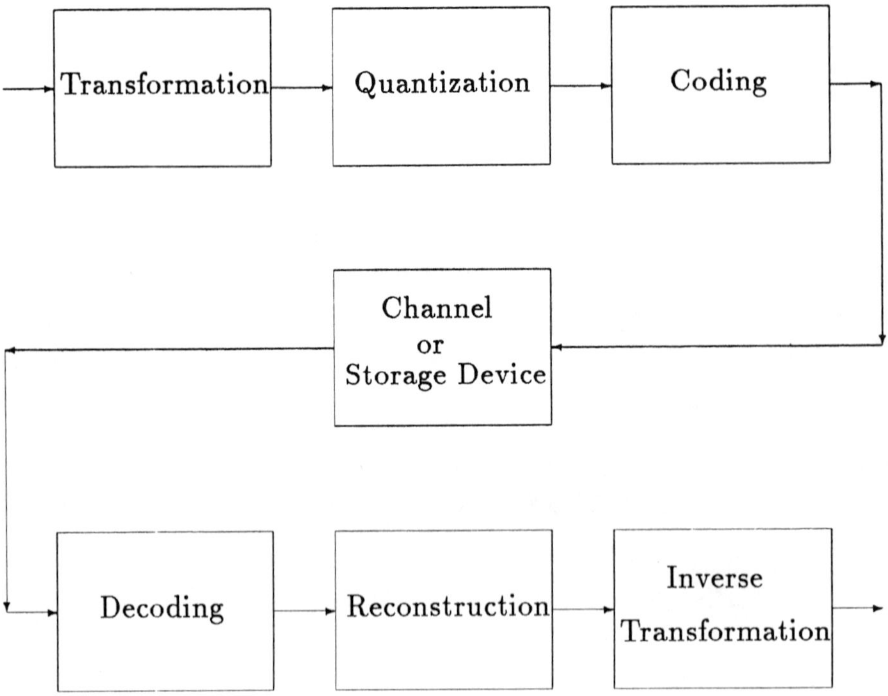

Figure 9.2: An image compression system based on transform coding.

9.3 Image Compression

Image compression allows the efficient transmission or storage of digital images with minimum distortion. A large portion of image compression systems employ transform coding techniques (Clarke, 1985). Figure 9.2 shows the general structure of a transform coding based image compression system. This system performs the coding of the input $\mathbf{x}$, in this case blocks of image pixels, into a specified number of bits. The main tradeoff in image compression is between the reduction of the redundancy existing in images and the quality of the reconstructed image. The quality of the reconstructed image can be measured by a quantitative distortion measure, such as the normalized mean-square error. However, the most reliable judge of the image quality is the human eye, which does not necessarily agree with mean-square error based quantitative measures.

The basic idea behind transform coding is that the image vectors $\mathbf{x}$ are substantially correlated. The objective of transform coding is the linear transformation of the input vectors $\mathbf{x}$ into less correlated, or ideally uncorrelated, vectors $\mathbf{y}$, which are called *transform coefficients*. Such an optimal transformation is provided by the *Karhunen-Loeve transform* or *Hotelling transform* (Clarke, 1985; Gersho and Gray, 1992). Assuming that $\mathbf{x}$ is zero-mean, $\mathbf{C} = \mathbf{C}_x = E\{\mathbf{x}\,\mathbf{x}^*\}$ is the covariance matrix of the input. Assume that $\lambda_1 \geq \lambda_2 \geq \ldots \geq \lambda_k \geq 0$ are the eigenvalues of $\mathbf{C}$ and let $\mathbf{u}_i$ be the eigenvector of $\mathbf{C}$, normalized to the unit norm, which corresponds to the eigenvalue λ_i. These eigenvectors are often referred to as the *principal components* of the input. The Karhunen-Loeve transform is defined as $\mathbf{T} = \mathbf{U}^*$, where $\mathbf{U}$ is the matrix whose columns are the eigenvectors $\mathbf{u}_i, i = 1, 2, \ldots, k$, that is, $\mathbf{U} = [\mathbf{u}_1\,\mathbf{u}_2\,\ldots\,\mathbf{u}_k]$. The transform coefficients can be obtained from the input vectors by $\mathbf{y} = \mathbf{T}\,\mathbf{x}$. Since $\mathbf{T}^{-1} = \mathbf{T}^*$, the input vectors $\mathbf{x}$ can be obtained from the transform coefficients by $\mathbf{x} = \mathbf{T}^*\,\mathbf{y}$. The covariance matrix of $\mathbf{y} = \mathbf{T}\,\mathbf{x} = \mathbf{U}^*\,\mathbf{x}$ is given by

$$\begin{aligned} \mathbf{C}_y &= E\{\mathbf{y}\,\mathbf{y}^*\} \\ &= E\{\mathbf{U}^*\,\mathbf{x}\,\mathbf{x}^*\,\mathbf{U}\} = \mathbf{U}^*\,\mathbf{C}_x\,\mathbf{U} = \mathrm{diag}(\lambda_1, \lambda_2, \ldots, \lambda_k) \end{aligned} \tag{9.3.1}$$

This latter equation indicates that the Karhunen-Loeve transform produces uncorrelated outputs $\mathbf{y}$. In addition, the variance of each transfom coefficient is the corresponding eigenvalue of $\mathbf{C}$. In image compression, the transform coefficients are evaluated on the basis of only the principal components of $\mathbf{C}$ which correspond to the larger eigenvalues and, therefore, carry most of the variance (Gersho and Gray, 1992). In practical applications, the Karhunen-Loeve transform is replaced by fixed orthogonal transforms. The discrete cosine transform is an orthogonal transform which is very widely used in applications (Clarke, 1985). Under certain assumptions regarding the statistics of the input, the discrete cosine transform coincides with the Karhunen-Loeve transform. The Hadamard transform is an example of a binary transform used in image compression (Gonzales and Wintz, 1977; Pratt, 1979).

Cottrell et al. suggested an alternative approach to image compression, which exploits the internal representations created by training a feed-forward neural network with analog output using Error Back Propagation (Cottrell et al., 1987). An image is typically represented by the density function $f(x, y)$, where x and y correspond to spatial coordinates within the image with intensity values between 0 and $2^K - 1$. In this approach, it was assumed that $K = 8$ and the intensity levels of the image were linearly transformed from 0-255 to 0-1. This image compression scheme employed a multilayered neural network with n_i inputs, $n_o = n_i$ analog output units and one layer of $n_h < n_i$ nonlinear hidden units. The nonlinearity associated with the hidden units in these experiments was the hyperbolic tangent, i.e., a monotone sigmoid function with a linear area around zero. This network was trained by the Error Back Propagation algorithm to map randomly selected portions of the image to the analog output of the network, i.e., to perform *auto-association.* This network is also called *self-supervised* because the output which corresponds to a certain input is not provided by an external teacher. Finally, this network is essentially an n to n encoder with analog inputs and outputs. After the training, the network generates a vector formed by the outputs of the hidden units for each portion of the image presented to it. In the context of transform coding, the outputs of the hidden units play the role of the transform coefficients while the matrix formed by the synaptic weights which connect the input and the hidden units plays the role of the transform matrix. In an image compression system, the outputs of the hidden units are quantized, coded and transmitted. The reconstruction of the image includes the decoding of the received binary signal, and the creation of a real-valued signal. The reconstructed image can simply be obtained from the resulting estimates of the outputs of the hidden units and the synaptic weights of the trained network which connect the hidden and the output units. In the context of transform coding, the matrix formed by the synaptic weights which connect the hidden and the output units plays the role of the inverse transform matrix.

The efficiency of the proposed image compression scheme was tested by several experiments (Cottrell, et al., 1987). In one of the experiments, the network was trained using blocks of the image containing 8×8 pixels, which produced input and output vectors of length 64. The network used in this experiment contained 16 hidden units. If b is the number of bits assigned to each output of the hidden units by the quantizer, the corresponding compression rate in this case is $16 \times b/64 = (b/4)\, bits/pixel$. This network required more than $100,000$ adaptation cycles to converge. The efficiency of this approach was also tested by modifying the size of the image blocks used in the training, the number of hidden units, and the quantization scheme used. Finally, the network trained using a certain image was tested on a different image with quite satisfactory results.

One of the most important questions raised by this image compression approach relates to how the network organizes the information presented to it by creating internal representations in the hidden layer. More specifically, experimental investigations were performed to determine if there exists a regularity in the outputs of the hidden units, thus indicating that the network performs principal component analysis of the input. However, the evaluation of the variance of each hidden unit indicated that the network does not perform exact principal component analysis (Cottrell et al., 1987).

The analysis and discussion which followed the publication of the image compression approach presented above is probably more important than the method itself. Motivated by this image compression approach, Boulard and Kamp studied a multi-layered neural network with analog output units which performs auto-association (Boulard and Kamp, 1988). If the network consists of linear hidden units, this study revealed the existence of an optimal solution which can be computed using singular value decomposition techniques and low-rank matrix approximation. If the hidden units are nonlinear and their nonlinearity can be approximated by a linear function for small values of the argument, the optimal parameters of the network can be found by simply extending the results obtained in the linear case (Boulard and Kamp, 1988). In the network used for image compression, the nonlinearity associated with the hidden units was

$\rho(x) = \tanh(x)$, which can be approximated for small values of x as $\tanh(x) \approx x$. In fact, it was experimentally verified that the hidden units of the network which performs auto-association behave as linear units during training (Cottrell, et al., 1987). The image compression approach also motivated the study of the landscape of the quadratic error function when it is used for training multi-layered neural networks with one layer of linear hidden units (Baldi, 1989; Baldi and Hornik, 1989). This study revealed that the quadratic error function has a unique local and global minimum corresponding to an orthogonal projection onto the subspace spanned by the first principal eigenvectors of the covariance matrix associated with the training patterns. All other critical point of this error function are saddle points (Baldi and Hornik, 1989). The existence of alternative more efficient techniques for training a network with analog output was the basis of a fair criticism of the Error Back Propagation algorithm. However, the generality of the singular value decomposition techniques is restricted by the assumption than the output of the network is analog, while the Error Back Propagation algorithm can train networks with nonlinear output units. The implication of this analysis goes beyond the criticism of the Error Back Propagation algorithm, which, despite all the surrounding myth, is just the result of the application of the first-order gradient descent method. An even more important outcome of the analysis described above is the relationship established between the internal representations created in multi-layered neural networks during training and principal components. Moreover, these studies substantiated earlier claims that the hidden layers behave like feature detectors of the input. In fact, the creation of feature detectors in multi-layered neural networks is not necessarily related to supervised learning. Linsker found the emergence of spatial opponents and orientation selective units in successive hidden layers of linear units trained without an external teacher using a modified Hebbian rule and random inputs (Linsker, 1986a; Linsker, 1986b; Linsker, 1988a). In a sense, such a system is an oversimplified version of the mammalian visual system, in which a spatially arranged array of retinal receptors becomes successively more and more transformed.

Some variations of the image compression approach proposed

by Cottrell et al. have been recently reported in the literature (Blanchet, 1990; Mougeot et al., 1990; Mougeot et al., 1991). Mougeot et al. argued that the convergence of the self-supervised training of multi-layered neural networks can be improved by adding certain constraints on the synaptic weights of the network. They also suggested that the quality of the compressed signal can be improved by minimizing some p-Holder norm, or L_p norm, instead of the quadratic error function. In fact, the quadratic error function corresponds to the L_2 norm. Mougeot et al. found that the use of the L_1 norm in the compression of a step-like one-dimensional signal provides an estimate of the compressed signal of a better quality than that obtained by minimizing the quadratic error function (Mougeot et al., 1990; Mougeot et al., 1991). However, the generality of this result must be supported by other experiments, since the L_1 norm performs better than the L_2 norm on step-like or spiky signals.

An alternative signal compression approach based on self-supervised training of a feed-forward neural network utilized preprocessing of the input data and a multi-layered neural network of an alternative architecture (Blanchet, 1990). The FFT of the speech signal used in this approach was initially evaluated and the resulting sequence was passed through a sixteen-channel filter bank, producing a sequence of 16-dimensional vectors. The elements of these 16-dimensional vectors were averaged and the result was considered to be the last coefficient of the 17-dimensional vectors which provided the inputs of the network. The network used by Cottrell et al. was modified by adding one layer of 20 nonlinear hidden units between the original hidden unit and the input and an identical layer between the original hidden layer and the output layer. Although the quality of the estimate provided by the compressed signal improved with the use of this network, there is a greater possibility that the network may be trapped in a local minimum during training (Blanchet, 1990).

The main criterion for the efficiency of these approaches is the relationship between the internal representations created by the self-supervised training of the multi-layered network used and the transform coefficients provided by the Karhunen-Loeve transform. The analysis of unsupervised training of feed-forward neural networks presented in section 2.2.7 indicates that image compression can be performed simply by single-layered neural networks trained without an external teacher, by either Sanger's or Oja's rule. Sanger presented an image compression experiment based on the unsupervised training of a single-layered neural network with $n_i = 64$ inputs and $n_o = 8$ linear output units according to his rule (Sanger, 1989a; Sanger, 1989b). The input vectors $\mathbf{x}$ were formed by 8×8 blocks of a 256×256 image with $2^8 = 256$ gray levels. After training the network, the weight vectors approach the directions of the n_o first eigenvectors of the covariance matrix of the inputs, ordered by decreasing eigenvalues. The output units provide the transform coefficients as

$$\hat{y}_i = \mathbf{w}_i^* \mathbf{x} \, \forall \, i = 1, 2, \ldots, n_o \tag{9.3.2}$$

The variance carried by each output unit is proportional to the corresponding eigenvalue. The difference in the amount of information carried by each output unit allows the assignment of a different number of bits to each output, depending on its importance. In Sanger's experiment, the outputs of the units were quantized by assigning five bits to the first two outputs, three bits to the third output and two bits to each of the other five outputs. Each block of $8 \times 8 = 64$ image pixels was represented by 23 bits, allowing the transmission of the image at the rate of $23/64 = 0.36 \, bits/pixel$. The image can be reconstructed from the outputs of the network by

$$\mathbf{x} = \sum_{i=1}^{n_o} \hat{y}_i \, \mathbf{w}_i \tag{9.3.3}$$

The same task could be performed by a single-layered neural network trained with no external teacher by Oja's rule. In this case, the weight vectors of the trained network span the same subspace spanned by the principal components but they are not directly related to them. As a result, the amount of information carried by the ordered outputs of the network is not necessarily decreasing. The use of Oja's rule could not allow the assignment of fewer bits to the less important outputs. Therefore, the use of Sanger's rule allows the transmission of the same amount of information at a lower compression rate.

Daugman utilized a 'neural' architecture to solve the minimization problem resulting when image compression is based on image representation using nonorthogonal transforms (Daugman, 1988a; Daugman, 1988b). A given image $X(i,j)$ may be represented by projecting it onto a selected set of vectors as

$$\hat{X}(i,j) = \sum_k a_k \, G_k(i,j) \tag{9.3.4}$$

The coefficients $\{a_k\}$ involved in this representation can be obtained by minimizing the error

$$\begin{aligned} E &= \frac{1}{2}\sum_i \sum_j [X(i,j) - \hat{X}(i,j)]^2 \\ &= \frac{1}{2}\sum_i \sum_j [X(i,j) - \sum_k a_k \, G_k(i,j)]^2 \end{aligned} \tag{9.3.5}$$

The minimization of (9.3.5) can be performed by solving the set of linear equations

$$\frac{\partial E}{\partial a_\ell} = -\sum_i \sum_j [X(i,j) - \sum_k a_k G_k(i,j)] G_\ell(i,j) = 0 \;\forall\, \ell \tag{9.3.6}$$

If the elementary functions $\{G_k(.)\}$ form a completely orthogonal set, the coefficients $\{a_\ell\}$ are given by a closed form expression as follows

$$a_\ell = \frac{\sum_i \sum_j X(i,j) G_\ell(i,j)}{\sum_i \sum_j G_\ell^2(i,j)} \tag{9.3.7}$$

Daugman used the set of nonorthogonal elementary functions provided by the two-dimensional Gabor transform (Daugman, 1988a; Daugman, 1988b). In this case, the coefficients $\{a_\ell\}$ can be obtained by solving the set of linear equations (9.3.6). However, such an approach is impractical, given the huge size of the matrix involved. Daugman suggested that the coefficients $\{a_\ell\}$ can be determined by employing an iterative method. More specifically, the minimization of E can be performed by updating the coefficients $\{a_\ell\}$ according to the rule $a_\ell \leftarrow a_\ell + \eta \Delta a_\ell$, where Δa_ℓ is obtained on the basis of the gradient descent method as

$$\Delta a_\ell = -\frac{\partial E}{\partial a_\ell} = \sum_i \sum_j G_\ell(i,j) X(i,j) - \sum_i \sum_j G_\ell(i,j) \sum_k a_k \, G_k(i,j) \tag{9.3.8}$$

Daugman suggested that the gradient descent method can be implemented by a neural network which consists of two layers of fixed synaptic weights and one layer containing adjustable synaptic weights (Daugman, 1988a; Daugman, 1988b). The synaptic weights of the first layer are the elementary functions which are multiplied with the image and form the term $\sum_i \sum_j G_\ell(i,j)X(i,j)$. The second layer consists of the coefficients $\{a_k\}$ which play the role of the adjustable synaptic weights of the network. The third layer is identical to the first layer and forms the terms $\sum_i \sum_j G_\ell(i,j) \sum_k a_k G_k(i,j)$. The change of the synaptic weights is evaluated by subtracting the feed-back signal, which is provided by the output of the third layer, from the feed-forward signal, which is provided by the output of the first layer. The adaptation of the synaptic weights is based on an unsupervised process. When the system reaches equilibrium, the resulting set of coefficients minimize the error function (9.3.5). The network proposed by Daugman is a special architecture, which was constructed for the implementation of the gradient descent method in this particular case (Daugman, 1988a; Daugman, 1988b). In addition to some analogies between the structure of this network and the human visual system pointed out by Daugman, an interesting feature of this approach is the use of an unsupervised scheme for image representation which eliminates its redundancy.

9.4 Recognition of Handwritten Signatures, Characters and Digits

Handwritten character recognition is among the most popular applications of neural networks. In fact, most of the early examples attempting to establish the usefulness of neural networks were based on character recognition. Over the years, interest switched from simple examples to the development of neural network systems of practical value. Such systems must be robust when subject to noise and other deformities of the input characters, so that they can be used in real-world applications. As a result, there is quite an extensive body of literature on this subject (Khotanzad and Lu, 1988;

Pawlicki, et al., 1988; Martin and Pittman, 1990; Skrzypek and Hoffman, 1991; Kim and Lee, 1991; Knerr et al., 1991; Shimada and Nishimura, 1991). The systems developed so far vary significantly in terms of their sophistication and efficiency. The basic problem in handwritten character recognition is the coding of the input information. Some approaches provide coded input information to simple network architectures. Other approaches use sophisticated network architectures that code the input information by creating implicit representations of the input. In these systems, certain parts of the network behave as feature detectors of the input. This section describes various neural network systems for handwritten character recognition in an attempt to illustrate the evolution of this research field.

9.4.1 Signature Recognition

Mighell et al. developed a neural network based system which could recognize true handwritten signatures from causal, i.e., non-professional, forgeries (Mighell et al., 1989). This is an important problem in many financial transactions. Although humans can easily recognize true signatures from causal forgeries, this could be very costly given the number of financial transactions occurring every day. The experiments used 80 true signatures of an individual and 66 forgeries. The forgeries were either computer generated or were signatures provided by other individuals. The training set consisted of 10 true signatures and 10 forgeries, while 70 true signatures and 56 forgeries were used to test the performance of the trained network. The signatures were scanned and thresholded to produce 128×64 binary images. In the first experiment, the 128×64 images were used as inputs to the networks. One of the networks used had 128×64 inputs plus one bias and one output unit. The network, trained by a Delta Rule algorithm, rejected 1 % of true signatures and accepted 4 % of the forgeries as true signatures. The addition of a hidden layer did not improve the overall performance of the system, indicating that this training set is linearly separable. This system is specialized to check the signature of only one person. A useful generalization would be a network trained to check the signa-

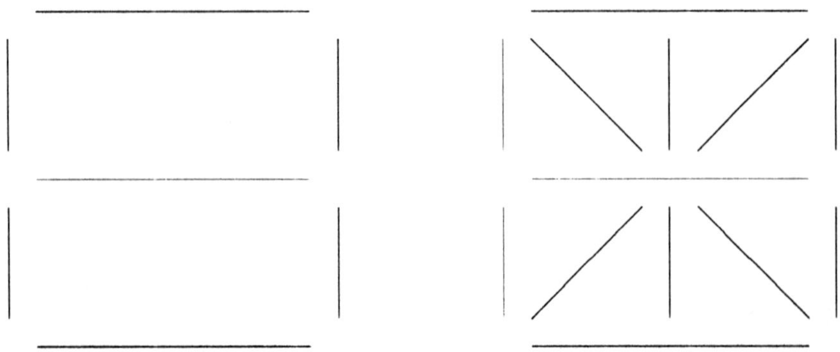

Figure 9.3: Horizontal, vertical and diagonal bars used for the representation of digits from 0 to 9 in two dimensions.

tures of several people, for example the customers of a bank. Such a task could probably require some kind of preprocessing of the input data.

9.4.2 Character Recognition

Burr developed a neural network based system for character and digit recognition using the *shadow coding* scheme to form the inputs of the network from the original data (Burr, 1988). Shadow coding is based on the projection of any character or digit onto a set of horizontal, vertical, and diagonal bars. Figure 9.3 shows a set of seven horizontal and vertical bars which can represent any digit from 0 to 9 in two dimensions. Since the system is developed for alphabetical characters as well as digits, some additional bars are needed. The representation of the characters "I" and "T" requires the two additional bars shown in Figure 9.3, and the representation of the characters "X" and "N" requires four diagonal bars also shown. The representation of each character or digit by the shadow code is achieved by projecting each point of the character/digit into the three closest horizontal, vertical, and diagonal bars. The projected shadow turns on a set of bits which are uniformly distributed along each bar. The "on" bits are counted, providing 13 numbers which are subsequently normalized between 0 and 1. Thus, each handwritten character or digit can be represented by a 13-dimensional vector which serves as the input of the network. A feed-forward neural network with one layer of hidden units was trained separately to recognize characters using the Error Back Propagation algorithm. The training set consisted of 104 character samples, with another set of 104 character samples used to test the performance of the network. The network consisting of 20 hidden units achieved the maximum success rate of 94 %. An increase in the number of hidden units resulted in a slight decrease of the success rate. The same network architecture was also trained using Error Back Propagation to recognize handwritten digits. In this case, the success reate achieved by the network with 5 hidden units was 95 %. The network with six hidden units achieved success rate of 97.5 %. Once again, a further increase in the number of hidden units had a negative effect on the performance of the network.

9.4.3 Zip Code Digit Recognition

One of the first commercial neural network applications was a system for zip code digit recognition developed for the U.S. Post Office (le Cun, Boser, et al., 1989; le Cun, Gardner, et al., 1989; le Cun, Jackel, et al., 1989; le Cun, Boser, et al., 1990; le Cun, Jackel, et al., 1990). le Cun et al. assumed that the problem of automatically separating the zip code digits had been solved and focused on automatic recognition of single digits. They used a database consisting of 9298 segmented numerals digitized from handwritten zip codes which appeared on actual U.S. mail. Another set of 3349 printed digits from 35 different fonts were also added to the handwritten digits. The training set consisted of 7291 handwritten digits and 2549 printed digits. The testing set was formed from the remaining 2007 handwritten and 700 printed digits. An aspect-ratio preserving linear transformation of each character provided characters fitting in a 16×16 pixel image. This transformation results in images with multiple gray levels, which are normalized to fall within the range -1 and $+1$.

An early version of this character recognition system contained constant synaptic weights in the input layers (Denker et al., 1989). In this early system, a convolutional method was used for locating features such as lines and edges, which are of particular importance in the character recognition process. The system developed by le Cun et al. is based on a multi-layered neural network with adaptive synaptic weights, which is trained using a variation of the Error Back Propagation algorithm (le Cun, Boser, et al., 1990). The input of the network used in the most recent system is a 16×16 image and the output consists of 10 units, each corresponding to a digit from 0 to 9. The network is trained in such a way that if the input of the network is the digit i, the ith unit of the output is "on", represented by $+1$, while the other output units are "off", represented by -1.

Figure 9.4 shows the general structure of the multi-layered neural network used in this application. Each character is presented to the network as a 16×16 image. In practice, the actual input was provided to the network as a 28×28 plane, in order to avoid problems when a kernel overlaps a boundary. The first hidden layer consists of 4 groups of 576 units organized in 24×24 feature maps in two dimensions. The input of each unit in a feature map is a 5×5 neighborhood on the image plane. All the units belonging to one of the four groups use the same set of 26 units, i.e., the 5×5 neighborhood plus a bias term. The second layer is an averaging/subsampling layer, consisting of 4 planes of size 12×12. Each unit of these 4 planes is fed by the outputs of a 2×2 unit neighborhood in the corresponding plane in the first layer of hidden units. Since all the synaptic weights are constrained to be equal, the second layer performs a local averaging and a 2 to 1 subsampling of the first layer in each direction. The third hidden layer consists of 12 groups of 64 units forming 8×8 feature maps in two dimensions. The function performed by the third hidden layer on the output of the second hidden layer is roughly similar to that performed by the first hidden layer on the input of the network. Finally, the fourth hidden layer is composed of 12 groups of 16 units forming 4×4 planes. This is once again an averaging/subsampling layer whose function is very similar to that of the second layer. The output

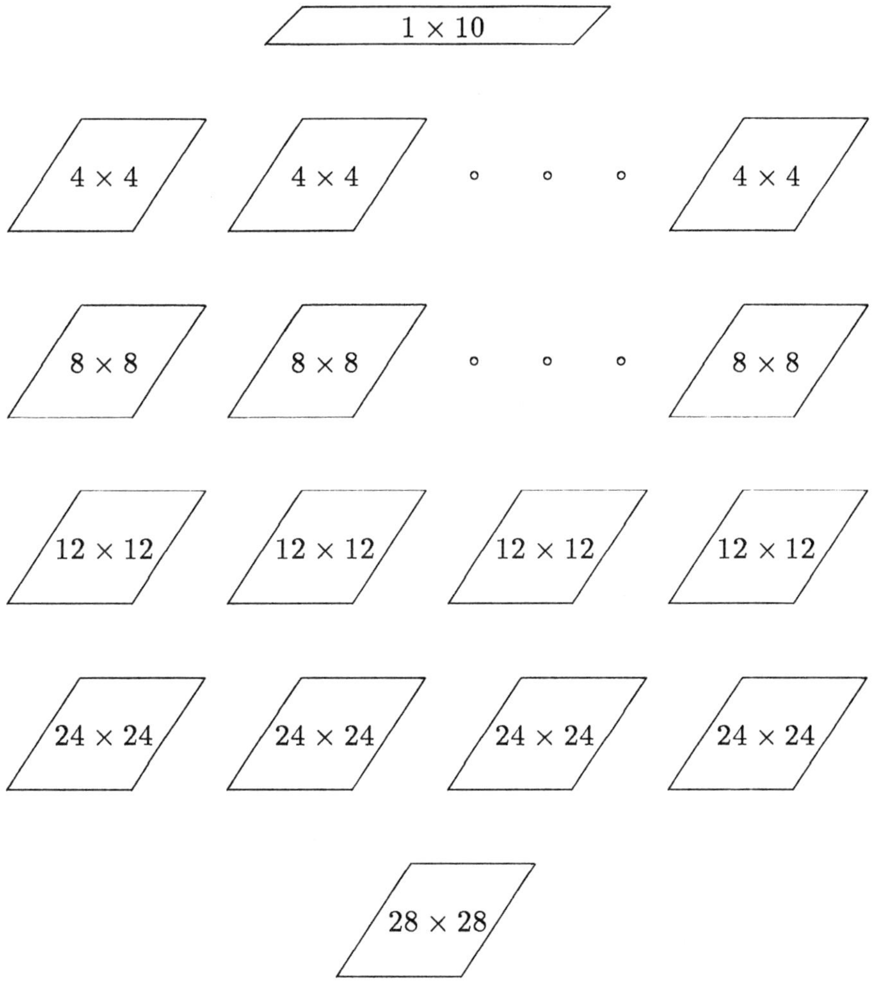

Figure 9.4: Multi-layered neural network architecture used in zip code digit recognition.

layer is fully connected to the fourth layer of hidden units. The final network consists of 4635 units, 98442 connections, and 2578 independent parameters, i.e., 4 times less parameters than the system reported earlier by the same team (le Cun, Boser, et al., 1989; le Cun, Jackel, et al., 1990). The reduction of the network parameters was performed by a technique proposed for the elimination of connections in multi-layered neural networks, known as optimal brain damage (Le Cun, Denker, et al., 1990).

The network was trained by a second-order variation of the Error Back Propagation algorithm. After 30 adaptation cycles the error rate of the network tested on the training set was 1.1 % by the error rate increased to 3.4 % when the network was tested on the testing set. The system was also evaluated by measuring the percentage of character rejections that are necessary in order to achieve an one percent error rate. It was found that the system tested on the entire set must reject 5.7 % of the characters in order to achieve an one percent error rate. The percentage of rejection increased to 9 % when the system was tested only on the handwritten characters included in the testing set. The performance of this system may be compared with that of an earlier system developed without elimination of connections. The error rate of this latter system tested on the training and testing sets was 0.14 % and 5 %, respectively. On the other hand, the system developed without elimination of connections required 12 % rejection in order to achieve 1 % error rate (le Cun, Boser, et al., 1989; le Cun, Jackel, et al., 1990).

9.4.4 The Cognitron, the Neocognitron and the Selective Attention Model

The *cognitron* is a self-organizing multi-layered neural network developed by Fukushima almost twenty years ago (Fukushima, 1975). This network learns to recognize patterns without a teacher. The self-organization of the cognitron is based on the following principle: Among the cells belonging to a certain area called "competition" area, only the one responding most strongly has its input connections reinforced. The reinforcement of each input connection to its maximum output cell is proportional to the intensity of the response of the cell from which the relevant connection leads. This principle is applied to both excitatory and inhibitory connections. Because of this self-organization principle, each cell extracts a different feature. Fukushima argued that the network trained on the basis of this principle is capable of exhibiting self-repair. The cognitron is formed by a number of similar layers connected in cascade. Each layer has the same number of excitatory and inhibitory cells. After the training of the cognitron, each excitatory cell extracts a feature of the stimulus presented to it during the training. The cell receives signals indicating the existence of the feature to be extracted through the excitatory connections. If an irrelevant feature is presented to the system, the inhibitory signal becomes stronger than the excitatory signal.

The disadvantage of the cognitron is its inability to recognize shifted or distorted patterns correctly (Fukushima, 1990). In order to overcome these problems, Fukushima developed the neocognitron in the early eighties (Fukushima, 1980; Fukushima et al., 1983; Fukushima, 1988). The neocognitron is trained by a teacher to recognize shifted or distorted patterns. In other words, the neocognitron has the ability to generalize, which is a highly desirable property in neural systems. The neocognitron is a hierarchical multilayered network consisting of many layers of cells. The input layer of the network consists of a two-dimensional array of receptive fields. Each of the succeeding stages has a layer of *S-cells*, followed by a layer of *C-cells*. The S-cells are feature extraction cells, which resemble, to some extent, the cells of the visual cortex. S-cells have the ability to extract features during training in the same way as the excitatory cells of the cognitron. After the learning phase, an S-cell is activated only if a particular feature is presented in a certain position of the input layer. The lower stages of the network extract local features, such as lines. The higher stages extract more global features, such as parts of the training pattern. C-cells resemble complex cells and allow positional errors in the features of the stimulus. The connections from S-cells to C-cells are fixed and invariable. Each C-cell receives signals from a group of S-cells which extract the same feature, but from slightly different positions. A C-cell is activated if at least one of these S-cells is active. This property guarantees that if the stimulus feature is shifted and another S-cell is activated, the same C-cell keeps responding. Since the changes in relative position of local features are tolerated by the C-cells, the S-cells in the succeeding stage can extract deformed global features. The layer of C-cells at the highest stage works as the recognition level. Each cell at this layer integrates the entire input pattern, while only one cell is active. This cell corresponds to the category of the input patterns. Some theoretical and numerical developments in the understanding of the feature extraction in the neocognitron were reported recently (Johnson et al., 1988). This study indicated that the feature extraction process is equivalent to a generalized nonlinear discriminant. The operation of the feature extraction process can be linked to the eigenvalues and eigenvectors of a matrix comprised of the excitatory and inhibitory convolution masks. Kim and Lee used a modified version of the neocognitron in handwritten Hangul recognition (Kim and Lee, 1991).

Fukushima recently developed the *selective attention model*, which is an improved version of the neocognitron (Fukushima, 1986; Fukushima, 1990; Fukushima et al., 1991). This system has the ability to recognize shifted and distorted patterns, even in the case where two or more patterns are presented to it simultaneously. The development of the selective attention model was achieved by adding top-down or backward connections to the bottom-up or forward connections that form the neocognitron. The forward signals in this system perform the function of pattern recognition. The forward path has almost the same structure and function as the neocognitron. The input patterns are presented to the lowest stage. The cells and connections in the backward path form the mirror image of the cell and connections in the forward path. The output of the recognition layer is returned to the lowest stage through the backward path. The function of the overall system is determined through the interaction of the forward and backward signals. The forward signals gate backward signal flow, while the backward signals facilitate the flow of forward signals. The backward signals perform the function of selective attention, pattern segmentation and associative recall.

9.5 Text to Speech Conversion

English text can be converted to speech through the use of a lookup table. Sejnowski and Rosenberg developed NETtalk, which is a neural network based system trained to convert English text to speech (Sejnowski and Rosenberg, 1987). This application demonstrated how a feed-forward neural network can be trained to transform *highest sensory representations* of letters into the *earliest articulatory representations* of the phonemes. The network used had seven groups of input units, each accepting a letter. The network was trained by sliding strings of seven letters. During training, the output of the network was the correct phoneme associated with the center letter of the seven letter window. The six letters adjacent to the center letter provide a partial context. Each letter is represented by a group of 29 dedicated units. The first 26 units correspond to the letters of the alphabet, while the other three correspond to punctuation and word boundaries. Only one of the 29 units is active for a given letter. The output of the network consists of 26 output units. The first 21 units represent 21 different articulatory features, while the other five represent stresses and syllable boundaries. Figure 9.5 shows the multi-layered neural network architecture used in the NETtalk.

A network consisting of 80 hidden units was trained using 1024 words pronounced by a child in first grade. The network learned 95 % of the correct phonemes after 50 adaptation cycles, and almost all secondary stresses and syllable boundaries after just 5 adaptation cycles. The performance of the trained network was tested on a set of 439 words from the same speaker with a success rate of 78 %. An important result was the evaluation of the network performance when the network is partly damaged. This was achieved by randomly changing the synaptic weights of the trained network. The performance of the network was only slightly affected by small changes of the synaptic weights. Larger changes of the synaptic weights resulted in a gradual degradation of the network performance. An even more important result is that the network was able to learn much faster after a moderate damage of the synaptic weights. In fact, this observation demonstrates the importance of

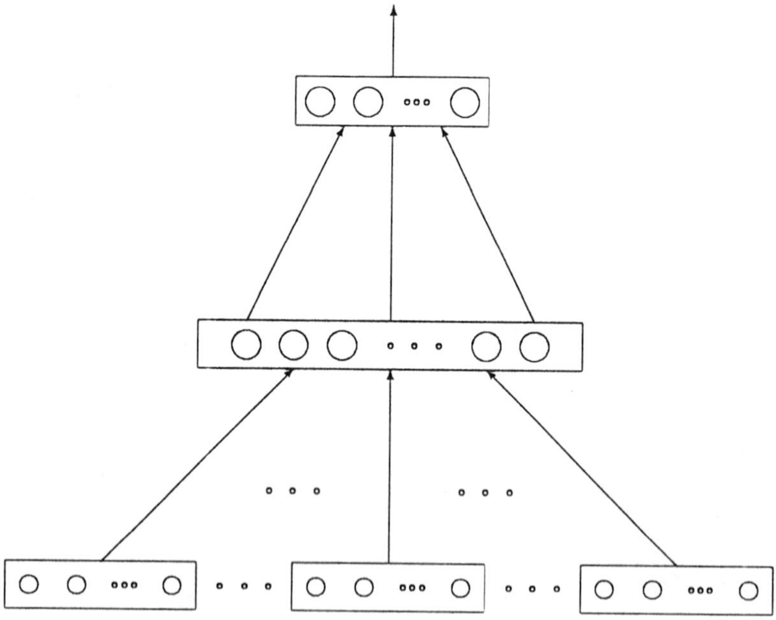

Figure 9.5: Multi-layered neural network architecture used in the NETtalk.

the algorithms presented in chapter 5, which determine the architecture of the network during training by eliminating hidden units and synaptic weights.

In another experiment, several networks with different numbers of hidden units were trained using the 1000 most commonly used words from the *Miriam Webster's Pocket Dictionary*. It was found that the learning rate increased with the number of hidden units. The generalization ability of the network with 120 hidden units was tested using 20,012 words contained in the dictionary, exhibiting an average success rate of 77 %. The same test was repeated but the synaptic weights of the network were continually updated as new words were presented to the network. The success rate reached 85 % after one pass through the dictionary and increased to 90 % after five passes. Clearly, the exposure of the network to more words increases its ability to correctly pronounce new words. Increasing the number of letters forming the input of the network from 7 to 11 slightly improved its performance.

Tenorio et al. reproduced the network structure used in NETtalk and carried out experiments to determine which characteristics of the network were responsible for which learning behavior and how closely it maps human speech development (Tenorio et al., 1988). These experiments indicate a strong dependence of the network mapping on the statistical characteristics of the input data. Tenorio et al. concluded that neural networks are satisfactory, provided that the application can be modeled by mapping and generalization of input characteristics (Tenorio, et al., 1988).

9.6 Classification Applications

The application of neural networks to classification problems is conceptually the most consistent with their structure and function. Considering a finite set of states or classes, the objective in classification applications is the assignment of a random sample to one of these states with minimum probability of error. Each sample is described by a set of parameters which form a vector, usually referred to as the feature vector. The development of such a classification system can be achieved by appropriately training a neural network in such a way that it provides an output corresponding to one of the classes, provided that the training sample used in forming its inputs belongs to this class. The ability of the neural network to correctly classify a test sample that is close in some sense to one of the training samples relates directly to its generalization ability. However, classical decision theory provides strong competition to neural networks as far as classification applications are concerned (Duda and Hart, 1973). The justification of the use of neural networks in classification applications depends on the existence of evidence that neural network classifiers are more efficient than alternative tools provided by decision theory. Another related issue is the variety of classification problems that can be treated using neural networks. Given the particular form of the inputs and outputs of neural networks, the application of neural networks to a broad class of classification problems strongly depends on the availability of a scheme transforming the data into admissible inputs of the neural network.

9.6.1 Classification of Sonar Signals

Gorman and Sejnowski developed a neural network based system for the classification of sonar targets (Gorman and Sejnowski, 1988a; Gorman and Sejnowski, 1988b). The task of the system was the classification of sonar return signals collected from a metal cylinder and a cylindrically shaped rock positioned lengthwise on a sandy ocean floor. The transmitted pulse was a wide-band linear FM chirp. The return signals were obtained from each target at various aspect angles. The preprocessing of the row data involved the evaluation of the spectral envelope of the sonar return signal. The input of the network consisted of 60 samples of the spectral envelope normalized between 0 and 1.

Gorman and Sejnowski presented the results of two experiments based on a set of 204 returns. In the first experiment, 16 of the returns were selected to form the testing set, while the other 192 returns formed the training set. This training set was used to train a network with 0, 2, 3, 6, 12, and 24 hidden units. The success rate achieved by the single-layered neutral network tested on the training set was 89.4 %. The test of the multi-layered neural networks used in this experiment indicated that the success rate increased monotonically with the number of hidden units. The highest success rate of 99.8 % was achieved by the networks consisting of 12 and 24 hidden units. The generalization ability of the networks used in this experiment was tested by feeding them processed return data not included in the training set. The test of the trained networks based on the testing set provided qualitatively similar but not identical results. The success rate increased monotonically from 77.1 % to 84.7 % as the number of hidden units increased from 0 to 12. However, the increase in the number of hidden units from 12 to 24 resulted in a slight decrease in the success rate. This experiment verifies that multi-layered neural networks perform better than single-layered neural networks. In addition, this experiment reveals that the increase in the number of hidden units does not necessarily improve the generalization ability of the network.

In an aspect-angle dependent experiment, the training and testing sets were selected in such a way that returns from each aspect angle were equivalently represented in the training and the testing sets (Gorman and Sejnowski, 1988a; Gorman and Sejnowski, 1988b). Both the training and testing sets used in this experiment contained 104 returns. Gorman and Sejnowski tested the performance of networks consisting of 0, 2, 3, 6, 12, and 24 hidden units. Once again, the success rate of the networks tested on the training set increased monotonically with the number of hidden units. The network with 24 hidden units achieved perfect classification. However, when the trained network was tested on the testing set the peak success rate of 84.7 % was achieved by the network with 12 hidden units. Once again, an increase in the number of hidden units from 12 to 24 resulted in a slight decrease in the success rate. The consistent difference in the performance of the trained networks tested on the training and testing data sets is an indication that when the number of hidden units increases above a certain threshold the trained networks tends to "memorize" the training set. This argument is consistent with the degradation of the generalization ability of the network observed in these experiments.

9.7 Medical Diagnosis

The development of automated systems for medical diagnosis is a significant challenge faced by physicians, engineers and computer scientists. Bayes theory provided the basis for the development of early systems for medical diagnosis based on symptoms and other individual items of patient information. Such systems require a data set sufficiently large in order to be considered a reliable statistical sample. Furthermore, these statistical approaches are based on the assumption that the symptoms are statistically independent. The validity of this assumption is strongly opposed by the medical community, on the basis of the argument that several symptoms often arise from the same organic cause (Bounds et al., 1988). An alternative approach for the development of automated medical diagnosis systems was the design of rule-based expert systems mimicking the interpretation of symptoms performed by human experts. Since the ability of human experts to perform medical diagnosis is a skill based on knowledge and enhanced by experience, the basic problem with the development of rule based expert systems is the difficulty of extracting explicit rules for medical diagnosis. The recent popularity of neural networks motivated the development of neural network based systems for medical diagnosis (Saito and Nakano, 1988; Bounds et al., 1988; Bounds et al., 1990; Harrison et al., 1991). This section outlines two neural network based systems developed for early diagnosis of heart attack, and the diagnosis of low back pain.

9.7.1 Early Diagnosis of Heart Attack

The diagnosis of heart attack, or acute myocardial infarction, is frequently performed by clinicians when patients suffering from chest pain are admitted to a hospital. Approximately 20-30 % of emergency hospital admissions are patients with possible heart attack, while almost half of the patients admitted to a coronary care unit suffer a heart attack. The patient's Electrocardiogram (ECG), which is readily available to the clinician, is not reliable because it often fails to show diagnostic changes. A standard test which gives a reliable diagnosis involves the measurement of enzyme and ECG changes over a period of 24 to 48 hours. The need for a system for reliable diagnosis shortly after the admission of the patient motivated the use of statistical techniques and expert systems for early heart attack diagnosis. However, none of these techniques or systems has gained widespread acceptance among the medical community.

Harrison et al. recently presented a system for early heart attack diagnosis based on a multi-layered neural network (Harrison et al., 1991). The network was trained on the basis of data collected from 300 consecutive emergency referrals with a complaint of chest pain. In order to train the network by a reliable teacher, heart attacks were diagnosed by the standard diagnosis method mentioned above. Half of the 300 cases formed the training set, while the remaining 150 cases formed the testing set. The data were used for training and testing a multi-layered neural network with 53 inputs, one binary output unit and one layer of 18 hidden units. The input of the network was formed by coding 38 different symptoms into 53-dimensional vectors. The output of the network was assigned to 1 if the patient was suffering from a heart attack and 0 otherwise. This network was trained by minimizing the quadratic error criterion and also by maximizing the relative entropy criterion presented in chapter 2. The algorithm based on the maximization of the relative entropy criterion converged faster than the conventional Error Back Propagation algorithm based on the minimization of the quadratic error criterion. The efficiency of this approach was evaluated by comparing the diagnosis provided by two trained networks and the diagnosis provided by a clinician. The criterion used in this evaluation comprised the *diagnostic accuracy*, i.e., the ratio of the number of correct diagnoses to the total number of cases, the *sensitivity*, i.e., the ratio of the correct positive diagnoses to the total number of patients with the disease, and the *specificity*, i.e., the ratio of the number of correct negative diagnoses to the total number of patients without the disease. The test of the trained networks indicated that the network trained by minimizing the quadratic error function achieved a slightly higher performance than that trained by maximizing the relative entropy criterion, while both networks achieved better overall performance than the clinician. The accuracy, sensitivity, and specificity achieved by the most efficient network was 0.88. Harrison et al. suggested that the decision regarding the patient's treatment could be made by a combination of a clinician's diagnosis and that provided by a neural network. Using such a mixed criterion, the sensitivity increased from 0.88 to 0.96.

9.7.2 Diagnosis of Low Back Pain

The diagnosis of the cause of back pain is a difficult task, since many symptoms are common in people having a serious spinal problem and people with less serious problems. Back pain is classified by physicians into Simple Low Back Pain (SLBP), Root Pain (ROOTP), Spinal Pathology (SPATH), and Abnormal Illness Behavior (AIB). Bounds et al. trained a neural network using data collected from 200 patients with low back pain (Bounds et al., 1988; Bounds et al., 1990). Half of these cases formed the training set while the remaining 100 formed the testing set. The reliability of the training and the testing sets was guaranteed by following these patients over a long period of time in order to ensure that they were representatives of one of the four classes. The inputs of the network were 50-dimensional vectors formed by coding the presence of various symptoms in the 200 patients. An input element was 1 if the patient had the corresponding symptom and 0 otherwise. If neither was true, the input was 0.5. Some analog inputs, valued from 0 to 1, were also used in cases where the binary coding was not possible. Bounds et al. tested a network with four output units, each corresponding to one of the classes, and a network with two output units whose outputs $(0,0)$, $(0,1)$, $(1,0)$, and $(1,1)$ were assigned to SLBP, AIB, ROOTP, and SPATH, respectively. The number of hidden units varied from 0, which corresponds to a single-layered neural network, to 50. The experiments indicated that the network with two output units achieves a higher success rate than the one with four output units. The performance of the networks tested improved slightly as the number of hidden units was increased from 0 to 30. However, a further increase of the number of hidden units degraded the efficiency of the network. The success rate of various networks was compared with that of three different sets of doctors and a fuzzy logic system developed for the same task. The neural networks with two output units achieved a higher success rate than all the sets of doctors. On the average, the network with two output units and one layer of 30 hidden units achieved an equivalent performance with the fuzzy logic system.

9.8 Prediction of Secondary Structures of Proteins

A basic problem in biochemistry is the determination of the structures and properties of proteins. Among other functions, proteins receive and send messages, protect the organism, and serve as enzymes in biochemical reactions. Each protein is formed from a primary sequence of amino-acids. The prediction of the three-dimensional conformation, or *tertiary structure,* of proteins is a particularly difficult problem. Biochemists study the more tractable problem of predicting the *secondary structure* of proteins, which is an arrangement of substructures known as α-helix, β-sheet, and coil. The characterization of a structure as coil implies that it is neither α-helix nor β-sheet. Any information regarding the secondary structure of a protein can be useful in determining its structural properties. The prediction of the secondary structure of a protein can be facilitated by the determination of a *homologous* protein of known structure. Statistical approaches and knowledge-based methods use available information from a database of known structures to predict the secondary structure of proteins for which no homologous structures are available. A summary of the results provided by such methods was reported recently (Mejia and Fogelman-Soulie, 1991). These methods achieved a success rate between 50 % and 60 %. The success rate is measured in this case by

$$Q = \frac{N_\alpha + N_\beta + N_{coil}}{N} \times 100 \qquad (9.8.1)$$

where N is the total number of residuals found, while N_α, N_β, and N_{coil} denote the number of residuals of type α-helix, β-sheet, and coil, respectively.

Qian and Sejnowski reported the first attempt to predict the secondary structure of proteins using multi-layered, feed-forward neural networks (Qian and Sejnowski, 1988a). The proteins were interpreted as words formed by combinations of an "alphabet" of 20 amino-acids. As a result, the network used in this approach was very similar with that used by Sejnowski and Rosenberg in the development of NETtalk. In this application the amino-acids play the role of letters while the secondary structure plays the role of phonemes. The input of the network is formed by sliding certain proteins in front of a window which reads 13 amino-acids. The input of the network is organized into 13 groups of 21 units. Each group of units receives a vector of 21 elements, only one of which is "on". The first 20 elements of this vector correspond to the amino-acids contained in the "alphabet" while the last represents space between sliding proteins. This kind of input formation is called *local coding.* The network is required to predict the substructure of the central amino-acid. The network consists of 3 output units, each corresponding to a substructure. The network also contains a hidden layer of nonlinear units. Other formations of the input and output are also possible. Such alternative formations could exploit some special biochemical properties of certain amino-acids (Qian and Sejnowski, 1988a).

A network consisting of 40 hidden units was trained by the Error Back Propagation algorithm using a set of real proteins. The peak performance of this network was 62.7 %. In another experiment, a network with 17 groups of 21 input units, 40 hidden units and 3 output units was initially trained using a set of real proteins. The network was subsequently pruned by arbitrarily setting to zero all the synaptic weights which were small in magnitude. It was found that 60 % of the synaptic weights were eliminated, and the performance of the network improved slightly. The role of the number of hidden units was revealed by the performance of a trained network with various numbers of hidden units. It was found that the success rate achieved by the trained network was not substantially affected by the number of hidden units. The peak success rate was achieved by the network with 40 hidden units. Nevertheless, the peak success rate was slightly higher than that achieved by a single-layered neural network. The role of the length of the input window was revealed by the performance of neural networks with input windows of length varying from 1 to 21. The peak performance was achieved by the network with a window of length 13. Qian and Sejnowski also attempted to improve the performance of the system by modifying the network architecture and the input representations (Qian and Sejnowski, 1988a).

At about the same time, Holley and Karplus reported a similar approach for the prediction of the secondary structure of proteins (Holley and Karplus, 1989). A feed-forward neural network with one layer of hidden units was trained using a string of amino-acids contained in certain proteins, while the input units were grouped into 17 groups of 21 units. The output of the network predicted the secondary structure of the amino-acid which was located at the center of the window. The network used in this approach had two output units taking the values 0 and 1. Among the four possible representations created by the output units, $(1,0)$, $(0,1)$, and $(0,0)$ were assigned to α-helix, β-sheet, and coil, respectively. Most of the results reported by Holley and Karplus were obtained by a network with one layer of hidden units. The training and testing sets used in these experiments were formed from a set of 64 proteins. The training set was formed by 48 proteins which contain 26 % α-helix, 20 % β-sheet, and 54 % coil. The remaining 14 proteins formed the testing set. The training and testing sets used in these experiments contained approximately the same percentage of secondary structures. The network with two hidden units was trained by the Error Back Propagation algorithm. When the trained network was tested on the training set, its success rate was 68.5 %. The success rate of the same network tested on the testing set decreased to 63.2 %. Holley and Karplus indicated that, with very few exceptions, this network provided better results than other statistical and knowledge-based methods existing in the literature. The highest success rate was achieved in these experiments by the network with an input window of 17 amino-acids. This contradicts the findings of Qian and Sejnowski, but the inconsistency of these results may be attributed to the different output representations used in the two approaches. The success rate of networks with various numbers of hidden units tested on the training set increased with the number of hidden units. When the same networks were tested on the testing set, the highest success rate was achieved by the network consisting of two hidden units. This is an indication that the increase of the number of hidden units facilitates the training of the network but degrades its generalization ability.

Mejia and Fogelman-Soulie recently reported their attempt to solve the same prediction problem using basically the same formulation and network architecture (Mejia and Fogelman-Soulie, 1991). The input of the network was provided in this approach by a window of 17 amino-acids. This choice was based on previous findings (Holley and Karplus, 1989) as well as biochemical evidence that this window size would cover most of the important residues. The network used consisted of 3 output units, and the number of units in the hidden layer was varied from 0 to 20. It was experimentally found that 2 to 3 hidden units were sufficient for this task. The success rate was slightly improved by a more structured initialization of the synaptic weights and the use of an asymmetric input window. If an asymmetric window was used, the output of the network did not predict the secondary structure of the amino-acid appearing in the center, if any, of the window. The success rate, after these modifications, was still lower than that reported elsewhere (Qian and Sejnowski, 1988a; Holley and Karplus, 1989). Mejia and Fogelman-Soulie attributed the inferiority of their results to the fact that Holley and Karplus included proteins in their test set which have homologies in the training set. On the other hand, they argued that Qian and Sejnowski achieved a higher success rate because of the size of the database that they used. Finally, they attempted to improve the network performance by pruning a network consisting of 20 hidden units after training it. The resulting improvement of the network performance again demonstrates the importance of learning algorithms which eliminate unnecessary hidden units and synaptic weights during training.

9.9 Weather Forecasting

A classical application of neural networks to weather forecasting was presented almost thirty years ago by Widrow and his team (Widrow and Smith, 1964). They used the Adaline to predict the occurrence of rainfall on the following day on the basis of fluctuations in the barometric pressure in the two preceding days. The percentage of successful predictions provided by the Adaline was comparable to that achieved by the official weather prediction agency, which employed a large set of parameters for weather forecasting.

9.9.1 Solar Flare Forecasting

Bradshaw et al. developed a neural network based system for solar flare forecasting (Bradshaw et al., 1989). This system was compared with a rule-based expert system which was developed for the same task and performed as well as a skilled human expert. A multi-layered neural network with one layer of hidden units was trained and tested using two sets of data of about 500 input/output pairs, with solar data as input and flare occurrence as output. The network was trained by the Error Back Propagation algorithm. The performance of the trained network was at least as good as the rule-based expert system. Bradshaw et al. emphasized that the development of the expert system required 700 rules and a man-year of work, while the neural network based system was developed in less than a week. Another important difference is that the expert system required about 5 minutes to process a single prediction while the network responded in only a few milliseconds.

9.9.2 Lighting Strike Prediction

Frankel et al. developed a neural network based system to predict lighting strikes at the Kennedy Space Center (Frankel et al., 1991). The early prediction of lighting strikes could prevent launches that might result in heavy financial losses or the even more devastating loss of human lives. Their approach was inspired by the argument made by Lapedes and Farber that neural networks may be used for predicting the future behavior of systems whose behavior appears to be "chaotic". A feed-forward neural network with one layer of hidden units was trained to predict the occurrence of lighting strikes on the basis of meteorological data including the wind, the electric field, and the wind divergence. The training and testing sets were formed by selecting data over a period of two weeks, which included days with and without thunderstorm activity. The data used for prediction were processed and combined, providing the 106 inputs of the network used in this approach. The output layer of the network consisted of 64 units, which correspond to 16 lighting strike areas each with 4 time intervals for prediction. The smallest network that provided satisfactory prediction performance consisted of only 6 hidden units. The neural network system was compared with the best prediction available. This comparison was based on a criterion comprising the *probability of detection*, the *false alarm rate*, and the *critical success index*. The network trained using only wind data achieved better performance than the alternative prediction system. The additional meteorological data used in training the network improved its performance. Frankel et al. suggested that the performance of the system could be further improved by training the network using other combinations of meteorological data.

9.10 Financial Predictions

A large portion of scientific research is devoted to the development of systems for prediction, such as predictions related to the availability of and/or demand for natural resources, weather forecasting, and financial predictions. Neural networks are suitable candidates for the development of systems predicting such events or developments, due to their nonlinear structure and generalization ability. The application of neural networks to prediction applications requires the determination of the parameters of the system under consideration, which are most likely affecting the events or developments of interest. Provided that such a set of parameters has been chosen, the neural network is trained using the history of the system under consideration. After its training, the neural network must be able to use the most recent parameters in order to predict future events or developments. The use of a neural network for prediction in a given system is based on the hypothesis that the future events or developments depend exclusively on the history of the system. Although this is frequently the case in many systems, there is no guarantee that this hypothesis is valid. In fact, the success of a neural-network-based predictor is also a test for the validity of this hypothesis. Among the possible applications of neural networks, their application to prediction is undoubtedly the "riskiest" in terms of success. However, the development of successful systems for prediction using neural networks is probably the most impressive in terms of the results.

9.10.1 The Efficient Market Hypothesis

A point of discussion among financial analysts and academics is the possibility of predicting the price of a stock from its history and other information available to the public. The dominant belief seems to be expressed by the *efficient market hypothesis,* which indicates roughly that the stock prices follow a random walk. The practical implication of this hypothesis is that the price of a stock cannot be reliably predicted on the basis of publicly available information including its history and the price of other stocks. White reported an interesting justification of this hypothesis based on common sense (White, 1988). In a profit-hunting environment, such as the stock market, everybody tries to exploit a rising profit opportunity. The apparent result is an increase in the price and, therefore, the disappearance of the profit opportunity. The validity of the efficient market hypothesis can be formally tested by using well-known statistical techniques. This test presumes the existence of a prediction model. Most of the financial prediction attempts are based on linear auto-regressive models. For such models, the existing evidence favors the efficient market hypothesis. The possibility of predicting future stock prices using nonlinear neural network models recently attracted the attention of several researchers. White used the daily IBM stock market return to test the hypothesis by using an auto-regressive linear model and a multi-layered, feed-forward neural network (White, 1988). The one-day stock return is defined by

$$r_t = \frac{p_t - p_{t-1} + d_t}{p_{t-1}} \tag{9.10.1}$$

where p_t is the price of the stock at closing on day t and d_t is the dividend paid on day t. Assume that the one-day stock return can be predicted by an auto-regressive model, that is,

$$r_t = w_0 + w_1\, r_{t-1} + w_2\, r_{t-2} + \ldots + w_p\, r_{t-p} + \varepsilon_t \tag{9.10.2}$$

where $w_i, i = 1, 2, \ldots, p$ are the model coefficients and ε_t represents random noise. The efficient market hypothesis implies that $w_1 = w_2 = \ldots = w_p = 0$. White indicated that the finding $w_1 = w_2 = \ldots = w_p = 0$ is consistent with either the efficient market hypothesis or the presence of nonlinear regularities in the time-series. In such a case, the use of a nonlinear neural network model would be a particularly useful consideration.

White tested the validity of the efficient market hypothesis by testing if $\mathrm{var}\{r_t\} = \mathrm{var}\{\varepsilon_t\}$. The criterion he used was the proximity of $R^2 = 1 - \mathrm{var}\{\varepsilon_t\}/\mathrm{var}\{r_t\}$ to zero. An estimate of R^2 evaluated using an auto-regressive model with $p = 5$, was $\hat{R}^2 = 0.008$. The proximity of $\hat{R}^2$ to zero is not enough for rejecting the efficient market hypothesis. White evaluated the same measure in the case where r_t is the output of a feed-forward neural network with one layer of hidden units. Since the convergence of the Error Back Propagation algorithm was very slow, this network was trained by a nonlinear least-squares method (White, 1988). The estimate of R^2 evaluated in their case was $\hat{R}^2 = 0.175$. Although this value is larger than that obtained in the case of a linear model, it is not high enough to provide a solid basis for the rejection of the efficient market hypothesis. White further tested the validity of the hypothesis by evaluating the correlation between r_t and the estimate $\hat{r}_t$ provided by the network. The correlation between r_t and $\hat{r}_t$ for a post-sampled period of 500 days was found to be -0.07, while the same correlation for a pre-sampled period of 500 days was 0.075. White concluded that the results provided by this network are not sufficient to reject the efficient market hypothesis.

9.10.2 Short Term Stock Market Prediction

The application of neural networks to stock market predictions is possibly the most difficult test of their potential use as predictors. There are many factors that affect the trading in a particular market, such as recent trading trends in this market and other major international markets, economic policy changes, trends and strength of the national economy, trends and developments in the international economy, increases or decreases in the interest rates, etc. On the other hand, company take-overs, acquisitions and mergers also affect everyday trading. In an unstable situation, such as a crash, the developments in the market are also affected by human psychology combined with the inflexibility of computerized trading. This discussion indicates that a financial market can be seen as a system with particularly complicated dynamics. The occurrence of crashes also indicates that this system is not always stable. For obvious reasons, the development of a system for short-term stock market predictions focuses only on periods of normal trading in the market.

The use of neural networks in the prediction of future trends of trading in the Toronto stock exchange (TSE) provides a sense of the reliability that can be achieved by neural network based financial predictors (Karayiannis, 1991a). The TSE is considered to be a very important indicator of economic trends and developments in Canada. It is well-known that developments in the New York stock exchange (NYSE) affect the trading trends not only in the TSE and other North American financial markets but also in overseas markets. Therefore, the development of a system for short-term predictions in the TSE should be based on the recent developments in the NYSE as well as the history of trading in this particular market. The first step towards the development of a system for short term stock market predictions is the quantitative description of the trading history in financial markets.

There is an index characterizing the trading in any financial market. For example, the Dow Jones (DJ) industrial average index is the widely observed index for the NYSE. On the other hand, the TSE is characterized by the TSE 300-stock composite index. The fluctuations of the index over a period of time, i.e., a day, a month, a year, provide a satisfactory description of the trading history in the corresponding financial market. The development of a system predicting the future fluctuations of the TSE index will be based on the recent trading history in the NYSE and the TSE, which is described by the fluctuation of the corresponding market indices. This formulation of the problem is based on the hypothesis that there exist some standard patterns of behavior in a certain financial market. In other words, the trading in a particular financial market exhibits some kind of "periodicity". According to the above discussion, the prediction of the TSE index fluctuation by such a system ignores many factors that in reality affect the climate of trading in the TSE. The major difficulty of developing a prediction system incorporating these factors relates to the lack of a code capable of translating these qualitative factors into suitable inputs for the neural network.

Let $I(k)$ be the value of the market index at the closing of the market on the kth day of trading. The change (increase or decrease) in the market index with respect to its value at the closing of the market on the $(k-1)$th day of trading can be represented by the "first-order derivative", defined as follows

$$i(k) = I(k) - I(k-1) \tag{9.10.3}$$

Clearly, $i(k) > 0$ if the market index is rising, while $i(k) < 0$ if the market index is falling. Certainly, $i(k)$ is a satisfactory indicator of the trend of trading in this market when only two successive days are considered. However, $i(k)$ alone provides no representative image of the recent trend of trading in this particular market. An even more reliable measure of the trend of trading can be provided by the "second-order derivative", defined as (Karayiannis, 1991a)

$$i'(k) = i(k) - i(k-1) = I(k) - 2I(k-1) + I(k-2) \tag{9.10.4}$$

Clearly, consideration of both measures $i(k)$ and $i'(k)$ provides an even more reliable representation of the recent market history than consideration of the measure $i(k)$ alone.

The development of the *neural network predictor (NNP)* in this application is based on the hypothesis that the fluctuation of the TSE index in the $(k+1)$th day of trading depends on the recent developments in the NYSE, described by $i_{DJ}(k)$ and $i'_{DJ}(k)$, and the TSE, described by the $i_{TSE}(k)$ and $i'_{TSE}(k)$. The output of the neural network is the sign of i_{TSE}, that is,

$$y_k = \text{sgn}(i_{TSE}(k+1)) \tag{9.10.5}$$

Clearly, $y_k = +1$ if the TSE index is rising, while $y_k = -1$ if the TSE index is falling. According to the assumptions made above, the corresponding inputs of the neural network are formed as follows

$$\mathbf{x}_k^* = f(i_{DJ}(k), i'_{DJ}(k), i_{TSE}(k), i'_{TSE}(k)) \tag{9.10.6}$$

where $f(.)$ is some function of the inputs. Assume that the NNP provides a prediction of the fluctuation of the TSE index on the (m^*+1)th day of trading. In addition, assume that the neural network is trained on the basis of the trading history in the NYSE and the TSE for the m most recent days. The above formulation indicates that the neural network is trained with respect to the associations $(y_k, \mathbf{x}_k)$, $k = m^*-1, m^*-2, \ldots, m^*-m$. After its training,

NNP	Correct Predictions	Success Rate
NNP I	21 out of 34	61.80 %
NNP II	24 out of 34	70.60 %
NNP III	20 out of 34	58.80 %
NNP IV	20 out of 34	58.80 %
NNP V	22 out of 34	64.70 %
NNP VI	20 out of 34	58.80 %

Table 9.2: Number of correct predictions and success rate achieved by various NNPs trained using the last 15 days of trading history.

the neural network is loaded with the information describing the fluctuation of the TSE index during the m most recent days of trading in conjunction with the recent developments in the NYSE and the TSE. The prediction of the fluctuation of the TSE index on the $(m^* + 1)$th day of trading is the output of the neural network when its input is $\mathbf{x}_{m^*}$. From the definition of $i(k)$ and $i'(k)$, the prediction for the $(m^* + 1)$th day of trading is based on developments in the NYSE and the TSE during the three most recent days m^*, $m^* - 1$, and $m^* - 2$.

A daily prediction of the TSE index was attempted by six distinct NNPs, using different schemes for coding the input data. These NNPs were tested for seven weeks (34 working days) from August 21, 1989 to October 6, 1989. The NNPs were trained using the market history during the 15 most recent days of trading. Table 9.2 shows the total results, that is, the number of correct predictions provided by each NNP and the corresponding success rate, when the NNPs were trained using the market history of the 15 most recent days. The highest success rate was 70.6 %, i.e., 24 correct predictions in 34 days, and was achieved by the NNP II.

According to these experiments, there are some days that all of the NNPs agree on their prediction for the trend of trading in the TSE the next day. However, there are also days that various NNPs provide conflicting predictions for the future fluctuations of the TSE index. This observation suggests that there exists a trade-off between the availability of predictions for every trading day and the reliability of these predictions. The alternative proposed here

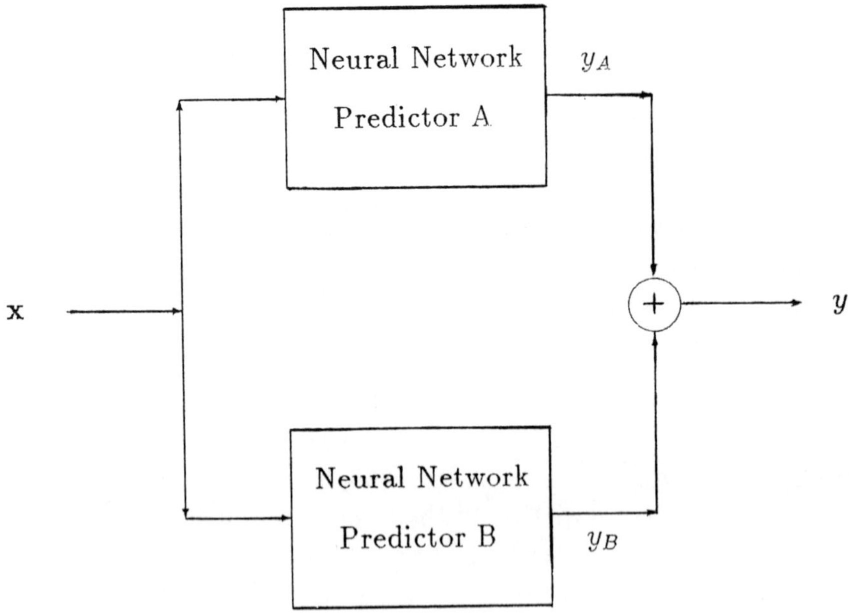

Figure 9.6: The structure of a generalized neural network predictor (GNNP).

is the development of a more "conservative" but more reliable system. Such a system provides no prediction when two selected NNPs provide conflicting predictions for the fluctuation of the TSE index for the next trading day. Figure 9.6 shows the structure of this system, called a *generalized neural network predictor (GNNP).* The two NNPs used in forming the GNNP are independently trained on the basis of the market history for the most recent days of trading. The input of the GNNP is based on the fluctuations of the market indices of the NYSE and the TSE described by the corresponding derivatives $i(k)$ and $i'(k)$. The output of the GNNP is the sum of the outputs of the two NNPs that form the GNNP. The possible outputs of the GNNP are -2, 0, and $+2$. The prediction for the TSE index is determined in terms of the output of the GNNP according to the following rule:

GNNP	Correct Predictions	Success Rate
GNNP I	16 out of 21	76.20 %
GNNP II	19 out of 27	70.40 %
GNNP III	14 out of 18	77.80 %
GNNP IV	21 out of 30	70.00 %
GNNP V	17 out of 24	70.80 %
GNNP VI	14 out of 19	73.70 %

Table 9.3: Number of correct predictions and success rate achieved by various GNNPs trained using the last 15 days of trading history.

- $y > 0$: The TSE index will be rising.
- $y = 0$: No prediction available.
- $y < 0$: The TSE index will be falling.

Five GNNPs were formed by combining the NNP II, the most successful one, with each one of the remaining NNPs. An additional GNNP, the GNNP VI, was formed by combining two NNPs with relatively high success rates, that is, the NNP I and the NNP V. The performance of these GNNPs was tested in the case that the corresponding NNPs were trained using the market history of the 15 most recent trading days. For each GNNP, Table 9.3 shows the number of days that a prediction was available, the number of correct predictions, and the corresponding success rate. Since the GNNPs do not necessarily provide a prediction for every day, the relative evaluation of various GNNPs is based on the number of days that each GNNP provides a prediction and the corresponding success rate. Table 9.3 indicates that these GNNPs achieve success rate higher than 70.0 %. In terms of the number of available predictions, the best GNNPs are the GNNP IV, providing predictions in 30 out of 34 days, and the GNNP II, providing predictions in 27 out of 34 days. Among the GNNPs tested in these experiments, the GNNP III achieves the highest success rate, 77.8 %, followed by the GNNP I with a success rate of 76.2 % and the GNNP VI with a success rate of 73.7 %. Once again, the overall performance of the GNNP I, providing predictions in 21 days with a success rate of

76.2 %, is higher than the one of the GNNP VI, providing predictions in 19 days with a success rate of 73.7 %. Therefore, the two best GNNPs are the GNNP III, providing predictions in 18 out of 34 days with a success rate of 78.8 %, and the GNNP I, providing predictions in 21 out of 34 days with a success rate of 76.2 %.

9.11 Other Applications

This section presents a review of the extensive literature on neural network applications in *speech processing, image processing* and *computer vision, pattern classification and recognition, system identification and control, robotics,* and other problems.

Neural networks have been applied to speech coding (Bengio et al., 1990), speech production (Komura and Tanaka, 1989), and speech recognition (Gold, 1986; Boulard and Wellekens, 1988; Burr, 1988; Homma et al., 1988; Rossen et al., 1988; Anderson et al., 1989; Franzini, 1989; Leung and Zue, 1989; Lippmann, 1989; Waibel, 1989; Boulard et al., 1990; Kowaleski and Strube, 1990; Lang et al., 1990; Lee and Lippmann, 1990; Boulard and Morgan, 1991; Morgan and Scofield, 1991; Tom and Tenorio, 1991).

Applications of neural networks in image processing and computer vision include image segmentation (Hurlbert and Poggio, 1989; Tunley, 1991; Gish and Blanz, 1990; Bilbro et al., 1988; Troudet and Tabatabai, 1988), image recognition (Cruz et al., 1990), texture analysis (Greenspan et al., 1991), computation of optical flow (Zhou and Chellappa, 1988; De Micheli et al., 1990), computation of motion (Koch et al., 1988), visual surface interpolation (Hutchinson and Koch, 1986), visual perception (Ogmen, 1991), and the random-dot stereograms problem, which is related to binocular depth perception (Qian and Sejnowski, 1989).

Neural networks have also been used in pattern classification and recognition (Silvermann et al., 1990; Yao et al., 1991; Yu et al., 1992), recognition of waveforms (Jansen, 1990), and object recognition (Wechsler and Zimmerman, 1987; Pawlicki, 1989; Zemel et al., 1989; Zemel et al., 1990).

Neural networks are currently used in system modeling and identification (Bhat et al., 1990; Narendra and Parthasarathy, 1990; Chu et al., 1990; Levin et al., 1991) and system control (Barto et al., 1983; Bavarian, 1988; Guez et al., 1988; Psaltis et al., 1988; Guez and Selinsky, 1988; Tolat and Widrow, 1988; Nguyen and Widrow, 1990; Chen, 1990; Jordan and Jacobs, 1990; Wieland, 1991; Moore and Naidu, 1991; Tulunay, 1991).

Neural networks are used in robot manipulator control (Guez and Ahmad, 1988; Josin et al., 1988; Kawato et al., 1988; Liu et al., 1988; Sobajic et al., 1988; Goldberg and Pearlmutter, 1989; Atkeson, 1990; Cruse and Bruwer, 1990; Handelman et al., 1990), in the design of controllers for collision-free movement of robot manipulators (Graf and LaLonde, 1988; Beckmann, 1990; Daunicht et al., 1990), in learning of visuomotor coordination of robot arms (Mel, 1988; Mel, 1989; Martinetz et al., 1990), and mobile robot control (Nagata et al., 1990).

Other applications of neural networks include natural language understanding (Nolfi and Parisi, 1988), linguistic problems (Ricotti et al., 1988), and the development of autonomous and expert systems (Shepanski and Macy, 1988; Tesauro and Sejnowski, 1988; Pomerleau, 1989; Goodman et al., 1989).

Chapter 10

Future Trends and Directions

Both the number of papers published in the scientific literature and the number of professional society meetings in the field of neural networks are growing rapidly. A number of technical disciplines are involved in the wide variety of university, government and industrial-based activities and studies. Psychologists, neurobiologists, neurophysiologists, mathematicians, physicists, computer scientists, and engineers are studying the issues of neural networks, formulating new theories, proposing new architectures, fast algorithms and implementations.

Some researchers are attempting to increase their understanding of how the brain works, which will enhance their ability to develop new types of neural networks. Others are studying biologically influenced neural networks, in order to achieve a better understanding of the brain. Computer scientists are combining conventional AI's symbolic and heuristic approach with neural networks, while still others are exploring the neural network implementation of fuzzy logic. Engineers and mathematicians are developing fast learning algorithms, while others are studying new architectures. Special electronic and optical devices are being developed for hardware implementations, and numerous promising applications are currently being explored.

Realizing the potential benefits of neural networks will require basic research to advance the understanding and technology on numerous fronts such as:

- Theory, including representations.
- Fast and efficient learning algorithms, including the issues of convergence and stability.
- Modular and highly interconnected architectures, including the study of overall system control.
- Implementation techniques (electronic, optical, biological).
- Applications.

Neural network research should assure good coupling to other branches of information technology, neuroscience, psychology and take advantages of conceptual breakthroughs in the difficult application areas, such as speech and character recognition, image processing and understanding, and vision.

From the point of view of theory, it is necessary to better understand the most elementary model of a neuron and to study the effects of different nonlinearities and weights on its performance. It is also necessary to develop new algorithms that are fast, robust in the presence of noise, and stable, and that guarantee convergence to a global minimum. Rigorous mathematical foundations must be developed to determine the characteristics of the training set and the network's ability to generalize from the training data.

From the point of view of architectures, there is a need to further investigate and better understand biological information processing. From such investigations, many new neural network architectures may be invented. There is also a need to develop techniques extending these networks to reasonably large scales, in order to solve real-world problems.

The future of the implementation of neural networks will benefit from current research in gallium arsenide (GaAs) and special-purpose charge-coupled devices (CCDS) which are increasing the number of interconnects-per-second. Also, continued developments in random-access memory (RAM) technology as well as the three-dimensional (3-D) chip technology are expected to increase current storage capabilities, while multiprocessing will extend the boundaries of simulation upward.

Other promising technologies include direct VLSI/VHSIC (very large scale integration / very high speed integrated circuits) which is a mature technology limited to a low density of interconnects due to its two-dimensional nature. Analog VLSI, is also developing rapidly and promises near-term results. Optical technology, is less developed and longer term than the silicon-based approaches, but it offers a much higher density of interconnects due to its three-dimensional nature. In such systems the information is stored holographically; another optical system reads the information in the first system and performs a comparison (associative memory) between the stored information and that which the second system attempts to recognize.

The diversity of successful neural network applications reported so far in the literature justifies and reinforces the recent interest in artificial neural networks. While current technology provides impressive solutions to the implementation of massive neural network systems, one of the most important challenges faced by the neural network community is the transformation of unsolved or partially solved problems into suitable candidates for neural network applications. The need for such a transformation suggests a fresh look at old problems in control, signal/image processing and other related disciplines. Conversely, new developments in other disciplines may open new directions in neural networks research. The authors believe that the last sentence of this book should be the following statement made by Mach in the start of this century, and also quoted in (Grossberg, 1988): "The foundations of science as a whole, and of physics in particular, await their next great elucidations from the side of biology, and especially from the analysis of the sensations...".

References

Abbott, L. F., and Kepler, T. B. 1989. Optimal learning in neural network memories. *Journal of Physics A,* **22,** L711-L717.

Abu-Mostafa, Y. S., and Jacques, J.-M. S. 1985. Information capacity of the Hopfield model. *IEEE Transactions on Information Theory,* **31,** 461-464.

Ackley, D. H., Hinton, G. E., and Sejnowski, T. J. 1985. A learning algorithm for Boltzmann machines. *Cognitive Science,* **9,** 147-169.

Albert, A. 1972. *Regression and the Moore-Penrose Pseudoinverse,* Academic Press, New York.

Almeida, L. B. 1987. A learning rule for asynchronous perceptrons with feed-back in a combinatorial environment. *Proceedings of the IEEE First International Conference on Neural Networks,* San Diego, California, pp. II-609 - II-618.

Almeida, L. B. 1988. Backpropagation in perceptrons with feedback. In *Neural Computers,* R. Eckmiller and Ch. v. d. Malsburg, pp. 199-208, Springer-Verlag, Berlin.

Amari, S. I. 1977. Neural theory of association and concept formation. *Biological Cybernetics,* **26,** 175-185.

Amari, S. I., and Maginu, K. 1988. Statistical neurodynamics of associative memory. *Neural Networks,* **1,** 63-73.

Anastassiou, D. 1988. Nonstandard A/D conversion based on symmetric neural networks. *Proceedings of the IEEE International Conference on Neural Networks,* San Diego, California, pp. I-181 - I-188.

Anderson, J. A. 1972. A simple neural network generating interactive memory. *Mathematical Biosciences,* **14,** 197-220.

Anderson, J. A. 1983. Cognitive and psychological computation with neural models. *IEEE Transactions on Systems, Man, and Cybernetics,* **13,** 799-815.

Anderson, J. A., and Rosenfeld, E., eds. 1988. *Neurocomputing: Foundations and Research,* MIT Press, Vol. I & II, Cambridge, Massachusetts.

Anderson, J. A., Silverstein, J. W., Ritz, S. R., and Jones, R. S. 1977. Distinctive features, categorical perception, and probability learning: Some applications of a neural model. *Psychological Review,* **84,** 413-451.

Anderson, S., Merrill, J., and Port, R. 1989. Dynamic speech categorization with recurrent networks. In *Proceedings of the 1988 Connectionist Models Summer School,* D. Touretzky, G. Hinton, and T. Sejnowski, eds., pp. 398-406, Morgan Kaufmann, San Mateo, California.

Angeniol, B., de la Croix Vaubois, G., and LeTexier, J.-Y. 1988. Self-organizing feature maps and the traveling salesman problem. *Neural Networks,* **1,** 289-293.

Anlauf, J. K., and Biehl, M. 1989. The AdaTron: An adaptive Perceptron algorithm. *Europhysics Letters,* **10,** 687-692.

Atkeson, C. G. 1990. Using local models to control movement. In *Advances in Neural Information Processing Systems 2,* D. S. Touretzky, ed., pp. 316-323, Morgan Kaufmann, San Mateo, California.

Baldi, P. 1989. Linear learning: Landscapes and algorithms. In *Advances in Neural Information Processing Systems 1,* D. S. Touretzky, ed., pp. 65-72, Morgan Kaufmann, San Mateo, California.

Baldi, P., and Hornik, K. 1989. Neural networks and principal component analysis: Learning from examples without local minima. *Neural Networks,* **2**, 53-58.

Banzhaf, W., and Haken, H. 1990. Learning in a competitive network. *Neural Networks,* **3**, 423-435.

Barto, A. G., Sutton, R. S., and Anderson, C. W. 1983. Neuronlike adaptive elements that can solve difficult learning control problems. *IEEE Transactions on Systems, Man, and Cybernetics,* **13**, 834-846.

Battiti, R. 1989. Accelerated Backpropagation learning: Two optimization methods. *Complex Systems,* **3**, 331-342.

Baum, E. B. 1986. Towards practical "neural" computation for combinatorial optimization problems. In *AIP Conference Proceedings,* J. Denker, ed., vol. 151, American Institute of Physics, New York.

Baum, E. B., and Haussler, D. 1989. What size net gives valid generalization? *Neural Computation,* **1**, 151-160.

Baum, E. B., and Wilczek, F. 1988. Supervised learning of probability distributions by neural networks. In *Neural Information Processing Systems,* D. Z. Anderson, ed., pp. 52-61, American Institute of Physics, New York.

Bavarian, B. 1988. Introduction to neural networks for intelligent control. *IEEE Control Systems Magazine,* April 1988, 3-7.

Becker, S., and le Cun, Y. 1989. Improving the convergence of Back Propagation learning with second order methods. In *Proceedings of the 1988 Connectionist Models Summer School,* D. Touretzky, G. Hinton, and T. Sejnowski, eds., pp. 29-37, Morgan Kaufmann, San Mateo, California.

Beckmann, J. 1990. Resistive network approach for obstacle avoidance in trajectory planning. In *Parallel Processing in Neural Systems and Computers,* R. Eckmiller, G. Hartmann, and G. Hauske, eds., pp. 405-408, Elsevier, North-Holland.

Bengio, Y., Cardin, R., Cosi, P., De Mori R., and Melro, E. 1990. Speech coding with multilayer networks. In *Neurocomputing: Algorithms, Architectures and Applications,* F. Fogelman-Soulie and J. Herault, eds., pp. 207-216, Springer-Verlag, Berlin.

Bhat, N. V., Minderman, P. A., McAvoy, T., and Wang, N. S. 1990. Modeling chemical process systems via neural computation. *IEEE Control Systems Magazine,* April 1990, 24-30.

Bilbro, G. L., White, M., and Snyder, W. 1988. Image segmentation with neurocomputers. In *Neural Computers,* R. Eckmiller and Ch. v. d. Malsburg, eds., pp. 71-79, Springer-Verlag, Berlin.

Blanchet, P. 1990. Data compression using multilayer perceptrons. In *Neurocomputing: Algorithms, Architectures and Applications,* F. Fogelman-Soulie and J. Herault, eds., pp. 237-240, Springer-Verlag, Berlin.

Block, H. D., Nilsson, N. J., and Duda, R. O. 1964. Determination and detection of features in patterns. In *Computer and Information Sciences,* J. T. Tou, R. H. Wilcox, eds., Spartan Books, Washington DC.

Boulard, H., and Kamp, Y. 1988. Auto-association by multilayer Perceptrons and singular value decomposition. *Biological Cybernetics,* **59,** 291-294.

Boulard, H., and Morgan, N. 1991. Merging multilayer Perceptrons and hidden Markov models: Some experiments in continuous speech recognition. In *Neural Networks: Advances and Applications,* E. Gelenbe, ed., pp. 215-239, Elsevier, North-Holland.

Boulard, H., Morgan, N., and Wellekens, C. J. 1990. Statistical inference in multilayer Perceptrons and hidden Markov models with applications in continuous speech recognition. In *Neurocomputing: Algorithms, Architectures and Applications,* F. Fogelman-Soulie and J. Herault, eds., pp. 219-226, Springer-Verlag, Berlin.

Boulard, H., and Wellekens, C. J. 1988. Multilayer Perceptrons and automatic speech recognition. *Proceedings of the IEEE International Conference on Neural Networks (ICNN '88),* San Diego, California, pp. II-407 - II-416.

Bounds, D. G. 1986. Numerical simulations of Boltzmann machines. In *AIP Conference Proceedings,* J. Denker, ed., Vol. 151, American Institute of Physics, New York.

Bounds, D. G., Lloyd, P. J., and Mathew, B. G. 1990. A comparison of neural network and other pattern recognition approaches to the diagnosis of low back disorders. *Neural Networks,* **3,** 583-591.

Bounds, D. G., Lloyd, P. J., Mathew, B., and Waddell, G. 1988. A multilayer perceptron network for the diagnosis of low back pain. *Proceedings of the IEEE International Conference on Neural Networks (ICNN '88),* San Diego, California, pp. IV-407 - IV-416.

Bradshaw, G., Fozzard, R., and Ceci, L. 1989. A connectionist expert system that actually works. In *Advances in Neural Information Processing Systems 1,* D. S. Touretzky, ed., Morgan Kaufmann, San Mateo, California.

Brandt, R. D., Wang, Y., Laub, A. J., and Mitra, S. K. 1988. Alternative networks for solving the traveling salesman problem and the list-matching problem. *Proceedings of the IEEE International Conference on Neural Networks (ICNN '88),* San Diego, California, pp. II-333 - II-340.

Bruck, J. and Goodman, J. W. 1988. On the power of neural networks for solving hard problems. In *Neural Information Pro-*

cessing Systems, D. Z. Anderson, ed., pp. 137-143, American Institute of Physics, New York.

Burr, D. J. 1988. Speech recognition experiments with Perceptrons. In *Neural Information Processing Systems,* D. Z. Anderson, ed., pp. 144-153, American Institute of Physics, New York.

Burr, D. J. 1988. Experiments on neural net recognition of spoken and written text. *IEEE Transactions on Acoustics, Speech, and Signal Processing,* **36,** 1162-1168.

Carpenter, G. A., and Grossberg, S. 1986. Adaptive resonance theory: Stable self-organization of neural recognition codes in response to arbitrary lists of input patterns. *Proceedings of the Eighth Annual Conference of the Cognitive Science Society,* Amherst, Massachusetts, pp. 45-62.

Carpenter, G. A., and Grossberg, S. 1987a. A massively parallel architecture for a self-organizing neural pattern recognition machine. *Computer Vision, Graphics, and Image Processing,* **37,** 54-115.

Carpenter, G. A., and Grossberg, S. 1987b. ART 2: Self-organization of stable category recognition codes for analog input patterns. *Applied Optics,* 4919-4930.

Carpenter, G. A., and Grossberg, S. 1988. The ART of adaptive pattern recognition by a self-organizing neural network. *IEEE Computer Magazine,* 77-88.

Carpenter, G. A., and Grossberg, S. 1989. ART 3: Hierarchical search using chemical transmitters in self-organizing pattern recognition architectures. *Neural Networks,* **3**, 129-152.

Chauvin, Y. 1989. A Back-propagation algorithm with optimal use of hidden units. In *Advances in Neural Information Processing Systems 1,* D. S. Touretzky, ed., pp. 519-526, Morgan Kaufmann, San Mateo, California.

Chen, F. C. 1990. Back-propagation neural networks for nonlinear self-tuning adaptive control. *IEEE Control Systems Magazine,* April 1990, 44-48.

Chen, H. H., Lee, Y. C., Maxwell, T., Sun, G. Z., Lee, H. Y., and Giles, C. L. 1986. Higher order correlation model for associative memory. In *AIP Conference Proceedings,* J. Denker, ed., Vol. 151, American Institute of Physics, New York.

Cheung, K. F., Oh, S., Marks, R. J., and Atlas, L. E. 1987. Neural net associative memories based on convex set projections. *Proceedings of the IEEE First International Conference on Neural Networks,* San Diego, California, pp. II-609 - II-618.

Chou, P. A. 1988. The capacity of the Kanerva associative memory is exponential. In *Neural Information Processing Systems,* D. Z. Anderson, ed., pp. 184-191, American Institute of Physics, New York.

Chu, S. R., Shoureshi, R., and Tenorio, M. 1990. Neural networks for system identification. *IEEE Control Systems Magazine,* April 1990, 31-35.

Clarke, R. J. 1985. *Transform Coding of Images,* Academic Press, Orlando, Florida.

Cohen, M. A., and Grossberg, S. 1983. Absolute stability of global pattern formation and parallel memory storage by competitive neural networks. *IEEE Transactions on Systems, Man, and Cybernetics,* **13,** 815-826.

Cottrell, G. W., Munro, P., and Zipser, D. 1987. Learning internal representations from gray-scale images: An example of extensional programming. In *Ninth Annual Conference of the Cognitive Science Society,* pp. 462-473, Erlbaum, Hillsdale.

Cowan, J. D. 1990. Neural networks: The early days. In *Advances in Neural Information Processing 2,* D. S. Touretzky, ed., pp. 828-841, Morgan Kaufmann, San Mateo, California.

Cruse, H., and Bruwer, M. 1990. A simple network controlling the movement of a three joint planar manipulator. In *Parallel Processing in Neural Systems and Computers,* R. Eckmiller, G. Hartman, and G. Hauske, eds., pp. 409-412, Elsevier, North-Holland.

Cruz, V., Cristobal, G., Michaux, T., and Barquin, S. 1990. Distortion invariant image recognition by Madaline and Back-propagation learning multi-networks. In *Neurocomputing: Algorithms, Architectures and Applications,* F. Fogelman-Soulie and J. Herault, eds., pp. 337-343, Springer-Verlag, Berlin.

Cybenko, G. 1989. Approximation by superpositions of a sigmoid function. *Mathematics of Control, Signals, and Systems,* **2**, 303-314.

Dahl, E. D. 1987. Accelerated learning using the generalized delta rule. *Proceedings of the IEEE First International Conference on Neural Networks,* San Diego, California, pp. II-523 - II 530.

Daunicht, W. J., Lades, M., Werntges, H., and Eckmiller, R. 1990. Inverse kinematics with obstacle avoidance implemented as a defanet. In *Parallel Processing in Neural Systems and Computers,* R. Eckmiller, G. Hartmann, and G. Hauske, eds., pp. 417-420, Elsevier, North-Holland.

Daugman, J. G. 1988a. Relaxation network for non-orthogonal image transforms. *Proceedings of the IEEE International Conference on Neural Networks (ICNN '88),* San Diego, California, pp. I-547 - I-560.

Daugman, J. G. 1988b. Complete discrete 2-D Gabor transforms by neural networks for image analysis and compression. *IEEE Transactions on Acoustics, Speech, and Signal Processing,* **36**, 1169-1179.

De Micheli, E., Radonich, G., Uras, S., and Torre, V. 1990. An algorithm for optical flow. *Neurocomputing: Algorithms, Ar-*

chitectures and Applications, F. Fogelman-Soulie and J. Herault, eds., pp. 345-356, Springer-Verlag, Berlin.

Denker, J. S., Gardner, W. R., Graf, H. P., Henderson, D., Howard, R. E., Hubbard, W., Jackel, L. D., Baird, H. S., and Guyon, I. 1989. Neural network recognizer for hand-written zip code digits. In *Advances in Neural Information Processing Systems 1,* D. S. Touretzky, ed., pp. 323-331, Morgan Kaufmann, San Mateo, California.

Duda, R. O., and Hart, P. E. 1973. *Pattern Classification and Scene Analysis,* John Wiley, New York.

Engel, J. 1988. Training feed-forward neural networks by simulated annealing. *Complex Systems,* **2**, 641-648.

Erlauson, R., and Abu-Mostafa, Y. 1989. Using an analog network for decoding. In *Proceedings of the 1988 Connectionist Models Summer School,* D. S. Touretzky, G. Hinton, and T. Sejnowski, eds., pp. 186-190, Morgan Kaufmann, San Mateo, California.

Fahlman, S. E. 1989. Fast-learning variations on Back-propagation: An empirical study. In *Proceedings of the 1988 Connectionist Models Summer School,* D. S. Touretzky, G. Hinton, and T. Sejnowski, eds., pp. 38-51, Morgan Kaufmann, San Mateo, California.

Fang, L., Wilson, W. H., and Li. T. 1990. A neural network for job sequencing. In *Parallel Processing in Neural Systems and Computers,* R. Eckmiller, G. Hartmann, and G. Hauske, eds., pp. 253-256, Elsevier Publishers, North Holland.

Foo, Y. P. S., and Takefuji, Y. 1988a. Stochastic neural networks for solving job-shop scheduling: Part 1. Problem representation. *Proceedings of the IEEE International Conference on Neural Networks (ICNN '88),* San Diego, California, pp. I-275 - I-282.

Foo, Y. P. S., and Takefuji, Y. 1988b. Stochastic neural networks for solving job-shop scheduling: Part 2. Architecture and simulations. *Proceedings of the IEEE International Conference on Neural Networks (ICNN '88),* San Diego, California, pp. I-283 - I-290.

Frankel, D., Schiller, I., Draper, J. S., and Barnes, A. A. 1991. Use of neural networks to predict lighting at Kennedy space center. *Proceedings of the International Joint Conference on Neural Networks (IJCNN '91),* Seattle, Washington, pp. I-319 - I-324.

Franzini, M. A. 1989. Learning to recognize spoken words: A study in connectionist speech recognition. In *Proceedings of the 1988 Connectionist Models Summer School,* D. Touretzky, G. Hinton, and T. Sejnowski, eds., pp. 407-416, Morgan Kaufmann, San Mateo, California.

Frean, M. 1990. The upstar algorithm: A method for constructing and training feed-forward neural networks. *Neural Computation,* **2**, 198-209.

Fukushima, K. 1975. Cognitron: A self-organizing multilayered neural network. *Biological Cybernetics,* **20**, 121-136.

Fukushima, K. 1980. Neocognitron: A self-organizing neural network model for a mechanism of pattern recognition unaffected by shift in position. *Biological Cybernetics,* **36**, 193-202.

Fukushima, K. 1986. A neural network model for selective attention in visual pattern recognition. *Biological Cybernetics,* **55**, 5-15.

Fukushima, K. 1988. A neural network for visual pattern recognition. *IEEE Computer Magazine,* March 1988, 65-75.

Fukushima, K. 1990. Neural network models for visual pattern recognition. In *Parallel Processing in Neural Systems and Computers,* R. Eckmiller, G. Hartmann, and G. Hauske, eds., pp. 351-356, Elsevier Publishers, North Holland.

Fukushima, K., Imagawa, T., and Ashida, E. 1991. Character recognition with selective attention. *Proceedings of the International Joint Conference on Neural Networks (IJCNN '91),* Seattle, Washington, pp. I-593 - I-598.

Fukushima, K., Miyake, S., and Ito, T. 1983. Neocognitron: A neural network model for a mechanism of visual pattern recognition. *IEEE Transactions on Systems, Man, and Cybernetics,* **13**, 826-834.

Gantmacher, F. R. 1977. *The Theory of Matrices,* Chelsea Publishing Company, New York.

Gersho, A., and Gray, R. M. 1992. *Vector Quantization and Signal Processing,* Kluwer Academic Publishers, Boston, Massachusetts.

Giles, C. L., Chen, D., Miller, C. B., Chen, H. H., Sun, G. Z., and Lee, Y. C. 1991. Second-order recurrent neural networks for grammatical inference. *Proceedings of the International Joint Conference on Neural Networks (IJCNN '91),* Seattle, Washington, pp. II-273 - II-281.

Giles, C. L., Sun, G. Z., Chen, H. H., Lee, Y. C., and Chen, D. 1990. Higher order recurrent networks and grammatical inference. In *Advances in Neural Information Processing Systems 2,* D. S. Touretzky, ed., pp. 380-387, Morgan Kaufmann, San Mateo, California.

Gish, S. L., and Blanz, W. E. 1990. Comparing the performance of connectionist and statistical classifiers on an image segmentation problem. In *Advances in Neural Information Processing Systems 2,* D. S. Touretzky, ed., pp. 614-621, Morgan Kaufmann, San Mateo, California.

Gold, B. 1986. Hopfield model applied to vowel and consonant discrimination. In *AIP Conference Proceedings,* J. Denker, ed., pp. 158-164, American Institute of Physics, New York.

Goldberg, K. Y., and Pearlmutter, B. A. 1989. Using backpropagation with temporal windows to learn the dynamics of the CMU Direct-Drive Arm II. In *Advances in Neural Information Processing Systems 1,* D. S. Touretzky, ed., pp. 356-363, Morgan Kaufmann, San Mateo, California.

Gonzales, R. C., and Wintz, P. 1977. *Digital Image Processing,* Addison-Wesley, Reading, Massachusetts.

Goodman, R. M., Miller, J. W., and Smyth, P. 1989. An information theoretic approach to rule-based connectionist expert systems. In *Advances in Neural Information Processing Systems 1,* D. S. Touretzky, ed., pp. 256-263, Morgan Kaufmann, San Mateo, California.

Gorman, R. P., and Sejnowski, T. J. 1988a. Learned classification of sonar targets using a massively parallel network. *IEEE Transactions on Acoustics, Speech, and Signal Processing,* **36,** 1135-1140.

Gorman, R. P., and Sejnowski, T. J. 1988b. Analysis of hidden units in a layered network trained to classify sonar targets. *Neural Networks,* **1,** 75-89.

Graf, D. H., and LaLonde, W. R. 1988. A neural controller for collision-free movement of general robot manipulators. *Proceedings of the IEEE International Conference on Neural Networks (ICNN '88),* San Diego, California, pp. I-77 - I-84.

Gray, R. M. 1984. Vector quantization. *IEEE ASSP Magazine,* April 1984, 4-29.

Greenspan, H., Goodman, R., and Chellappa, R. 1991. Texture analysis via unsupervised and supervised learning. *Proceedings of the International Joint Conference on Neural Networks (IJCNN' 91),* Seattle, Washington, pp. I-639 - I-644.

Grimson, W. E. L. 1981. *From Images to Surfaces: A Computational Study of the Human Visual System,* MIT Press, Cambridge, Massachusetts.

Grossberg, S. 1987a. *The Adaptive Brain I: Cognition, Learning, Reinforcement and Rhythm,* Elsevier, North Holland.

Grossberg, S. 1987b. *The Adaptive Brain II: Speech, Language and Motor Control,* Elsevier, North Holland.

Grossberg, S. 1987c. Competitive learning: From interactive activation to adaptive resonance. *Cognitive Science,* **11,** 23-63.

Grossberg, S. 1988a. *Neural Networks and Natural Intelligence,* MIT Press, Cambridge, Massachusetts.

Grossberg, S. 1988b. Nonlinear neural networks: Principles, mechanisms, and architectures. *Neural Networks,* **1,** 17-61.

Grossman, T., Meir, R., and Domany, E. 1988. Learning by choice of internal representations. *Complex Systems,* **2,** 555-575.

Guez, A., and Ahmad, Z. 1988. Solution to the inverse kinematics problem in robotics by neural networks. *Proceedings of the IEEE International Conference on Neural Networks (ICNN '88),* San Diego, California, pp. II-617 - II-624.

Guez, A., Eilbert, J. L., and Kam, M. 1988. Neural network architecture for control. *IEEE Control Systems Magazine,* April 1988, 22-25.

Guez, A., and Selinsky, J. 1988. A neuromorphic controller with a human teacher. *Proceedings of the IEEE International Conference on Neural Networks (ICNN '88),* San Diego, California, pp. II-595 - II-602.

Handelman, D. A., Lane, S. H., and Gelfand, J. J. 1990. Integrating neural networks and knowledge-based systems for intelligent control. *IEEE Control Systems Magazine,* April 1990, 77-87.

Hanson, S. J., and Pratt, L. Y. 1989. Comparing biases for minimal network construction with Back-propagation. In *Advances in Neural Information Processing Systems 1,* D. S. Touretzky, ed., pp. 177-185, Morgan Kaufmann, San Mateo, California.

Harrison, R. F., Marshall, S. J., and Kennedy, R. L. 1991. The early diagnosis of heart attacks: A neurocomputational approach. *Proceedings of the International Joint Conference on Neural Networks (IJCNN '91),* Seattle, Washington, pp. I-1 - I-5.

Hebb, D. O. 1949. *The Organization of Behavior,* J. Wiley, New York.

Hedge, S. U., Sweet, J. L., and Levy, W. B. 1988. Determination of parameters in a Hopfield/Tank computational network. *Proceedings of the IEEE International Conference on Neural Networks (ICNN-88),* San Diego, California, pp. I-291 - I-298.

Hemani A., and Postula, A. 1990. Scheduling by self-organization. *Proceedings of the International Joint Conference on Neural Networks (IJCNN '90),* Washington, DC, pp. II-543 - II-546.

Hertz, J., Krogh, A., and Palmer, R. G. 1991. *Introduction to the Theory of Neural Computation*, Addison-Wesley, Redwood City, California.

Hinton, G. E. 1986. Learning distributed representations of concepts. In *Proceedings of the Eighth Annual Conference of the Cognitive Science Society,* Amherst, Massachusetts, pp. 1-12, Erlbaum, Hillsdale.

Hinton, G. E., and Sejnowski, T. J. 1986. Learning and relearning in Boltzmann machines. In *Parallel Distributed Processing,* vol. 1, D. E. Rumelhart, and J. L McClelland, eds., MIT Press, Cambridge, Massachusetts.

Hirose, Y., Yamashita, K., and Hijiya, S. 1991. Back-propagation algorithm which varies the number of hidden units. *Neural Networks,* **4,** 61-66.

Holley, L. H., and Karplus, M. 1989. Protein structure prediction with a neural network. *Proceedings of the National Academy of Sciences U.S.A.,* **86,** 152-156.

Homma, T., Atlas, L. E., and Marks, R. J. 1988. An artificial neural network for spatio-temporal bipolar patterns: Application to phoneme classification. In *Neural Information Processing Systems,* D. Z. Anderson, ed., pp. 31-40, American Institute of Physics, New York.

Hopfield, J. J. 1982. Neural networks and physical systems with emergent collective computational abilities. *Proceedings of the National Academy of Sciences U.S.A.,* **79,** 2554-2558.

Hopfield, J. J. 1984. Neurons with graded response have collective computational properties like those of two-state neurons. *Proceedings of the National Academy of Sciences U.S.A.,* **81,** 3088-3092.

Hopfield, J. J. 1987. Learning algorithms and probability distributions in feed-forward and feed-back networks. *Proceedings of National Academy of Sciences, USA,* **84,** 8429-8433.

Hopfield, J. J., and Tank, D. W. 1985. Neural computation of decisions in optimization problems. *Biological Cybernetics,* **52,** 141-152.

Hurlbert, A., and Poggio, T. 1989. A network for image segmentation using color. In *Advances in Neural Information Processing Systems 1,* D. S. Touretzky, ed., pp. 297-304, Morgan Kaufmann, San Mateo, California.

Hush, D. R., and Salas, J. M. 1988. Improving the learning rate of back-propagation with the gradient reuse algorithm. *Proceedings of the IEEE International Conference on Neural Networks (ICNN '88),* San Diego, California, pp. I-441 - I-447.

Hutchinson, J. M., and Koch, C. 1986. Simple analog and hybrid networks for surface interpolation. In *AIP Conference Proceedings,* J. Denker, ed., vol. 151, pp. 235-240, American Institute of Physics, New York.

Iyer, K. V. K., Ogmen, H., and Koc, C. K. 1992. Landscape reshaping algorithm for additive neural networks with application to graph mapping problems. *Electronics Letters,* **28**, 109-111.

Jacobs, R. A. 1988. Increased rate of convergence through learning rate adaptation. *Neural Networks,* **1**, 295-307.

Jang, J. S., Lee, S. Y., and Shin, S. Y. 1988. An optimization network for matrix inversion. In *Neural Information Processing Systems,* D. Z. Anderson, ed., pp. 397-401, American Institute of Physics, New York.

Jansen, B. H. 1990. Artificial neural nets for K-complex detection. *IEEE Engineering in Medicine and Biology,* September 1990, 50-52.

Johnson, K., Daniell, C., and Burman, J. 1988. Feature extraction in the neocognitron. *Proceedings of the IEEE International Conference on Neural Networks (ICNN '88),* San Diego, California, pp. I-117 - I-126.

Jordan, M. I., and Jacobs, R. A. 1990. Learning to control an unstable system with forward modeling. In *Advances in Neural Information Processing Systems 2,* D. S. Touretzky, ed., pp. 324-331, Morgan Kaufmann, San Mateo, California.

Josin, G., Charney, D., and White, D. 1988. Robot control using neural networks. *Proceedings of the IEEE International Conference on Neural Networks (ICNN '88),* San Diego, California, pp. II-625 - II-631.

Kamimura, R. 1991. Recurrent neural network to acquire the grammatical competence. *International Joint Conference on Neural Networks,* Seattle, Washington, pp. I-361 - I-366.

Kanerva, P. 1988. *Sparse Distributed Memory,* MIT Press, Cambridge, Massachusetts.

Karayiannis, N. B. 1991a. *Artificial Neural Networks: Learning Algorithms, Performance Evaluation, and Applications,* Ph.D. dissertation, University of Toronto, Toronto, Canada.

Karayiannis, N. B. 1991b. ALADIN: Algorithms for Learning and Architecture DetermINation. *University of Houston Technical Report No. 91-12.*

Karayiannis, N. B. 1992a. Fast learning algorithms for feed-forward neural networks with analog and binary output. *University of Houston Technical Report No. 92-02.*

Karayiannis, N. B. 1992b. ALADIN: Algorithms for Learning and Architecture DetermINation. *Proceedings of the International Joint Conference on Neural Networks (IJCNN '92),* Baltimore, Maryland, pp. I-601 - I-606.

Karayiannis, N. B., and Venetsanopoulos, A. N. 1988. The correlational associative memory realizes Hebbian learning. *Abstracts of the First Annual INNS Meeting,* Boston, Massachusetts, pp. 187.

Karayiannis, N. B., and Venetsanopoulos, A. N. 1989. Regularization theory in image restoration: The regularizing operator approach. *Optical Engineering,* **28**, 761-780.

Karayiannis, N. B., and Venetsanopoulos, A. N. 1990a. On the dynamics of neural networks realizing associative memories of first and higher order. *Network: Computation in Neural Systems*, **1**, 345-364.

Karayiannis, N. B., and Venetsanopoulos, A. N. 1990b. Recursive least-squares learning algorithms for single-layered neural networks. *Proceedings IASTED International Conference on Modeling, Simulation, and Optimization,* Montreal, Canada, pp. 162-165.

Karayiannis, N. B., and Venetsanopoulos, A. N. 1990c. Efficient learning algorithms for single-layered neural networks. *Parallel Processing in Neural Systems and Computers,* R. Eckmiller, G. Hartmann, and G. Hauske, eds., pp. 173-176, Elsevier Publishers, North Holland.

Karayiannis, N. B., and Venetsanopoulos, A. N. 1990d. Application of neural networks to environmental protection. *Proceedings of the International Neural Network Conference,* Paris, France, pp. 334-337.

Karayiannis, N. B., and Venetsanopoulos, A. N. 1990e. Regularization theory in image restoration: The stabilizing functional approach. *IEEE Transactions on Acoustics, Speech, and Signal Processing,* **38,** 1155-1179.

Karayiannis, N. B., and Venetsanopoulos, A. N. 1991. Fast learning algorithms for neural networks. In *Artificial Neural Networks,* T. Kohonen, K. Makisara, O. Simula, J. Kangas, eds., pp. 1141-1144, Elsevier Publishers, North Holland.

Karayiannis, N. B., and Venetsanopoulos, A. N. 1992a. Image interpolation based on variational principles. *Signal Processing,* **25,** 259-288.

Karayiannis, N. B., and Venetsanopoulos, A. N. 1992b. Fast learning algorithms for neural networks. *IEEE Transactions on Circuits and Systems,* in press.

Karayiannis, N. B., and Venetsanopoulos, A. N. 1992c. Efficient LEarning Algorithms for Neural NEtworks (ELEANNE). submitted to *IEEE Transactions on Systems, Man, and Cybernetics,* under review.

Karayiannis, N. B., and Venetsanopoulos, A. N. 1992d. On the performance of single-layered neural networks. *Biological Cybernetics,* in press.

Karayiannis, N. B., and Venetsanopoulos, A. N. 1992e. High-order neural networks and neural networks with composite key patterns. submitted for publication in *Network: Computation in Neural Systems,* under review.

Karayiannis, N. B., and Venetsanopoulos, A. N. 1992f. Neural networks in decision making. submitted to *Neurocomputing,* under review.

Kawato, M., Uno, Y., Isobe, M., and Suzuki, R. 1988. Hierarchical neural network model for voluntary movement with application to robotics. *IEEE Control Systems Magazine,* April 1988, 8-16.

Keeler, J. D. 1988. Capacity for patterns and sequences in Kanerva's SDM as compared to other associative memory models. In *Neural Information Processing Systems,* D. Z. Anderson, ed., pp. 412-421, American Institute of Physics, New York.

Khotanzad, A., and Lu, J. H. 1988. Distortion invariant character recognition by a multi-layer perceptron and back propagation learning. *Proceedings of the IEEE International Conference on Neural Networks (ICNN '88),* San Diego, California, pp. I-625 - I-632.

Kim, E. J., and Lee, Y. 1991. Handwritten Hangul recognition using a modified neocognitron. *Neural Networks,* **4**, 743-750.

Knerr, S., Personnaz, L., and Dreyfus, G. 1991. A new approach to the design of neural network classifiers and its application to the automatic recognition of handwritten digits. *Proceedings of the International Joint Conference on Neural Networks (IJCNN '91),* Seattle, Washington, pp. I-91 - I-96.

Koch, C., Luo, J., Mead, C., and Hutchinson, J. 1988. Computing motion using resistive networks. In *Neural Information Processing Systems,* D. Z. Anderson, ed., pp. 422-431, American Institute of Physics, New York.

Kohonen, T. 1972. Correlation matrix memories. *IEEE Transactions on Computers,* **21**, 353-359.

Kohonen, T. 1974. An adaptive associated memory principle. *IEEE Transactions on Computers,* 444-445.

Kohonen, T. 1987. *Content-addressable memories,* 2nd Edition, Springer-Verlag, Berlin.

Kohonen, T. 1988a. An introduction to neural networks. *Neural Networks,* **1,** 3-16.

Kohonen, T. 1988b. The "neural" phonetic typewriter. *IEEE Computer Magazine,* 11-22.

Kohonen, T. 1989 *Self-Organization and Associative Memory,* 3rd Edition, Springer-Verlag, Berlin.

Kohonen, T. 1990. The self-organizing map. *Proceedings of the IEEE,* **78,** 1464-1480.

Kohonen, T., Barna, G., and Chrisley, R. 1988. Statistical pattern recognition with neural networks: Benchmarking studies. *Proceedings of the IEEE International Conference on Neural Networks (ICNN '88),* San Diego, California, pp. I-61 - I-68.

Kohonen, T., and Ruohonen, M. 1973. Representation of associated data by matrix operators. *IEEE Transactions on Computers,* **22,** 701-702.

Kollias, S., and Anastassiou, D. 1989. An adaptive least squares algorithm for the efficient training of artificial neural networks. *IEEE Transactions on Circuits and Systems,* **36,** 1092-1989.

Komura, M., and Tanaka, A. 1989. Speech production using a neural network with a cooperative learning mechanism. In *Advances in Neural Information Processing Systems 1,* D. S. Touretzky, ed. pp. 232-239, Morgan Kaufmann, San Mateo, California.

Kowalewski, F., and Strube, H. W. 1990. Word recognition with a recurrent neural network. In *Parallel Processing in Neural Systems and Computers,* R. Eckmiller, G. Hartmann, and G. Hauske, eds., p.p. 391-394, Elsevier, North-Holland.

Kramer, A., and Sangiovanni-Vincentelli, A. 1989. Efficient parallel learning algorithms for neural networks. In *Advances in Neural Information Processing Systems 1,* D. S. Touretzky, ed., pp. 40-48, Morgan Kaufmann, San Mateo, California.

Krogh, A., and Hertz, J. A. 1990. Hebbian learning of principal components. *Parallel Processing in Neural Systems and Computers,* R. Eckmiller, G. Hartmann, and G. Hauske, eds., pp. 183-186, Elsevier Publishers, North Holland.

Krogh, A., Thorbergsson, G. I., and Hertz, J. A. 1990. A cost function for internal representations. In *Advances in Neural Information Processing Systems 2,* D. S. Touretzky, ed., pp. 733-740, Morgan Kaufmann, San Mateo, California.

Kung, S. Y., and Hu, Y. H. 1991. A Fobenius approximation reduction method (FARM) for determining optimal number of hidden units. *Proceedings of the International Joint Conference on Neural Networks (IJCNN '91),* Seattle, Washington, pp. II-163 - II-168.

Lampinen, J., and Oja, E. 1990a. Distortion tolerant feature extraction with Gabor functions and topological coding. *Proceedings of International Neural Network Conference,* Paris, France, pp. 301-304.

Lampinen, J., and Oja, E. 1990b. Fast computation of Kohonen self-organization. In *Neurocomputing: Algorithms, Architectures and Applications,* F. Fogelman-Soulie and J. Herault, eds., pp. 65-74, Springer-Verlag, Berlin.

Lang, K. J., Waibel, A. H., and Hinton, G. E. 1990. A time-delay neural network architecture for isolated word recognition. *Neural Networks,* **3,** 23-43.

Lapedes, A., and Farber, R. 1988. How neural networks work. In *Neural Information Processing Systems,* D. Z. Anderson, ed., pp. 442-456, American Institute of Physics, New York.

Lau, C., ed. 1992. *Neural Networks: Theoretical Foundations and Analysis,* IEEE Press, New York.

le Cun, Y., Boser, B., Denker, J. S., Henderson, D., Howard, R. E., and Hubbard, W., and Jackel, L. D. 1989. Backpropagation applied to handwritten zip code recognition. *Neural Computation,* 1, 541-551.

le Cun, Y., Gardner, W. R., Graf., H. P., Henderson, D., Howard, R. E., and Hubbard, W., Jackel, L. D., Baird, H. S., and Guyon, I. 1989. Neural network recognizer for hand-written zip code digits. In *Advances in Neural Information Processing Systems 1,* D. S. Touretzky, ed., pp. 323-331, Morgan Kaufmann, San Mateo, California.

le Cun, Y., Jackel, L. D., Boser, B., Denker, J. S., Graf, H. P., Guyon, I., Henderson, D., Howard, R. E., and Hubbard, W. 1989. Handwritten digit recognition: Application of neural chips and automatic learning. *IEEE Communications Magazine,* 41-46.

le Cun, Y., Boser, B., Denker, J. S., Henderson, D., Howard, R. E., and Hubbard, W., and Jackel, L. D. 1990. Handwritten digit recognition with a Back-Propagation Network. In *Advances in Neural Information Processing Systems 2,* D. S. Touretzky, ed., pp. 396-404, Morgan Kaufmann, San Mateo, California.

le Cun, Y., Denker, J. S., and Solla, S. A. 1990. Optimal brain damage. In *Advances in Neural Information Processing Systems 2,* D. S. Touretzky, ed., pp. 598-605, Morgan Kaufmann, San Mateo, California.

le Cun, Y., Jackel, L. D., Boser, B., Denker, J. S., Graf, H. P., Guyon, I., Henderson, D., Howard, R. E., and Hubbard, W. 1990. Handwritten digit recognition: Application of neural chips and automatic learning. In *Neurocomputing: Algorithms, Architectures, and Applications,* F. Fogelman-Soulie and J. Herault, eds., pp. 303-318, Springer-Verlag, Berlin.

Lee, Y., and Lippmann, R. P. 1990. Practical characteristics of neural network and conventional pattern classifiers on artificial and speech problems. In *Advances in Neural Information Processing Systems 2,* D. S. Touretzky, ed., pp. 168-177, Morgan Kaufmann, San Mateo, California.

Leung, H. C., and Zue, V. W. 1989. Applications of error back-propagation to phonetic classification. In *Advances in Neural*

Information Processing Systems 1, D. S. Touretzky, ed. pp. 206-214, Morgan Kaufmann, San Mateo, California.

Levin, E., Gewirtzman, R., and Inbar, G. F. 1991. Neural network architecture for adaptive system modeling and control. *Neural Networks,* **4,** 185-191.

Levine, D. S. 1990. *Introduction to Neural and Cognitive Modeling,* Lawrence Elbaum Associates, Publishers, Hillsdale, New Jersey.

Linde, Y., Buzo, A., and Gray, R. M. 1980. An algorithm for vector quantization. *IEEE Transactions on Communication,* **28,** 84-95.

Linsker, R. 1986a. From basic network principles to neural architecture: Emergence of spatial-opponent cells. *Proceedings of the National Academy of Sciences U.S.A.,* **83,** 7508-7512.

Linsker, R. 1986b. From basic network principles to neural architecture: Emergence of orientation-selective cells. *Proceedings of the National Academy of Sciences U.S.A.,* **83,** 8390-8394.

Linsker, R. 1986c. From basic network principles to neural architecture: Emergence of orientation columns. *Proceedings of the National Academy of Sciences U.S.A.,* **83,** 8779-8783.

Linsker, R. 1988a. Self-organization in a perceptual network. *IEEE Computer Magazine,* 105-117.

Linsker, R. 1988b. Toward an organizing principle for a layered perceptual network. In *Neural Information Processing Systems,* D. Z. Anderson, ed., pp. 485-494, American Institute of Physics, New York.

Lippmann, R. P. 1987. An introduction to computing with neural nets. *IEEE Acoustics, Speech and Signal Processing Magazine,* April 1987, 4-22.

Lippmann, R. P. 1989a. Pattern classification using neural networks. *IEEE Communications Magazine,* 47-64.

Lippmann, R. P. 1989b. Review of neural networks for speech recognition. *Neural Computation,* **1,** 1-38.

Liu, H., Iberall, T., and Bekey, G. A. 1988. Building a generic architecture for robot hand control. *Proceedings of the IEEE International Conference on Neural Networks (ICNN '88),* San Diego, California, pp. II-567 - II-573.

Marchand, M., Colea, M., and Rujan, P. 1990. A convergence theorem for sequential learning in two-layer Perceptrons. *Europhysics Letters,* **11**, 487-492.

Martin, G. L., and Pittman, J. A. 1990. Recognizing handwritten letters and digits. In *Advances in Neural Information Processing Systems 2,* D. S. Touretzky, ed., pp. 405-410, Morgan Kaufmann, San Mateo, California.

Martinetz, J., Ritter, J., and Schulten, K. J. 1990a. Learning of visuomotor-coordination of a robot arm with redundant degrees of freedom. In *Parallel Processing in Neural Systems and Computers,* R. Eckmiller, G. Hartmann, and G. Hauske, eds., pp. 431-434, Elsevier, North-Holland.

Martinetz, J., Ritter, J., and Schulten, K. J. 1990b. Three-dimensional neural net for learning visuomotor coordination of a robot arm. *IEEE Transactions on Neural Networks,* **1,** 131-136.

McCormack, M. D. 1991. Neural computing in geophysics. *Geophysics: The Leading Edge of Exploration*, 11-15.

McCulloch, W. S., and Pitts, W. 1943. A logical calculus of the ideas immanent in nervous activity. *Bulletin of Mathematical Biophysics,* **5,** 115-133.

McEliece, R. J., Posner, E. C., Rodemich, E. R., and Venkatesh, S. S. 1987. The capacity of the Hopfield associative memory. *IEEE Transactions on Information Theory,* **33,** 461-482.

Mejia, C., and Fogelman-Soulie, F. 1990. Incorporating knowledge in multi-layer networks: The example of protein secondary structure prediction. In *Neurocomputing: Algorithms, Architectures, and Applications,* F. Fogelman-Soulie and J. Herault, eds., pp. 3-13, Springer-Verlag, Berlin.

Mel, B. W. 1987. MURPHY: A robot that learns by doing. In *Neural Information Processing Systems,* D. Z. Anderson, ed., pp. 544-553, American Institute of Physics, New York.

Mel, B. W. 1989. Further explorations in visually-guided reaching: Making MURPHY smarter. In *Advances in Neural Information Processing Systems 1,* D. S. Touretzky, ed., pp. 348-355, Morgan Kaufmann, San Mateo, California.

Mezard, M., and Nadal, J. P. 1989. Learning in feed-forward neural networks: The tiling algorithm. *Journal of Physics A,* **22**, 2191-2204.

Mighell, D. A., Wilkinson, T. S., and Googman, J. W. 1989. Backpropagation and its application to handwritten signature verification. In *Advances in Neural Information Processing Systems 1,* D. S. Touretzky, ed., pp. 340-347, Morgan Kaufmann, San Mateo, California.

Minsky, M. L., and Papert, S. A. 1969. *Perceptrons: An Introduction to Computational Geometry,* MIT Press, Cambridge, Massachusetts.

Moore, B. 1989. ART 1 and pattern clustering. In *Proceedings of the 1988 Connectionist Models Summer School,* D. Touretzky, G. Hinton, and T. Sejnowski, eds., pp. 174-185, Morgan Kaufmann, San Mateo, California.

Moore, K. L., and Naidu, S. 1991. Linear quadratic regulation using neural networks. *Proceedings of the International Joint Conference on Neural Networks (IJCNN '91),* Seattle, Washington, pp. II-735 - II-739.

Morgan, A. P., and Scofield, C. L. 1991. *Neural networks and speech processing,* Kluwer Academic Publishers, Boston, Massachusetts.

Mougeot, M., Azencott, R., and Angeniol, B. 1990. A study of image compression with backpropagation. In *Neurocomputing: Algorithms, Architectures and Applications,* F. Fogelman-Soulie and J. Herault, eds., pp. 333-336, Springer-Verlag, Berlin.

Mougeot, M., Azencott, R., and Angeniol, B. 1991. Image compression with back propagation: Improvement of the visual restoration using different cost functions. In *Neural Networks,* **4,** 467-476.

Mozer, M. C., and Smolensky, P. 1989. Skeletonization: A technique for trimming the fat from a network via relevance assessment. In *Advances in Neural Information Processing Systems 1,* D. S. Touretzky, ed., pp. 107-115, Morgan Kaufmann, San Mateo, California.

Nagata, S., Sekiguchi, M., and Asakawa, K. 1990. Mobile robot control by a structured hierarchical neural network. *IEEE Control Systems Magazine,* April 1990, 69-76.

Nakano, K. 1972. Associatron - A model of associative memory. *IEEE Trans. on Systems, Man, and Cybernetics,* **2,** 381-388.

Nakano, K., Sakaguchi, Y., Isotani, R., and Ohmari, T. 1988. Self-organizing system obtaining communication ability. *Biological Cybernetics,* **58,** 417-425.

Narendra, K. S., and Parthasarathy, K. 1990. Identification and control of dynamical systems using neural networks. *IEEE Transactions on Neural Networks,* **1,** 4-27.

Nasrabadi N. M., and Feng, Y. 1988. Vector quantization of images based upon the Kohonen self-organizing feature maps. *Proceedings of the IEEE International Conference on Neural Networks (ICNN '88),* San Diego, California, pp. I-101 - I-108.

Nguyen, D. H., and Widrow, B. 1990. Neural networks for self-learning control systems. *IEEE Control Systems Magazine,* April 1990, 18-23.

Noble, B., and Daniel, J. W. 1977. *Applied Linear Algebra,* Prentice-Hall, Englewood Cliffs, New Jersey.

Nolfi, S., and Parisi, D. 1988. Learning to understand sentences in a connectionist network. *Proceedings of the IEEE International Conference on Neural Networks (ICNN '88),* San Diego, California, pp. II-215 - II-219.

Ogmen, H. 1991. A continuous-time global theory of visual perception. *Proceedings of the International Joint Conference on Neural Networks (IJCNN '91),* Seattle, Washington, pp. I-627 - I-632.

Oja, E. 1982. A simplified neuron model as a principal component analyzer. *Journal of Mathematical Biology,* **15**, 267-273.

Oja, E. 1983. *Subspace Methods of Pattern Recognition,* Research Studies Press and J. Wiley, Letchworth, England.

Oja, E. 1989. Nonlinear networks, principal components, and subspaces. *International Journal of Neural Systems,* **1**, 61-68.

Oja, E. 1992. Principal components, minor components, and linear neural networks. *Neural Networks,* in press.

Oja, E., Ogawa, H., and Wangviwattana, J. 1991. Learning in nonlinear constrained Hebbian networks. In *Artificial Neural Networks,* T. Kohonen, K. Makisara, O. Simula, J. Kangas, eds., pp. 385-390, Elsevier Publishers, North Holland.

Parker, D. B. 1987. Optimal algorithms for adaptive networks: Second order Back Propagation, second order direct Propagation, and second order Hebbian learning. *Proceedings of the IEEE First International Conference on Neural Networks,* San Diego, California, Vol. II, pp. 593-600.

Pawlicki, T. 1989. NORA: Neural-network object recognition architecture. In *Proceedings of the 1988 Connectionist Models Summer School,* D. Touretzky, G. Hinton, and T. Sejnowski, eds., pp. 444-451, Morgan Kaufmann, San Mateo, California.

Pawlicki, T. F., Lee, D. S., Hull, J. J., and Srihani, S. N. 1988. Neural network models and their application to handwritten digit recognition. *Proceedings of the IEEE International Conference on Neural Networks (ICNN '88),* San Diego, California, pp. I-63 - I-70.

Pearlmutter, B. A. 1989. Learning state space trajectories in recurrent neural networks. In *Proceedings of the 1988 Connectionist Models Summer School,* D. Touretzky, G. Hinton, and T. Sejnowski, eds., pp. 113-117, Morgan Kaufmann, San Mateo, California.

Personnaz, L., Guyon I., and Dreyfus, G. 1985. Information storage and retrieval in spin-glass like neural networks. *J. Physique Lett.,* **46,** L-359 - L-365.

Peterson, C., and Anderson, J. R. 1987. A mean field theory learning algorithm for neural networks. *Complex Systems,* **1,** 995-1019.

Pineda, F. J. 1987. Generalization of back-propagation to recurrent neural networks. *Physical Review Letters,* **59,** 2229-2232.

Pineda, F. J. 1988a. Generalization of back-propagation to recurrent and higher order neural networks. In *Neural Information Processing Systems,* D. Z. Anderson, ed., pp. 602-611, American Institute of Physics, New York.

Pineda, F. J. 1988b. Dynamics and architecture for neural computation. *Journal of Complexity,* **4,** 216-245.

Pineda, F. J. 1989. Recurrent backpropagation and the dynamical approach to adaptive neural computation. *Neural Computation,* **1,** 161-172.

Platt, J. C., and Hopfield, J. J. 1986. Analog decoding using neural networks. In *AIP Conference Proceedings,* J. S. Denker, ed., pp. 364-369, American Institute of Physics, New York.

Poggio, T. 1975. On optimal nonlinear associative recall. *Biological Cybernetics,* **19,** 201-209.

Pomerleau, D. A. 1989. ALVINN: An autonomous land vehicle in a neural network. In *Advances in Neural Information Processing Systems 1,* D. S. Touretzky, ed., pp. 305-313, Morgan Kaufmann, San Mateo, California.

Pratt, W. K. 1979. *Image Transmission Techniques,* Academic Press, New York.

Psaltis, D., and Park, C. H. 1986. Nonlinear discriminant functions and associative memories. In *AIP Conference Proceedings,* J. Denker, ed., Vol. 151, American Institute of Physics, New York.

Psaltis, D., Park, C. H., and Hong, J. 1988. Higher order associative memories and their optical implementations. *Neural Networks,* **1,** 143-163.

Psaltis, D., Sideris, A., and Yamamura, A. A. 1988. A multilayered neural network controller. *IEEE Control Systems Magazine,* April 1988, 17-21.

Puskorius, G. V., and Feldcamp, L. A. 1988. Decoupled extended Kalman filter training of feed-forward layered networks. *Proceedings of the International Joint Conference on Neural Networks (IJCNN '91),* Seattle, Washington, pp. I-771 - I-777.

Qian, N., and Sejnowski, T. J. 1988a. Predicting the secondary structure of globular proteins using neural network models. *Journal of Molecular Biology,* **202,** 865-884.

Qian, N., and Sejnowski, T. J. 1988b. Learning to solve random-dot stereograms of dense transparent surfaces with recurrent back-propagation. In *Proceedings of the 1988 Connectionist Models Summer School,* D. Touretzky, G. Hinton, and T. Sejnowski, eds., pp. 435-443, Morgan Kaufmann, San Mateo, California.

Ramanujam, J., and Sadayappan, P. 1988. Optimization by neural networks. *Proceedings of the IEEE International Conference on Neural Networks (ICNN '88),* San Diego, California, pp. I-325 - I-332.

Ramanujam, J., and Sadayappan, P. 1989. Parameter identification for constrained optimization using neural networks. In *Proceedings of the 1988 Connectionist Models Summer School,* D. S. Touretzky, G. Hinton, and T. Sejnowski, eds., pp. 154-161, Morgan Kaufmann, San Mateo, California.

Rao, C. R., and Mitra, S. K. 1971. *Generalized Inverse of Matrices and its Applications,* J. Wiley, New York, 1971.

Ricotti, L. P., Ragazzini, S., and Martinelli, G. 1988. Learning of word stress in a sub-optimal second order back-propagation network. *Proceedings of the IEEE International Conference on Neural Networks (ICNN '88),* San Diego, California, pp. I-355 - I-361.

Ritter, H. J., Martinez, T. M., and Schulten, K. J. 1989. Topology conserving maps for learning visuo-motor coordination. *Neural Networks,* **2,** 159-168.

Rodrigues, J. S., and Almeida, L. B. 1991. Improving the learning speed in topological maps of patterns. In *Neural Networks: Advances and Applications,* E. Gelenbe, ed., pp. 63-78, Elsevier, North-Holland.

Rogers, S. K., and Kabrisky, M. 1991. *An Introduction to Biological and Artificial Neural Networks for Pattern Recognition,* SPIE Optical Engineering Press, Bellingham.

Rohwer, R., and Forrest, B. 1987. Training time-dependence in neural networks. *Proceedings of the IEEE First International*

Conference on Neural Networks, San Diego, California, pp. II-701 - II-708.

Rosenblatt, F. 1962. *Principles of Neurodynamics: Perceptrons and Theory of Brain Mechanisms,* Spartan Books, Washington, DC.

Rossen, M. L., Tajchman, G. N., Bush, M. A., and Anderson, J. A. 1988. Training methods for a connectionist model of consonant-vowel syllable recognition. *Proceedings of the IEEE International Conference on Neural Networks (ICNN '88),* San Diego, California, pp. I-239 - I-246.

Ruck, D. W., Rogers, S. K., Kabrisky, M., Oxley, M. E., and Suter, B. W. 1990. The multilayer Perceptron as an approximation to Bayes optimal discriminant function. *IEEE Transactions on Neural Networks,* **1**, 296-298.

Rumelhart, D. E., Hinton, G. E., and Williams, R. J. 1986. Learning internal representations by error propagation. In *Parallel Distributed Processing,* D. E. Rumelhart and J. L. McClelland, eds., Vol. I, pp. 318-362. MIT Press, Cambridge, Massachusetts.

Rumelhart, D. E., McClelland, J. L., and the PDP Group 1986. *Parallel Distributed Processing,* MIT Press, Vol. I & II, Cambridge, Massachusetts.

Saito, K., and Nakano, R. 1988. Medical diagnostic expert system based on PDP model. *Proceedings of the IEEE International Conference on Neural Networks (ICNN '88),* San Diego, California.

Sanchez-Sinencio, E., and Lau, C., eds. 1992. *Artificial Neural Networks: Paradigms, Applications, and Hardware Implementations,* IEEE Press, New York.

Sanger, T. D. 1989a. Optimal unsupervised learning in a single-layer linear feed-forward neural network. *Neural Networks,* **2**, 459-473.

Sanger, T. D. 1989b. An optimality principle for unsupervised learning. In *Advances in Neural Information Processing Systems 1,* D. S. Touretzky, ed., pp. 11-19, Morgan Kaufmann, San Mateo, California.

Sanger, T. D. 1990. Analysis of the two-dimensional receptive fields learned by generalized Hebbian algorithm in response to random input. *Biological Cybernetics,* **63,** 221-228.

Scalero, R. S., and Tepedelenioglu, N. 1992. A fast new algorithm for training feedforward neural networks. *IEEE Transactions on Signal Processing,* **40,** 202-210.

Scalettar, R., and Zee, A. 1988. Emergence of grandmother memory in feed forward networks: Learning with noise and forgetfulness. In *Connectionist Models and Their Implementations: Readings from Cognitive Science,* D. Waltz and J. A. Feldman, eds., pp. 309-332, Albex, Norwood.

Segee, B. E., and Carter, M. J. 1991. Fault tolerance of pruned multilayer networks. *Proceedings of the International Joint Conference on Neural Networks (IJCNN '91),* Seattle, Washington, pp. II-444 - II-453.

Sejnowski, T. J. 1986. Higher-order Boltzmann machines. In *Neural Networks for Computing,* J. S. Denker, ed., pp. 398-403, American Institute of Physics, New York.

Sejnowski, T. J., and Rosenberg, C. R. 1987. Parallel networks that learn to pronounce English text. *Complex Systems,* **1,** 145-168.

Shawe-Taylor, J. S., and Cohen, D. A. 1990. Linear programming algorithm for neural networks. *Neural Networks,* **3,** 575-582.

Shepanski, J. F., and Macy, S. A. 1988. Teaching artificial neural systems to drive: Manual training techniques for autonomous systems. In *Neural Information Processing Systems,* D. Z. Anderson, pp. 693-700, American Institute of Physics, New York.

Shimada, T., and Nishimura, K. 1991. A new self-organizing method and its applications to handwritten digit recognition. *Proceedings of the International Joint Conference on Neural Networks (IJCNN '91),* Seattle, Washington, pp. I-275 - I-281.

Sietsma, J., and Dow, R. J. F. 1988. Neural net pruning - why and how. *Proceedings of the IEEE International Conference on Neural Networks (ICNN '88),* San Diego, California, pp. I-325 - I-333.

Sietsma, J., and Dow, R. J. F. 1991. Creating artificial neural networks that generalize. *Neural Networks,* **4,** 67-79.

Silva, F. M., and Almeida, L. B. 1990. Acceleration techniques for the Backpropagation algorithm. In *Proceedings of EURASIP Workshop on Neural Networks,* L. B. Almeida and C. J. Wellekens, eds., Springer-Verlag, Berlin.

Silverman, R. H., and Noetzel, A. S. 1990. Image processing and pattern recognition in ultrasonograms by backpropagation. *Neural Networks,* **3**, 593-603.

Simard, P. Y., Ottaway, M. B., and Ballard, D. H. 1989. Analysis of recurrent backpropagation. In *Proceedings of the 1988 Connectionist Models Summer School,* D. Touretzky, G. Hinton, and T. Sejnowski, eds., pp. 103-112, Morgan Kaufmann, San Mateo, California.

Simpson, P. K. 1990. *Artificial Neural Systems: Foundations, Paradigms, Applications, and Implementations,* Pergamon Press, New York.

Singhal, S., and Wu, L. 1989. Training multilayer Perceptrons with the Extended Kalman Filter. In *Advances in Neural Information Processing Systems 1,* D. S. Touretzky, ed., pp. 29-37, Morgan Kaufmann, San Mateo, California.

Skrzypek, J., and Hoffman, J. 1991. Visual recognition of script characters and neural network architectures. In *Neural Net-*

works: Advances and Applications, E. Gelenbe, ed., pp. 109-144, Elsevier, North-Holland.

Sobajic, D. J., Lu, J. J., and Pao, Y. H. 1988. Intelligent control of the Intelledex 6057 robot manipulator. *Proceedings of the IEEE International Conference on Neural Networks (ICNN '88),* San Diego, California, pp. II-633 - II-640.

Solla, S. A., Levin, E., and Fleisher, M. 1988. Accelerated learning in layered neural networks. *Complex Systems,* **2**, 625-640.

Tagliarini, G. A., and Page, E. W. 1988. A neural-network solution to the concentrator assignment problem. In *Neural Information Processing Systems,* D. Z. Anderson, ed., pp. 775-782, American Institute of Physics, New York.

Tank, D. W., and Hopfield, J. J. 1986. Simple "neural" optimization networks: An A/D converter, a signal detection circuit, and a linear programming circuit. *IEEE Transactions on Circuits and Systems,* **33**, 533-541.

Tenorio, M. F., Tom, M. D., and Schwartz, R. G. 1988. Adaptive networks as a model for human speech development. *Proceedings of the IEEE International Conference on Neural Networks (ICNN '88),* San Diego, California.

Terzopoulos, D. 1984. Multiresolution algorithms in computational vision. In *Image Understanding,* S. Ullman and W. Richards, eds., pp. 225-262, Albex, Norwood.

Tesauro, G., and Sejnowski, T. J. 1988. A 'neural' network that learns to play backgammon. In *Neural Information Processing Systems,* D. Z. Anderson, ed., pp. 794-803, American Institute of Physics, New York.

Tikhonov, A. N., and Arsenin, V. Y. 1977. *Solutions of ill-posed problems,* J. Wiley, New York.

Tolat, V. V., and Widrow, B. 1988. An adaptive "broom balancer" with visual inputs. *Proceedings of the IEEE International*

Conference on Neural Networks (ICNN '88), San Diego, California, pp. II-641 - II-647.

Tom, M. D., and Tenorio, M. F. 1991. Short utterance recognition using a network with minimum training. *Neural Networks,* **4,** 711-722.

Troudet, T., and Tabatabai, A. 1988. An adaptive neural net approach to the segmentation of mixed gray-level and binary pictures. *Proceedings of the IEEE International Conference on Neural Networks (ICNN '88),* San Diego, California, pp. I-585 - I-592.

Tulunay, E. 1991. Introduction to neural networks and their application to process control. In *Neural Networks: Advances and Applications,* E. Gelenbe, ed., pp. 241-273, Elsevier, North-Holland.

Tunley, H. 1991. Dynamic image segmentation and optic flow extraction. *Proceedings of the International Joint Conference on Neural Networks (IJCNN '91),* Seattle, Washington, pp. I-599 - I-604.

Van den Bout, D. E., and Miller, T. K. 1988. A traveling salesman objective function that works. *Proceedings of the IEEE International Conference on Neural Networks (ICNN '88),* San Diego, California, pp. II-299 - II-303.

Van den Bout, D. E., and Miller III, T. K. 1989. Improving the performance of the Hopfield-Tank neural network through normalization and annealing. *Biological Cybernetics,* **62,** 129-139.

Vogl, T. P., Mangis, J. K., Rigler, A. K., Zink, W. T., and Alton, D. L. 1988. Accelerating the convergence of the Back-Propagation method. *Biological Cybernetics,* **59,** 257-263.

Waibel, A. 1989. Consonant recognition by modular construction of large phonemic time-delay neural networks. In *Advances in Neural Information Processing Systems 1,* D. S. Touretzky, ed. pp. 215-223, Morgan Kaufmann, San Mateo, California.

Wasserman, P. D. 1989. *Neural Computing: Theory and Practice,* Van Nostrand Reinhold, New York.

Watrous, R. L. 1987. Learning algorithms for connectionist networks: Applied gradient methods of nonlinear optimization. *Proceedings of the IEEE First International Conference on Neural Networks,* San Diego, California, pp. II-619 - II-627.

Wechsler, H., and Zimmerman, G. L. 1987. Fault-tolerant recognition using DAM's. *Proceedings of the IEEE First International Conference on Neural Networks,* San Diego, California, pp. II-719 - II-726.

Werbos, P. 1974. Beyond regression: New tools for prediction and analysis in behavioral sciences. *Ph.D. dissertation,* Harvard University, Cambridge, Massachusetts.

White, H. 1988. Economic prediction using neural networks: The case of IBM daily stock returns. *Proceedings of the IEEE International Conference on Neural Networks (ICNN '88),* San Diego, California, pp. II-451 - II-459.

Widrow, B. 1962. Generalization and information storage in networks of Adaline 'neurons'. In *Self-Organizing Systems,* M. Yovitz, G. Jacobi, and G. Goldstein, eds., pp. 435-461, Spartan Books, Washington, DC.

Widrow, B. 1987. The original adaptive neural net broom-balancer. *Proceedings of International Symposium on Circuits and Systems,* Philadelphia, pp. 351-357.

Widrow, B., Glover, J., Kaunitz, J., Williams, C., and Hearn, R. 1975. Adaptive noise canceling: Principles and Applications. *Proceedings of the IEEE,* **63,** 1692-1716.

Widrow B., and Lehr, M. A. 1990. 30 years of adaptive neural networks: Perceptron, Madaline, and Backpropagation. *Proceedings of the IEEE,* **78,** 1415-1442.

Widrow, B., and Smith, F. W. 1964. Pattern-recognizing control systems. In *Computer and Information Sciences,* J. T. Tou, R. H. Wilcox, eds., Spartan Books, Washington DC.

Widrow, B., and Stearns, S. D. 1985. *Adaptive Signal Processing,* Prentice-Hall, Englewood Cliffs, New Jersey.

Widrow, B., and Winter, R. G. 1988. Neural nets for adaptive filtering and adaptive pattern recognition. *IEEE Computer,* March 1988, 25-39.

Widrow, B., Winter, R. G., and Baxter, R. A. 1988. Layered neural nets for pattern recognition. *IEEE Transactions on Acoustics, Speech, and Signal Processing,* **36,** 1109-1118.

Wieland, A. P. 1991. Evolving neural network controllers for unstable systems. *Proceedings of the International Joint Conference on Neural Networks (IJCNN '91),* Seattle, Washington, pp. II-667 - II-673.

Wilson, G. V., and Pawley, G. S. 1988. On the stability of the traveling salesman problem algorithm of Hopfield and Tank. *Biological Cybernetics,* **58,** 63-70.

Xu, X., and Tsai, W. T. 1991. Effective neural algorithms for the traveling salesman problem. *Neural Networks,* **4,** 193-205.

Yao, Y., Freeman, W. J., Burke, B., and Yang, Q. 1991. Pattern recognition by a distributed neural network: An industrial application. *Neural Networks,* **4,** 103-121.

Yu, P., Anastassopoulos, V., and Venetsanopoulos, A. N. 1992. Pattern classification and recognition based on morphology and neural networks. *Canadian Journal of Electrical and Computer Engineering,* **17,** 58-64.

Yuille, A. L., Kammen, D. M., and Cohen, D. S. 1989. Quadrature and the development of orientation selective cortical cells by Hebb rules. *Biological Cybernetics,* **61,** 183-194.

Zemel, R. S., Mozer, M. C., and Hinton, G. E. 1989. TRAFFIC: A model of object recognition based on transformations of feature instances. In *Proceedings of the 1988 Connectionist Models Summer School,* D. Touretzky, G. Hinton, and T. Sejnowski, eds., pp. 452-461, Morgan Kaufmann, San Mateo, California.

Zemel, R. S., Mozer, M. C., and Hinton, G. E. 1990. TRAFFIC: Recognizing objects using hierarchical reference frame transformations. In *Advances in Neural Information Processing Systems 2,* D. S. Touretzky, ed., pp. 266-273, Morgan Kaufmann, San Mateo, California.

Zhou, Y. T., Chellappa, R. 1988. Computation of optical flow using a neural network. *Proceedings of the IEEE International Conference on Neural Networks (ICNN '88),* San Diego, California, pp. II-71 - II-78.

Zhou, Y. T., Chellappa, R., Vaid, A., and Jenkins, B. K. 1988. Image restoration using a neural network. *IEEE Transactions on Acoustics, Speech, and Signal Processing,* **36**, 1141-1151.

Subject Index

2/3 Rule, 75

α-helix, 355, 357

β-sheet, 355, 357

Σ S Exchange transformation, 49, 50, 52

A/D converter, 319

Abnormal illness behavior, 353

Absolute error, 147

Absolute error criterion, 146, 147

Absolute temperature, 59

Active hidden unit, 201, 205, 207, 210, 212, 213, 214, 215

Actual probability, 60, 82

Actual response, 26, 229

Acute myocardial infraction, 353

Adaline, 3, 11

Adaptation cycle, 27, 31, 36, 98, 107-109, 111, 112, 116, 117, 120, 122, 123, 124, 125-131, 145-147, 150, 159, 160, 163, 165, 166, 168, 171-174, 177-185, 189 199, 200, 201-208, 209-213, 310, 333, 344

Adaptation gain, 69

Adaptation rule, 88, 99, 142

Adaptation stage, 111

Adaptive algorithm, 264

Adaptive filtering, 30

Adaptive learning, 142

Adaptive parameter, 68

Adaptive pattern recognition, 75

Adaptive process, 68

Adaptive resonance theory (ART), 6, 11, 73, 75

Adaptive signal processing, 3

AdaTron, 142

Additive noise, 327

ALADIN, 7, 195, 198, 203-205, 208, 211, 212

Amino-acid, 355-356

Amplifier, 53

Analog input, 332

Analog output, 14, 24, 28-31, 89, 90, 95, 98, 99, 102, 107, 118, 132, 146, 152, 163-166, 176-178, 220, 221, 247-249, 332, 333

Applications of neural networks, 87, 301
Architectural modification, 211
Architecture determination, 195, 215
Arithmetic operation, 36, 101, 112, 122, 126, 130, 131, 176, 183, 184, 310
ART 1, 6, 73-75
ART 2, 75
ART 3, 75
Articulatory feature, 347
Articulatory representation, 347
Artificial intelligence (AI), 4
Artificial neural network, 2, 3
Artificial neuron, 2, 12-14, 16-18
Associated data, 265, 266
Associated pattern, 220, 256, 260, 264, 270, 277, 280
Association, 26, 29, 35, 90-93, 95, 96, 98, 100, 104, 105, 107, 108, 111, 112, 113, 114, 120, 121, 128, 133, 134, 138, 149, 154, 155, 157, 159, 163, 168-170, 172-176, 178, 192, 199, 203-206, 263, 269, 278, 279, 281, 286, 287, 289, 292, 295, 296
Associative learning, 280
Associative recall, 280, 346
Asymptotic convergence, 238, 239, 242
Asymptotic equilibrium, 13, 46
Attentional subsystem, 74
Attribute, 280
Auto-association, 126, 176, 333
Auto-regressive model, 361
Auto-associative update error, 92, 93
Automated system, 351
Autonomous system, 73, 369
Auxiliary network, 63, 65
Auxiliary waste pattern, 306
Averaging layer, 344
Back propagation learning, 88
Backward connection, 346
Backward path, 346
Backward signal, 346
Barometric pressure, 358
Bayes optimal discriminant function, 27
Bayes theory, 351
Bayesian decision theory, 73
Benchmark, 130
Bias term, 165, 197, 327, 330
Binary output, 14, 27, 31, 42, 89, 98, 99, 102, 104, 118, 119, 128, 132, 146-148, 152, 153, 162, 164-167, 171, 178-181, 183-185, 199, 220, 221, 247-249
Binary representation, 319

Binary signal, 332

Binocular depth perception, 368

Biochemical reactions, 355

Biological neuron, 13

Boltzmann's constant, 59

Boltzmann machine, 6, 11, 58, 59-62, 82, 143

Boltzmann-Gibbs distribution, 59, 60, 325

Bottom-up pattern, 75

Brain-state-in-a-box model, 11, 49, 51

C-cell, 346

Capacitor, 53

Capacity, 57, 219, 227, 229, 234, 236, 240-242, 244, 246, 249, 260, 287, 290-293, 294, 296

Capacity bound, 236-240, 242, 294, 296, 297

Capacity ratio, 236-242, 244-246, 289-291, 294-296

Cell membrane, 13

Center cell, 70

Central processing unit (CPU), 2

Character recognition, 8, 317 318, 339, 340, 342

Character recognition system, 317

Characteristic number, 308, 309, 313

Check verification system, 317

Chemical pollutant, 304, 305, 306, 307, 310, 312, 313, 314

Classification application, 349

Classification of sonar signals, 8, 318, 350

Classification problem, 349, 350

Classification system, 350

Classification task, 214, 215

Codebook design, 71

Codebook vector, 72

Coding scheme, 303

Cognitron, 344

Coil, 355, 357

Collision-free movement, 369

Column space, 91, 133

Column vector, 28, 81, 93, 132, 221, 265, 267, 269, 282

Competitive dynamical system, 48

Competitive learning, 67

Complex cell, 346

Composite key pattern, 7, 279, 280, 282, 284, 290, 292, 294

Compressed signal, 126, 335, 336

Compression rate, 333, 336

Computation of motion, 368

Computation of optical flow, 368

Computational burden, 91, 119, 123

Computational complexity, 6, 89, 145

Computational effort, 62,

Computational requirement, 36, 88, 112, 122, 126, 128, 264

Computer simulation, 71

Computer vision, 317, 326, 368

Computing devices, 56

Concentrator assignment, 319

Concurrent processor, 319

Conditional probability, 83

Confidence interval, 202, 205, 207

Conjunctive interaction, 62

Constraint, 325, 326

Constrained minimization, 325, 326, 328

Constrained search, 197

Content addressable memory (CAM), 11, 47, 56, 57

Continuous function, 26, 28, 41, 99, 145, 152, 195, 247

Control of robot arm, 71

Conventional inverse, 92

Convergence, 16, 30, 35, 88, 89, 99, 131, 132, 142, 143, 145, 164, 166, 179

Convergence properties, 89, 112

Convergence rate, 36, 126, 127, 142, 146

Convex function, 26, 145

Convolution mask, 346

Convolutional method, 342

Cooperative dynamical system, 48

Correlated patterns, 224, 230

Correlation, 224, 230, 231, 242, 271, 273, 275, 276, 286, 287

Correlation matrix, 271, 284, 286

Correlational learning, 29, 56

Cost function, 36, 37

Covariance matrix, 40, 331, 334, 336

Critical success index, 359

Deactivation acceptance test, 208

Decay of activity, 13, 15, 53, 327

Decay term, 15, 16, 39, 42

Decision boundaries, 72

Decision making, 8, 303, 304, 305, 307, 315

Decision theory, 349

Decoupled extended Kalman filter, 89

Decreasing function, 230

Delta Rule, 9, 29, 30, 76, 110, 112, 115, 116, 142, 168, 171, 247, 339

Density function, 332

Desired output, 26, 247

Desired probability, 60, 82

Diagnosis of heart attack, 318, 352, 353

Diagnosis of low back pain, 318, 352, 353

Diagnostic accuracy, 353

Diagonal elements, 88, 224, 225, 275, 283, 320

Diagonal matrix, 226

Differentiable everywhere function, 26, 28, 41, 51, 99, 145, 152, 200, 247

Differential equation, 13, 18-20, 23, 47, 52, 63, 65

Digit recognition, 317, 318, 340

Digital computer, 2, 56

Direct inversion, 326

Direction cosine, 57, 148, 229, 255, 271, 285

Discrepancy measure, 26, 60, 82, 216

Discrete cosine transform, 331

Distinct path, 323

Dominant eigenvalue, 224-227, 231, 232, 251, 273, 274, 277, 286, 287

Dow Jones (DJ) index, 363

Dynamic behavior, 13, 18, 19, 49, 51, 63, 65

Dynamical equation, 50, 51, 85

Dynamical model, 13

Dynamical system, 47, 65

Echo cancelation, 3

Ecosystem, 305, 306, 307, 310

Efficient adaptation, 119, 174

Efficient learning, 318

Efficient learning algorithm, 87, 89, 109, 122, 131, 132, 141, 182, 299, 303

Efficient market hypothesis, 318, 361, 363

ELEANNE, 6, 143

ELEANNE 1, 90, 93, 94, 132

ELEANNE 2, 94, 97, 98, 102, 103, 132, 134, 249

ELEANNE 3, 104-107, 111, 112, 115, 132, 135, 248

ELEANNE 4, 108, 109, 115, 132, 248

ELEANNE 5, 111, 113, 114, 118, 122, 128-132, 138, 183, 184, 310

ELEANNE 6, 115, 117, 118, 120, 122-125, 128-132, 184, 185, 310

ELEANNE 7, 118, 120, 121-132, 185

Electric field, 359

Electrical circuit, 53

Electrical hardware, 55

Electrocardiogram (ECG), 352
Elementary function, 337, 338
Energy function, 39, 43
Environmental pollution, 304
Environmental problem, 304, 305
Environmental protection, 300, 303, 304
Equilibrium, 13, 14, 16, 18-20, 23, 24, 40, 43, 49, 51, 62, 63, 65, 85, 338
Equilibrium state, 13, 16, 23, 65, 86, 321, 329
Equilization, 3
Error back propagation, 3, 4, 11, 33-37, 62, 78, 88, 89, 110, 115, 118, 122-131, 142-144, 162, 164, 166, 176-182, 196, 198, 205, 208, 332, 334, 341, 344, 353, 357, 359, 362
Error curve, 185
Error function, 37, 142
Error rate, 344, 345
Estimate of the gradient, 104
Estimation error, 145
Euclidean norm, 39, 42, 197
Euclidean distance, 68, 69, 321
Exclusive OR, 35, 37, 128, 129, 179
Excitatory cell, 345, 346
Excitatory connection, 53, 345
Excitatory signal, 53, 345
Expected output, 27, 32, 33, 65, 98, 110, 143-147, 152, 198
Expected response, 151
Expert system, 303, 304, 305, 352, 359, 360, 369
Exponential function, 69
Extended Kalman algorithm, 89
Extensive search, 321
External bias, 64
External input, 321
External teacher, 26, 32, 37, 67, 332, 336, 337
False alarm rate, 359
Fast ALADIN, 198, 205-207, 209-214
Fast algorithm, 183
Fast back propagation, 6, 162-164, 166, 172, 176, 177, 178-183, 188, 191, 207
Fast delta rule, 6, 153, 154, 185
Fast ELEANNE 3, 155, 157, 158, 167, 168, 187
Fast ELEANNE 4, 158, 160, 161, 167, 171, 172
Fast ELEANNE 5, 167, 169, 170, 183, 192
Fast ELEANNE 6, 171-174, 184
Fast ELEANNE 7, 174, 175, 185
Fast learning, 318

Fast learning algorithm, 141, 144, 145, 151, 161, 176, 185, 199, 299, 303

Feature detector, 334, 339

Feature extraction, 345

Feature map, 74, 342

Feature vector, 349

Feed-back neural model, 5, 11

Feed-back neural network, 5, 9, 11, 47, 51, 62

Feed-back signal, 338

Feed-back system, 46

Feed-back term, 320, 327

Feed-forward neural network, 6, 7, 9, 11, 12, 26, 27, 35-38, 42, 59, 62, 64, 88, 89, 118, 120, 128, 143-145, 147, 153, 165, 174, 179, 186, 195-199, 201, 207, 209, 212, 214-215, 310, 332, 335, 336, 341, 355, 356, 359, 361, 362

Feed-forward signal, 338

FFT, 335

Filter bank, 335

Financial market, 363

Financial prediction, 8, 360, 364

Financial transaction, 339

Finite approximation, 227-230, 234, 235, 240, 251-254, 256

Finite problem, 321

Finite series, 230

Finite-element method, 88

First-order derivative, 365

First-order estimate, 266

First-order learning algorithm, 88

First-order neural network, 9, 20, 223, 236, 237, 242, 259, 260, 264, 271, 274, 276-279, 290, 294, 296-298

Fixed point, 67

Flexible algorithm, 91

Fobenius approximation method, 196

Fobenius norm, 196

Forgetting term, 39

Formulation of the problem, 300

Forward connection, 346

Forward path, 346

Forward signal, 346

Free-running system, 61, 84

Fuzzy logic, 318, 354

Gabor transform, 337

Gain control module, 75

Gaussian distribution, 202

Gaussian random variable, 235, 236, 239-241

General Methodology, 300, 302

Generalization, 2, 35, 37, 195, 219

Generalization ability, 7, 35, 143, 165, 195, 214, 215, 219, 224, 227, 228, 231, 232, 234, 243, 244, 249, 276, 278, 286, 348-351, 357, 360

Generalization properties, 318

Generalized characteristic number, 309, 313, 314, 315

Generalized criterion, 7, 36, 146, 147, 151, 162, 165, 186

Generalized criterion for training, 6, 198

Generalized error criterion, 147

Generalized inverse, 29, 91, 92, 97, 133, 222, 265

Generalized neural network predictor (GNNP), 366, 367

Generalized objective function, 145, 147, 166, 178, 199

Generalized rule, 70

Generalized training criterion, 144

Generalized update equation, 70

Global feature, 346

Global minimum, 334

Global ordering, 69

Gradient, 38, 42, 77, 100, 101, 142, 155, 156

Gradient approximation, 102

Gradient descent, 35, 36, 38, 41, 42, 62, 63, 66, 88, 142

Gradient descent method, 6, 38, 41, 42, 61, 63, 66, 76, 84, 88, 99, 113, 116, 139, 153, 168, 185, 192, 334, 338

Gradient direction, 143

Gradient reuse algorithm, 142

Gramian, 223

Graph mapping, 319

Graph partitioning problem, 319

Gray level, 329, 336, 341

Greville formula, 91

Hadamard product, 283, 284

Hadamard transform, 331

Hamming distance, 146, 147, 249, 308

Handwritten character, 344

Handwritten digits, 341, 348

Handwritten signature, 339

Handwritten zip code, 341

Hard-limiter, 99, 152, 247

Hard-limiting nonlinearity, 14, 23, 55, 98, 99, 152, 200, 247, 250, 261, 327, 328

Heart attack, 352

Hebbian learning, 11, 39

Hebbian learning rule, 38, 42

Hebbian rule, 5, 38, 39, 41, 79, 81, 334

Hebbian term, 40, 42

Hessian matrix, 88, 99, 100, 102-104, 106, 107, 112, 120, 127, 128, 130, 135-138, 155, 156, 158, 187-189

Hetero-associative update error, 92

Heuristic technique, 142

Heuristics, 196, 198, 321

Hidden layer, 23, 24, 32, 35, 58, 110, 126, 195, 196, 200, 205, 207, 208, 261, 333, 335, 336, 342, 348, 356, 358

Hidden state, 83

Hidden unit, 11, 12, 22-25, 32, 37, 58-60, 63, 64, 78, 82, 109-111, 115, 118-120, 123, 126, 128-131, 161, 162, 165, 166, 169, 171, 172, 174, 176, 179, 191, 195-198, 199-204, 205-213, 214, 215, 216, 217, 261, 269, 310, 332, 333, 334, 341, 344, 349, 351, 353, 355, 357, 358, 360, 363

High activity, 13

High-order Boltzmann machine, 62

High-order neural network, 7, 9, 20, 259, 260, 263, 264, 270, 285, 293, 297

Homologous protein, 355

Homologous structure, 355

Homology, 358

Hopfield model, 5, 47, 52-54, 57, 58, 327, 328

Hopfield network, 47, 223, 319, 320, 328, 329

Hotelling transform, 331

Human expert, 352

Human function, 151

Human like operation, 87

Human speech development, 349

Hybrid implementation, 301

Hybrid scheme, 112

Hyperbolic tangent, 15, 99, 153, 200, 247, 332

Hypercube, 55

Identification problem, 89

Identity matrix, 96, 223, 270, 273, 283

Ill-posed problem, 318, 326

Image compression, 8, 71, 118, 126, 330, 332-334, 336, 337

Image formation process, 327

Image interpolation, 326

Image processing, 317, 326, 368

Image recognition, 368

Image recognition system, 87, 141

Image representation, 329, 337, 339

Image restoration, 326, 327

Image segmentation, 369

Implicit representation, 339

Independent random variable, 288, 289, 292
Infinite expansion, 232
Infinite norm, 227, 275
Infinite series, 225-227, 274
Information retrieval, 2
Inhibitory cell, 345
Inhibitory connection, 48, 53, 345
Inhibitory signal, 75, 345
Initial estimate, 34, 138
Initial state, 46, 55, 56, 57, 321, 323
Inner product, 68, 222
Input characteristics, 349
Input layer, 12, 22, 24, 57, 346
Input pattern, 12, 74, 90, 93, 106, 110, 112, 133, 134, 161, 167, 168, 233, 234, 243-246, 249, 261, 263, 264, 269, 276-278, 281, 285, 286, 288, 292, 295, 346
Input signal, 13, 17, 67
Input unit, 64, 195, 203, 208
Input vector, 26, 29, 35, 38, 43, 69, 73, 215, 220, 336
Instability, 96
Interference term, 229, 235
Internal representation, 32, 36, 37, 332-334, 335
Intrinsic function, 47
Inverse matrix, 86
Inverse problem, 326
Inverse transform matrix, 332
Iterative method, 337
Job sequencing, 319
Job-shop sequencing problem, 319
Kanerva's associative memory, 57
Karhunen-Loeve transform, 45, 330-332, 335
Key data, 265, 266, 279
Key pattern, 220, 223-225, 227, 229, 230, 234, 249, 250, 256, 260, 263, 264, 267, 269-271, 273-281, 284, 286, 287, 290, 292, 294, 296
Key-word, 56
Khatri-Rao product, 267
Knowledge-based methods, 356, 358
Kohonen feature map, 6, 11
Kronecker delta, 44, 51, 77, 293
Kronecker product, 261, 262, 266, 269, 279-282, 285, 287, 291, 293
L_p norm, 335
Language understanding, 369
Large-scale problem, 35-37
Large-scale system, 317, 318
Lateral interaction, 68

Learning, 9

Learning algorithm, 27, 29, 30, 32, 36, 37, 62, 83, 87-90, 94, 99, 102, 104, 111, 126, 127, 131, 132, 142, 143, 145, 146, 150, 176, 185, 200, 247, 260, 263, 264

Learning behavior, 349

Learning function, 41, 42, 46

Learning matrix, 3

Learning phase, 12, 37

Learning process, 62

Learning rate, 30, 33-35, 66, 100, 128, 130, 138, 142, 143, 155, 177-183, 349

Learning rule, 12, 38, 41, 42, 44, 45

Learning scheme, 9, 29, 56, 75, 76, 142, 219, 235, 246, 249, 264, 270, 298

Learning step, 69, 71

Learning speed, 197

Learning strategy, 151

Learning task, 66

Learning vector quantization (LVQ), 6, 11, 71

Learning vector quantization 1 (LVQ1), 72, 73

Learning vector quantization 2 (LVQ2), 73

Learning vector quantization 3 (LVQ3), 73

Least disturbance principle, 307

Least-squares formulation, 94

Least-squares method, 362

Least-squares sense, 28, 29, 91, 95, 221, 265, 267, 269, 282

Least-squares solution, 97

Least-squares technique, 28, 90, 102, 103, 221

Legitimate path, 324

Legitimate tour, 323

Lexicographic ordering, 328

Lighting strike, 359

Lighting strike prediction, 359

Limestone, 214

Linear combination, 93, 132, 134

Linear discriminant function, 18, 259

Linear equation, 337, 338

Linear function, 13, 18, 51, 52, 69

Linear mapping, 221, 265

Linear model, 327

Linear neuron, 14

Linear programming, 36, 319

Linear space, 91

Linear term, 266

Linear transformation, 341

Linear unit, 42, 44, 79, 81

Linearly dependent, 250

Linearly independent, 81, 223
Linearly nonseparable problem, 35
Linearly separable, 18, 339
Linguistic problems, 369
Lithology, 214
Local coding, 356
Local feature, 345
Local minimum, 30, 35, 47, 55-57, 66, 150, 197, 198, 334, 336
Logarithmic error scale, 180
Logarithmic scale, 210
Logistic function, 99
Long-term memory (LTM), 74
Lookup table, 347
Low activity, 13, 14
Low-rank matrix approximation, 333
Lower bound, 195
Lower network, 32, 111, 112, 113, 116, 118, 120, 138, 139, 161, 167-169, 171, 172, 176, 192
Lyapunov function, 46, 48, 52, 53, 55, 319, 322, 327, 329
Massively parallel architecture, 3
Matrix equation, 29, 282
Matrix inverse, 174, 222
Matrix inversion, 63, 91, 319
Matrix inversion lemma, 95, 96, 103, 106, 109, 156, 160
Matrix norm, 226
Matrix of synaptic weights, 24, 28, 29, 46, 81, 90, 91-93, 95-97, 100, 103, 106, 107, 109, 111, 115, 116, 118, 132-135, 153, 158, 160, 168, 169, 172, 221, 222, 228-236, 239, 240, 248, 256, 265, 271, 284, 288, 320
Matrix product, 261
Maximum a posteriori (MAP) estimation, 327
McCulloch-Pitts model, 11, 49, 52
Mean field annealing process, 62
Mean-square error, 331
Measure of relevance, 198
Measure of similarity, 68, 229, 255
Medical diagnosis, 8, 351
Memory state, 55, 56
Minimal network architecture, 210
Minimization problem, 337
Minimum distortion, 330
Minimum norm generalized inverse, 133
Mobile robot control, 369
Monotone function, 41, 47

Multi-layered neural network, 5-7, 11, 12, 21-24, 32, 34, 35, 59, 88, 89, 109-111, 115, 118-120, 122, 123, 126, 130, 132, 142, 144, 161, 162, 167, 169, 171, 176, 180-185, 192, 195, 196, 198, 207, 209, 217, 310, 333-335, 342-344, 346, 348, 350, 353, 355, 358, 361

n to *n* encoder, 35, 63, 130, 131, 180-185, 210, 332

Necessary condition, 304

Neocognitron, 345-347

Neumann-type expansion, 225

Neural impulse, 13

Neural network application, 141, 317, 318

Neural network architecture, 9, 11, 18, 19, 75, 76, 299, 300, 301, 303, 318

Neural network model, 47

Neural network predictor (NNP), 365, 367

Neural network system, 141, 318

Neural network with composite key patterns, 260, 279-282, 285-287, 290, 291, 294-296, 298

Neuron activity, 13, 15

NETtalk, 318, 347, 348, 355

Network architecture, 216

Network complexity, 197

Network construction algorithm, 196

Network performance, 247, 349

New York stock exchange (NYSE), 364-367

Newton method, 88

Newton-Raphson algorithm, 99

Noise cancelation, 3

Noise component, 235, 241

Noise level, 244

Noise term, 229, 230, 235, 236, 239, 288-290, 292, 294

Nondecreasing function, 13, 15

Nonlinear discriminant, 346

Nonlinear discriminant function, 18, 259

Nonlinear dynamic system, 89

Nonlinear dynamics, 303

Nonlinear function, 13, 16, 265

Nonlinear mapping, 265, 266

Nonlinear operator, 26, 220

Nonlinear processing element, 11

Nonlinear structure, 360

Nonlinear units, 11, 41, 356

Nonsingular matrix, 82, 96, 102, 103, 222

Normal equation, 90, 95

Normal threshold, 305, 307, 312

Normalized Hamming distance, 231, 233, 234, 243-245, 276-278, 286, 295

Normalized Hebbian rule, 39

Normalized mean-square error, 331

Novelty, 93

Novelty filter, 93, 94

NP-complete problem, 55, 318, 319, 321

Numerical integration technique, 56

Object recognition, 71, 368

Observed image, 327

Objective function, 30, 33, 47, 55, 78, 88, 95, 98, 101, 102, 106, 110, 113, 115, 116, 120, 138, 139, 144-148, 152, 155, 158, 162, 166, 169, 171, 172, 177, 185, 188, 192, 197, 198, 205, 208, 247, 320-323, 325, 329

Oja's learning rule, 39, 81

Oja's rule, 40, 43, 46, 79, 336

Optical hardware, 55

Optical implementation, 270, 297

Optimal associative recall, 221, 264

Optimal brain damage, 196, 344

Optimal estimate, 28, 29, 95, 97, 266

Optimal learning algorithm, 142

Optimal least-squares estimate, 97, 103, 104

Optimal least-squares method, 220

Optimal least-squares technique, 220, 249

Optimal least-squares training, 7, 29, 93, 131, 220, 223, 260, 264-266, 270, 271, 275, 297

Optimal matrix of synaptic weights, 91, 92, 98, 223, 226-234, 241, 242, 267, 270, 273-276, 278, 282, 284, 285

Optimal nonlinear association, 266

Optimal solution, 28, 29, 90, 91, 95, 221, 223, 265, 266, 267, 270, 282, 333

Optimal training, 92

Optimally trained neural network, 229, 231, 233, 234, 242-247, 249

Optimization, 8, 47, 318

Optimization application, 56

Optimization problem, 56, 71, 102, 103, 143, 144, 317, 319

Optimization strategy, 88, 102, 103, 132, 142, 143, 145, 155, 248

Optimization technique, 56

Ordered map, 71

Ordering of a feature map, 71
Ordering process, 69, 71
Orientation selective unit, 335
Orienting subsystem, 74, 75
Original object, 327
Orthogonal projection, 334
Orthogonal set, 337
Orthogonal transform, 331, 337
Orthogonalization effect, 229, 230, 275, 276
Orthonormal vectors, 44, 46
Oscillation, 178, 181, 182
Oscillatory behavior, 202
Outer-product rule, 7, 29, 57, 98, 141, 219, 220, 223, 224, 226, 231, 233, 236-238, 240-242, 246, 247, 249, 250, 260, 264, 270-278, 287, 288, 290, 292, 294, 296-298
Output activity, 20, 22
Output error, 163, 167
Output layer, 12, 23, 24, 57, 261, 335
Output nonlinearity, 41, 46, 165, 247
Output pattern, 12, 90, 144, 199, 215
Output unit, 24, 31, 37, 64, 90, 105, 108, 114, 119, 128, 130, 157, 159, 170, 179, 195, 203, 210, 212, 214, 333, 353, 354
Output vector, 26, 215, 220
p-Holder norm, 335
Parity problem, 35
Partial derivative, 216, 217
Partition function, 60
Partitioned matrix, 91, 95
Pattern classification, 4, 317, 368
Pattern recognition, 2, 317, 346, 368
Pattern recognition application, 224
Pattern recognition problem, 63
Pattern recognition system, 11
Pattern segmentation, 346
Penalizing error, 151
Perceptron, 3, 4, 11, 35
Perceptron learning rule, 37
Perfect association, 133
Perfect classification, 351
Perfect recall, 231, 234, 244, 246, 247, 277, 286, 295
Performance, 282
Performance characteristics, 301
Performance criterion, 197
Performance evaluation, 219, 220, 261, 303
Performance of neural networks, 29, 219, 341, 342
Permutation matrix, 321, 322

Point spread matrix, 327

Pollution pattern, 305, 306, 307, 312

Pollution source, 300, 304, 305, 306, 307, 308, 309, 310, 311, 312, 313, 314

Positive definite function, 26, 47, 145

Positive definite matrix, 81, 96, 102, 103

Positive semi-definite matrix, 96, 103, 104, 137, 138, 156

Prediction application, 360

Prediction model, 362

Prediction problem, 357

Prediction system, 360

Preprocessing of the input data, 340

Principal component, 40, 44, 45, 126, 330, 331, 335, 337

Principal component analysis, 38, 44-46, 81, 333

Principal eigenvector, 334

Probability of detection, 359

Probability of error, 349

Probability of perfect recall, 237, 238

Process control, 68

Pruning approach, 196

Quadratic error criterion, 7, 27, 142, 143, 144, 147, 148, 165, 353, 354

Quadratic error, 26, 27, 34

Quadratic error function, 33, 65, 66, 166, 221, 334, 335, 353

Quadratic function, 19, 166, 326-328

Quadratic functional, 327, 329

Quadratic term, 266

Quantitative correlation measure, 224, 273

Quantitative distortion measure, 330

Quantization scheme, 333

Random-dot stereogram, 66, 368

Random input, 334

Random noise, 361

Random number generator, 34, 126, 128, 130, 176, 179, 182

Random variable, 236, 239

Random vector, 57

Rate of convergence, 31, 130, 142

Rayleigh quotient, 226

Real-valued signal, 332

Real-world application, 141, 317, 318

Receptive field, 345

Recognition level, 346

Recognition of waveforms, 369

Reconstracted image, 330

Rectangular matrix, 29

Recurrent back propagation, 6, 11, 62, 65, 66, 84
Recurrent network, 63, 66
Recurrent neural network, 62-64, 66, 259
Recursion, 103
Recursive algorithm, 91, 92, 104, 136
Recursive evaluation, 91-97, 106, 119, 174
Recursive inversion, 127, 128, 131
Recursive least-squares algorithm, 6, 29, 94
Recursive least-squares learning algorithm, 89, 90, 131, 132
Redundant hidden unit, 202
Regular grammar, 66
Regularization theory, 326
Regularized solution, 326, 327
Regularizing parameter, 327
Relative effect, 303, 307, 308, 313, 314
Relative entropy, 61, 143, 166
Relative entropy criterion, 27, 34, 143, 165, 166, 353
Relative sensitivity measure, 201
Relaxation, 143
Reset wave, 74, 75
Resistor, 52
Restored image, 329
Rewarding success, 151
Robot arms, 369
Robot manipulator, 369
Robot manipulator control, 369
Robotics, 67, 317, 368
Robotics application, 71
Robust principal component analysis, 46
Robustness to noise, 219
Root pain (ROOTP), 353
Rule-based expert system, 352, 358
S-cell, 346
Saddle points, 333
Sample mean, 202, 208
Sample standard deviation, 202, 203, 208
Sandstone, 214
Sanger's learning rule, 45, 46
Sanger's rule, 44, 45, 336, 337
Saturation level, 165
Saturation limit, 51
Scaling problem, 56
Scaling properties, 318
Second-order algorithm, 6, 153
Second-order Boltzmann machine, 62
Second-order derivative, 88, 197, 364

Second-order learning algorithm, 88, 89, 110, 248
Second-order method, 143
Second-order neural network, 20, 261, 263, 264, 279, 286, 297
Second-order partial derivative, 99, 120, 188
Second-order recurrent network, 66
Secondary structures of proteins, 8, 318, 352, 355-357
Selective attention, 346
Selective attention model, 345, 346
Self-organization, 2, 11, 67, 345
Self-organization principle, 345
Self-organization process, 72
Self-organizing feature map, 67, 68, 71
Self-organizing learning, 67
Self-organizing neural architecture, 67, 73
Self-organizing neural network, 9, 67
Self-repair, 345
Self-supervised neural network, 332
Self-supervised training, 335
Sensitivity, 353
Sensitivity analysis, 209
Sensory representation, 347
Sensitivity measure, 201, 202, 203, 205, 209
Sequential adaptation, 107
Shadow coding, 340
Shale, 214
Short-term memory (STM), 75
Short-term stock market prediction, 318
Shortest path, 321-325
Sigma-pi unit, 259
Sigmoid function, 52, 53, 332
Sigmoid nonlinearity, 42, 55, 63
Signal compression, 335
Signature recognition, 339
Signum function, 27, 49, 98, 152, 200, 221, 247
Similarity transformation, 226
Simple low back pain (SLBP), 354
Simulated annealing, 37, 62, 66, 325
Simulated temperature, 326
Single-layered neural network, 5-7, 11, 18, 21, 24, 27-30, 43-44, 81, 89, 90, 92-94, 98, 102, 104, 111, 122, 126, 132, 144, 152, 167, 219-221, 232, 237, 242, 247-249, 261, 263, 269, 281, 298, 310, 336, 350, 354, 356

Singular value decomposition, 333
Smooth approximation, 152
Skeletonization, 197
Soft-limiter, 15
Soft-limiting nonlinearity, 31, 99, 152, 247
Solar flare forecasting, 358
Spatial opponents, 334
Spatial ordering, 69
Spatial resolution, 69
Specificity, 354
Spectral envelope, 350
Speech processing, 62, 317, 368
Speech production, 368
Speech recognition systems, 87, 141
Speech recognition, 62, 72, 368
Speech signal, 335
Speed of convergence, 237
Spiky signal, 335
Spin glass models of neural networks, 142
Spinal pathology (SPATH), 353
Stability, 48, 97, 103, 142, 326
Stability-plasticity dilemma, 73
Stabilizing functional, 326
Stabilizing functional approach, 326
Stable algorithm, 104
Stable estimate, 327
Stable fixed point, 64, 66
Stable state, 42, 47, 55-57, 65, 322
State space, 55
State-of-the-art paradigm, 141, 317
Statistical approach, 355
Statistical characteristics, 349
Statistical decision theory, 71
Statistical sample, 351
Statistical technique, 353, 362
Statistically independent, 362
Stereo disparity, 141
Stimulus, 280, 297
SPDE image model, 328
Stochastic unit, 11, 58
Stock market, 362
Stock market prediction, 363
Storage requirement, 112
Stored association, 93, 134
Stored memory, 56
Stored pattern, 57, 221, 230, 231, 233, 234, 236, 237, 242-246, 248, 255, 295
Strongly correlated patterns, 225, 273, 274, 276
Subsampling layer, 342
Success rate, 341, 349, 351, 352, 355, 357, 358, 359, 366, 368, 369

Supervised learning, 5, 9, 12, 26, 37, 335

Supervised training, 26

Syllable boundary, 347

Symmetric matrix, 96, 135, 309

Synaptic weight, 11-13, 18, 22, 24, 26, 27, 30-34, 36-39, 42-44, 49, 53, 55-58, 60, 61, 63, 66, 67, 77-79, 89, 98-100, 104, 107, 110-113, 115, 116, 118, 120, 126, 130, 138-140, 142, 143, 145, 153, 155, 161, 162, 165, 167-169, 171, 172, 174, 176, 179, 182, 186, 189-194, 197, 199, 203, 205, 209, 216, 217, 221, 240, 247, 261, 264, 328, 330, 333, 335, 338, 342, 344, 348, 349, 357

System control, 317, 368, 369

System design, 301, 305

System identification, 368, 369

System implementation, 301, 310

System modeling, 369

System realization, 301, 310

System verification, 301, 310

Taylor expansion, 79

Temperature, 60, 62

Temporal behavior, 66

Temporal trajectory, 66

Temporary association, 113, 116, 168

Temporary input pattern, 113, 115, 116, 118-120, 168, 169, 172, 174, 176

Temporary oscillation, 129

Tertiary structure, 355

Test pattern, 231, 277

Test sample, 215, 350

Testing set, 344, 351, 354, 357, 360

Text to speech conversion, 8, 347

Texture analysis, 368

Thermal equilibrium, 59

Thinking machine, 87

Time constant, 53

Time evolution, 15

Time requirements, 36, 88, 264

Tolerance parameter, 203, 211, 212, 213, 214, 215

Top-down pattern, 75

Topological neighborhood, 69, 70

Toronto stock exchange (TSE), 363, 364, 365, 366

Toronto stock exchange (TSE) index, 364-367

Total energy, 59

Trainable control device, 87, 141

Training, 261

Training error, 197

Training of neural networks, 30, 143

Training phase, 261

Training problem, 94, 104, 143

Training process, 32, 73, 151, 188, 200-203, 211, 216

Training scheme, 249, 297, 301, 303

Training set, 12, 26, 27, 35, 73, 74, 179, 195, 200, 203, 212, 215, 219, 220, 246, 339, 341, 351, 354, 357, 360

Training task, 126, 127, 128, 129, 176, 182, 184, 202, 212

Trajectories, 48

Transfer operator, 93, 94

Transform coding, 318, 332, 333

Transform coding technique, 330

Transform coefficient, 330-332, 336

Transform matrix, 332

Traveling salesman problem, 55, 319, 321-324

Two-dimensional feature map, 67, 68

Uncorrelated patterns, 223, 225, 229, 231, 270, 271

Unconstrained minimization, 325, 326

Unlearning, 61

Unstable fixed point, 39

Unstable recursion, 97

Unsupervised learning, 9, 37, 38, 67

Unsupervised process, 338

Unsupervised scheme, 339

Unsupervised training, 5, 37, 38, 44, 336

Update equation, 30, 31, 33, 34, 38-42, 44, 46, 61, 70, 76-79, 88, 91, 92, 100, 103, 106, 118, 132-135, 138, 139, 140 153, 156, 158, 162, 167, 169, 172, 186, 189, 190-194 203, 208, 209, 216, 217

Update error, 93

Update rule, 42, 69, 72

Upper bound, 195, 220, 226, 227, 293

Upper network, 111, 112, 115, 116, 118, 161, 167-169, 171, 172

Variance, 235, 236, 239, 240, 288, 289, 292, 293

Vector quantization, 71, 72

Vigilance parameter, 75

Visible state, 83

Visible units, 11, 58-61, 82, 83

Visual cortex, 345

Visual perception, 368

Visual surface interpolation, 368

Visual surface reconstruction, 326

Waste pattern, 306, 307, 308, 312

Weakly correlated patterns, 225, 227, 233, 273, 274

Weather forecasting, 8, 318, 358

Weight decay approach, 196

Weight matrix, 44

Weight space, 36

Weight vector, 39, 40, 44, 45, 68, 69, 70-72, 337

Well-logging data, 214

Wind divergence, 359

Zero-mean random variable, 235, 236, 239-241, 288

Zip code digit recognition, 341, 342

Author Index

Abbott, L. F., 142
Abu-Mostafa, Y. S., 57, 319
Ackley, D. H., 58, 60-62
Ahmad, Z., 369
Albert, A., 29, 222
Almeida, L. B., 63, 142
Amari, S. I., 219, 234, 236, 237, 290, 294
Anastassiou, D., 89, 128, 143, 319
Anderson, J. A., 4, 29, 38, 49, 50, 56, 223
Anderson, J. R., 62
Anderson, S., 368
Angeniol, B., 71
Anlauf, J. K., 142
Arsenin, V. Y., 326
Atkeson, C. G., 369
Baldi, P., 334
Banzhaf, W., 38
Barto, A. G., 369
Battiti, R., 143
Baum, E. B., 27, 143, 195, 321, 322
Bavarian, B., 369
Becker, S., 88, 143
Beckmann, J., 369
Bengio, Y., 368
Bhat, N. V., 368
Biehl, M., 142
Bilbro, G. L., 368
Blanchet, P., 335
Blanz, W. E., 368
Block, H. D., 310, 312
Boser, B., 342, 344
Boulard, H., 118, 126, 333, 368
Bounds, D. G., 60, 62, 352, 354
Bradshaw, G., 358
Brandt, R. D., 319, 324
Bruck, J., 321
Bruwer, M., 369
Burr, D. J., 340, 368
Carpenter, G. A., 73-75
Chauvin, Y., 197
Chellappa, R., 368
Chen, F. C., 369
Chen, H. H., 20, 259

Cheung, K. F., 93
Chou, P. A., 57
Chu, S. R., 368
Clarke, R. J., 330, 331
Cohen, D. A., 36
Cohen, M. A., 47-49
Cottrell, G. W., 118, 126, 332-334
Cowan, J. D., 3
Cruse, H., 369
Cruz, V., 368
Cybenko, G., 195
Dahl, E. D., 142
Daniel, J. W., 95, 226
Daunicht, W. J., 369
Daugman, J. G., 337, 338
De Micheli, E., 368
Denker, J. S., 342
Dow, R. J. F., 196
Duda, R. O., 73, 349
Engel, J., 37
Erlauson, R., 319
Fahlman, S. E., 143
Fang, L., 319
Farber, R., 195
Feldkamp, L. A., 89
Feng, Y., 71
Fogelman-Soulie, F. 355, 357, 358
Foo, Y. P. S., 319
Forrest, B., 63
Frankel, D., 359, 360
Franzini, M. A., 368
Frean, M., 196
Fukushima, K., 344-346
Gantmacher, F. R., 223
Gardner, W. R., 341
Gersho, A., 71, 331
Giles, C. L., 66, 259
Gish, S. L., 368
Gold, B., 368
Goldberg, K. Y., 369
Gonzales, R. C., 332
Goodman, J. W., 321
Goodman, R. M., 369
Gorman, R. P., 350, 351
Graf, D. H., 71, 369
Gray, R. M., 71, 331
Greenspan, H., 368
Grimson, W. E. L., 326
Grossberg, S., 3, 4, 47-52, 73-75, 373
Grossman, T., 37
Guez, A., 369
Haken, H., 38
Handelman, D. A., 369
Hanson, S. J., 197
Harrison, R. F., 352, 353

Hart, P. E., 73, 349
Haussler, D., 195
Hebb, D. O., 38
Hedge, S. U., 319, 324
Hemani A., 71
Hertz, J., 4, 27, 35, 38-40, 43-45, 59, 62, 118, 128, 130, 146, 195, 197
Hinton, G. E., 3, 58, 60, 61, 197
Hirose, Y., 198
Hoffman, J., 339
Holley, L. H., 356-358
Homma, T., 368
Hopfield, J. J., 3, 27, 46, 51-58, 143, 223, 319-323
Hornik, K., 334
Hu, Y. H., 196
Hurlbert, A., 368
Hush, D. R., 142
Hutchinson, J. M., 326, 368
Iyer, K. V. K., 319
Jackel, L. D., 341, 344
Jacobs, R. A., 142, 370
Jacques, J.-M. S., 57
Jang, J. S., 319
Jansen, B. H., 368
Johnson, K., 346
Jordan, M. I., 369
Josin, G., 369
Kabrisky, M., 4
Kamimura, R., 66
Kamp, Y., 118, 126, 333
Kanerva, P., 57
Karayiannis, N. B., 89-91, 94, 96, 100, 106, 110, 122, 126, 145, 146, 148, 152, 153, 156, 158, 160, 165, 167, 176, 198-201, 207, 223, 225, 238, 259, 260, 264, 270, 274, 300, 314, 326, 328, 363
Karplus, M., 356-358
Kawato, M., 369
Keeler, J. D., 57
Kepler, T. B., 142
Khotanzad, A., 338
Kim, E. J., 339, 346
Knerr, S., 339
Koch, C., 327, 368
Kohonen, T., 3, 4, 13, 29, 56, 62, 67-73, 90, 91, 93, 221, 223, 265
Kollias, S., 88, 128, 143
Komura, M., 368
Kowalewski, F., 368
Kramer, A., 197
Krogh, A., 36, 37, 43, 44
Kung, S. Y., 196
LaLonde, W. R., 71, 369
Lampinen, J., 71
Lang, K. J., 368

Lapedes, A., 195

Lau, C., 4

le Cun, Y., 88, 143, 196, 197, 341, 342, 344

Lee, Y., 339, 346, 368

Lehr, M. A., 12, 18, 21, 30, 247

Leung, H. C., 368

Levin, E., 369

Levine, D. S., 4

Linde, Y., 71

Linsker, R., 38, 42, 334

Lippmann, R. P., 3, 12, 62, 368

Liu, H., 369

Lu, J. H., 338

Macy, S. A., 369

Maginu, K., 219, 234, 236, 237, 290, 294

Marchand, M., 196

Martin, G. L., 339

Martinetz, J., 71, 369

McClelland, J. L., 3, 4, 259

McCormack, M. D., 214

McCulloch, W. S., 3, 49

McEliece, R. J., 57, 219, 236, 237, 239, 294, 296

Mejia, C., 355, 357, 358

Mel, B. W., 369

Mezard, M., 196

Mighell, D. A., 339

Miller, T. K., 319, 325

Minsky, M. L., 3

Mitra, S. K., 29, 91, 133, 222, 225, 267

Moore, B., 74

Moore, K. L., 369

Morgan, A. P., 368

Morgan, N., 368

Mougeot, M., 335

Mozer, M. C., 197

Nadal, J. P., 196

Nagata, S., 369

Naidu, S., 369

Nakano, K., 29, 56, 223, 280, 352

Narendra, K. S., 368

Nasrabadi N. M., 71

Nguyen, D. H., 369

Nishimura, K., 339

Noble, B., 95, 226

Nolfi, S., 369

Ogmen, H., 368

Oja, E., 38-44, 46, 71, 81

Page, E. W., 319

Papert, S. A., 3

Parisi, D., 369

Park, C. H., 18, 259, 270

Parker, D. B., 88, 143

Parthasarathy, K., 368

Pawley, G. S., 319, 324
Pawlicki, T., 339, 368
Pearlmutter, B. A., 66, 369
Personnaz, L., 222
Peterson, C., 62
Pineda, F. J., 63, 65
Pittman, J. A., 339
Pitts, W., 3, 49
Platt, J. C., 319
Poggio, T., 265, 267, 270, 368
Pomerleau, D. A., 369
Postula, A., 71
Pratt, L. Y., 197
Pratt, W. K., 332
Psaltis, D., 18, 20, 259, 270, 271, 290, 293, 369
Puskorius, G. V., 89
Qian, N., 66, 356-358, 368
Ramanujam, J., 319
Rao, C. R., 29, 91, 133, 222, 225, 267
Ricotti, L. P., 369
Ritter, H. J., 71
Rogers, S. K., 4
Rohwer, R., 63
Rosenblatt, F., 3, 11
Rosenberg, C. R., 347
Rosenfeld, E., 4
Rossen, M. L., 368
Ruck, D. W., 27
Rumelhart, D. E., 3, 4, 12, 21, 33, 35, 99, 110, 118, 128, 130, 162, 203, 259
Ruohonen, M., 29, 90, 221, 265
Sadayappan, P., 319
Saito, K., 352
Salas, J. M., 142
Sanchez-Sinencio, E., 4
Sanger, T. D., 38, 44-46, 336
Sangiovanni-Vincentelli, A., 197
Scalero, R. S., 143
Scalettar, R., 197
Scofield, C. L., 368
Sejnowski, T. J., 58, 60-62, 66, 347, 350, 351, 356-358, 368, 369
Selinsky, J., 369
Shawe-Taylor, J. S., 36
Shepanski, J. F., 369
Shimada, T., 339
Sietsma, J., 196
Silva, F. M., 142
Silverman, R. H., 368
Simard, P. Y., 66
Simpson, P. K., 4
Singhal, S., 89, 143
Skrzypek, J., 339
Smith, F. W., 358
Smolensky, P., 197

Sobajic, D. J., 369
Solla, S. A., 27, 143, 165
Stearns, S. D., 3, 30
Strube, H. W., 368
Tabatabai, A., 368
Tagliarini, G. A., 319
Takefuji, Y., 319
Tanaka, A., 368
Tank, D. W., 55, 319-324
Tenorio, M. F., 349, 368
Tepedelenioglu, N., 143
Terzopoulos, D., 326
Tesauro, G., 369
Tikhonov, A. N., 326
Tolat, V. V., 369
Tom, M. D., 368
Troudet, T., 368
Tsai, W. T., 319
Tulunay, E., 369
Tunley, H., 368
Van den Bout, D. E., 319, 325
Venetsanopoulos, A. N., 90, 91, 94, 100, 106, 126, 145, 146, 148, 153, 160, 167, 176, 199, 223, 259, 260, 300, 326, 328
Vogl, T. P., 143
Waibel, A., 368
Wasserman, P. D., 4
Watrous, R. L., 88
Wechsler, H., 368
Werbos, P., 33
Wellekens, C. J., 368
White, H., 361, 362
Widrow, B., 3, 11, 18, 21, 29, 30, 99, 247, 358, 369
Wieland, A. P., 369
Wilczek, F., 27, 143
Williams, R. J., 3
Wilson, G. V., 319, 323, 324
Winter, R. G., 30
Wintz, P., 332
Wu, L., 89, 143
Xu, X., 319
Yao, Y., 368
Yu, P., 368
Yuille, A. L., 38, 42
Zee, A., 197
Zemel, R. S., 368
Zhou, Y. T., 328, 329, 368
Zimmerman, G. L., 368
Zue, V. W., 368